Y0-ABS-372

PREPARATIVE

INORGANIC

REACTIONS

Volume 6

PREPARATIVE INORGANIC REACTIONS

Volume 6

Editor

WILLIAM L. JOLLY

Department of Chemistry
University of California
and
Inorganic Materials Research Division
Lawrence Radiation Laboratory
Berkeley, California

1971
WILEY-INTERSCIENCE
a division of John Wiley & Sons, Inc.
New York · London · Sydney · Toronto

Library of Congress Catalog Card Number: 64-17052

ISBN 0-471-44688-2

Printed in the United States of America

10 9 8 7 6 5 4 3 2 1

CONTENTS

PREPARATIVE
INORGANIC
REACTIONS

Volume 6

Complexes of Macrocyclic Ligands

LEONARD F. LINDOY* and DARYLE H. BUSCH

Evans Chemical Laboratory
Ohio State University, Columbus

* Present address: The James Cook University of N. Queensland, Townsville, Queensland, Australia

I. INTRODUCTION

Although metal complexes of organic macrocyclic ligands have been known for more than 50 years, it is only during the past decade that a large number of new macrocyclic-ligand complexes have been prepared and investigated. Early examples were found in natural products—for example, porphyrin and corrin ring derivatives—or were confined to the phthalocyanines. Many of the new complexes exhibit unusual properties, and the associated theoretical implications have provided a stimulus for much of the recent activity in this area. The appearance of several reviews covering various aspects of the coordination chemistry of macrocyclic ligands reflects the increased attention being given to these compounds.[1–5] The possibility of using synthetic macrocycles as models for more intricate biological macrocyclic systems has long been recognized. Although of limited significance in the past, such investigations will undoubtedly increase interest in synthetic procedures for obtaining new macrocycles. The synthesis of macrocyclic ligands often seems to be aided by the presence of metal ions. In this connection a number of coordination template effects have been recognized.

The macrocyclic complexes of only a small number of metal ions, mostly first-row transition metal ions in their usual oxidation states, have been studied in any detail. The majority of the macrocyclic ligands reported to date are quadridentates containing four nitrogen donors, although a few macrocycles containing oxygen or sulfur donors are also known. No publications have appeared concerning ligands incorporating other donor atoms—for example, phosphorus or arsenic. In addition, examples of pentadentate, sexadentate, and higher polydentate macrocycles are rare, and such ligands have been used to prepare only a relatively small number of metal complexes. It is apparent that there remain immense opportunities for new synthetic studies in these little explored areas.

Although a general discussion of synthetic methods is given here, emphasis has been placed on those considerations which are unique to the preparation of macrocyclic complexes. Unfortunately the principles underlying some of the synthetic procedures are yet little understood, and indeed, in addition to its contribution to fundamental understanding, the elucidation of the mechanisms of the formation of macrocyclic complexes will undoubtedly be of great benefit in the design of new syntheses.

II. GENERAL CONSIDERATIONS

Apart from the usual considerations attending the design and use of organic ligands, there are additional concerns associated with the cyclic nature of macrocycles when used as chelating agents. Foremost among these is the constraint imposed on the positions of the donor atoms in a cyclic complex. This constraint may be twofold, affecting, first, the relative positions of the donor atoms with respect to one another and, second, the positions of the donor atoms with respect to the central metal ion and other ligands in the same complex. These effects are often manifested by a restriction of the possible conformations of the macrocycle and of the configurations of its donor atoms on coordination.

In many of the macrocyclic compounds to be discussed the in-plane ligand field strength, due to the donors of the ring, has been shown to be somewhat greater than expected on the basis of the kinds of donors present. In these cases the increase in ligand field strength may well be a consequence of a forced shortening of the metal-donor bond distances, although other factors are undoubtedly important. It has been calculated that the ligand field strength is quite sensitive to small changes in the metal-donor distances.[6] In addition to this the capacity of the donor electron orbitals for effective overlap with the appropriate metal orbitals is a function of the angular dispositions of the donors. Hence the character of the metal-donor bonds in such a complex is expected to be affected by the size of the overall macrocyclic ring, as well as by the sizes of the various individual chelate rings. A survey of the quadridentate macrocycles reported so far indicates that ring sizes of 13 to 16 members are most common. Molecular models confirm that, for those complexes in which the macrocycle completely encircles the metal ion, such ring sizes are favorable, provided that the donor atoms are so positioned that they are able to form five-, six-, or seven-membered chelate rings upon coordination. These observations are restricted mainly to the complexes of a few first-row transition metal ions, and in fact other ring sizes may also be suitable in complexes of metals having widely different ionic radii. In addition quadridentate macrocycles of smaller ring sizes may easily be accommodated by the metal ion by coordinating in a folded (nonplanar) form.

As a result of steric interactions between hydrogens and other ring substituents, the possible conformations of many macrocyclic ligands are restricted, and the geometry assumed by the donor atoms may be determined by such interactions. For example, the distribution of substituents permits or prevents the folding of certain rings, thereby controlling alternate modes of chelation.

In the design of macrocyclic chelating agents the steric and electronic consequences of unsaturation in the ring system are important. A large number of the macrocycles prepared so far are highly conjugated and yield complexes that exhibit extensive electron delocalization. This delocalization often contributes in a major way to the unusual properties associated with these compounds. In addition, extensive conjugation usually dictates an essentially planar ligand system. If the macrocycle is aromatic, the resulting complexes may also show additional stability.

The synthesis of an organic macrocycle in the absence of a metal ion is very often a low-yield reaction as a result of the competing linear polymerization reaction, which may dominate the cyclization process. Classically the yield of cyclized product is sometimes improved by effecting the ring-closing step at high dilution. More dramatically, however, the addition of a metal ion during or before this step can often result in a striking improvement in the yield of the required macrocycle as its metal complex. In these cases the improvement in yield is a consequence of one or more metal-ion effects that are collectively referred to as *coordination template effects*. If the directive influence of the metal ion controls the steric course of a sequence of stepwise reactions, the process is called the *kinetic template effect*. If the metal ion perturbs an existing equilibrium in the organic system to produce the required product in large yield, as its metal complex, the process is called the *thermodynamic template effect*. In simple terms the latter amounts to an inversion of the usual function of metal ion and ligand in sequestration processes. Usually a ligand is used to sequester a metal and remove it effectively from a competing equilibrium system. In the case of the thermodynamic coordination template effect a ligand is sequestered and removed from equilibrium by a metal ion. Examples of each effect are known in the formation of macrocyclic ligands in the presence of metal ions, and these are discussed in the appropriate following sections.

In certain cases the metal ion may serve to improve the yield of recoverable products substantially in other ways; for example, the metal ion may selectively chelate with the single product most suitable for chelation and facilitate its separation without affecting the mechanism of the formation reaction or its equilibrium position. This is the simplest application of sequestration to ligand synthesis.

III. THE CLASSICAL MACROCYCLES

The classical macrocycles are those which incorporate rings identical or closely related to the various ring systems found in nature. No attempt has been made to be comprehensive in discussing synthetic procedures for these compounds. Rather it is the aim of what follows to provide, by means of a few representative examples for each class, some insight into the kinds of procedure used to obtain this large family of macrocycles. Although the choice of examples is somewhat arbitrary, some emphasis has been placed on those synthetic procedures which involve metal ions. Biosynthetic methods have been treated elsewhere,[7] and it is inappropriate to discuss them here.

A. Phthalocyanine Syntheses

The phthalocyanine ring PcH_2 (**I**), although closely related to the biologically important porphyrins, does not occur in nature. Because of their intense colors and chemical inertness the metal phthalocyanines are of great

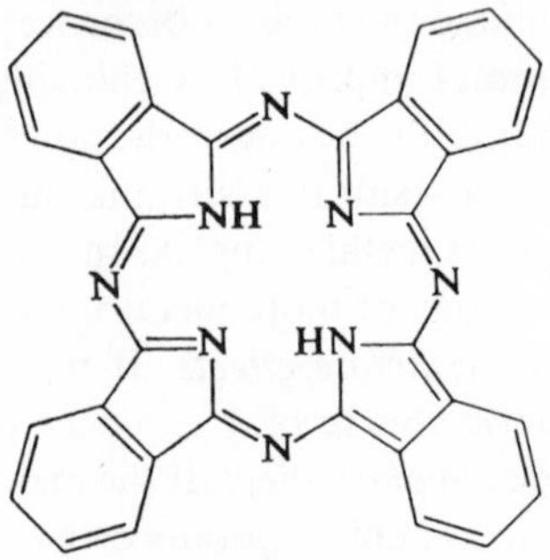

N NH N N N HN N N

I

industrial importance as pigments and dyeing agents. As a consequence this class of compound has been studied intensively for more than 30 years, resulting in an immense number of publications and patents in which synthetic details are given. Only those syntheses which can be performed conveniently on a laboratory scale are considered here. Aspects of these have also been discussed in two other general reviews of the phthalocyanines.[3,8]

Phthalocyanine complexes of a very large number of metals have been reported, but there is no general method that can be used to prepare all the known complexes.

The free phthalocyanine ligand (**I**) can be prepared directly in the absence of metal ions by cyclization reactions, in the presence of a catalyst, involving *o*-cyanobenzamide (**II**)[9] or phthalonitrile (**III**).[10] Alternatively the free

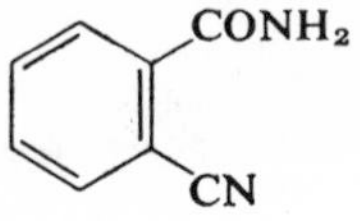

ligand can often be obtained from labile metallic derivatives of **I** by treatment with acids or water.[8–10]

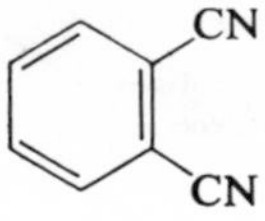

Many of the phthalocyanine metal complexes can be prepared from the free ligand by reaction with metal halides in a solvent, such as boiling quinoline or benzophenone.[9] Several of the complexes can also be obtained directly from less stable metal derivatives, such as the alkali metal salts, by metathesis. The use of the dilithium salt is particularly attractive in this regard, for, unlike **I**, it is readily soluble in common organic solvents.[11]

In Figure 1 some of the more common routes to phthalocyanines from phthalonitrile and its derivatives are illustrated. Nevertheless it is emphasized that many variations of these methods are also used. For example, the reaction of phthalonitrile with metal hydrides or metal oxides also yields phthalocyanine complexes in some instances. No attempt has been made here to list all the variations, but instead the reader is referred to the book by Moser and Thomas,[8] which lists the preparations of specific metal phthalocyanines individually.

When the cyclization reactions are carried out in the presence of a metal halide, there is often a strong tendency for halogenation of the aromatic rings of the product to occur. This tendency can be reduced by the addition of a base—for example, urea—or the use of a basic solvent for the reaction.

Many of the phthalocyanine complexes can be purified by sublimation at about 400°; nevertheless some complexes do not sublime, and these are generally purified by recrystallization or chromatography.

It is difficult to decide the role of template reactions in many of the cyclization reactions that yield phthalocyanine complexes, since in several cases the free macrocycle is also formed in the absence of a metallic element.

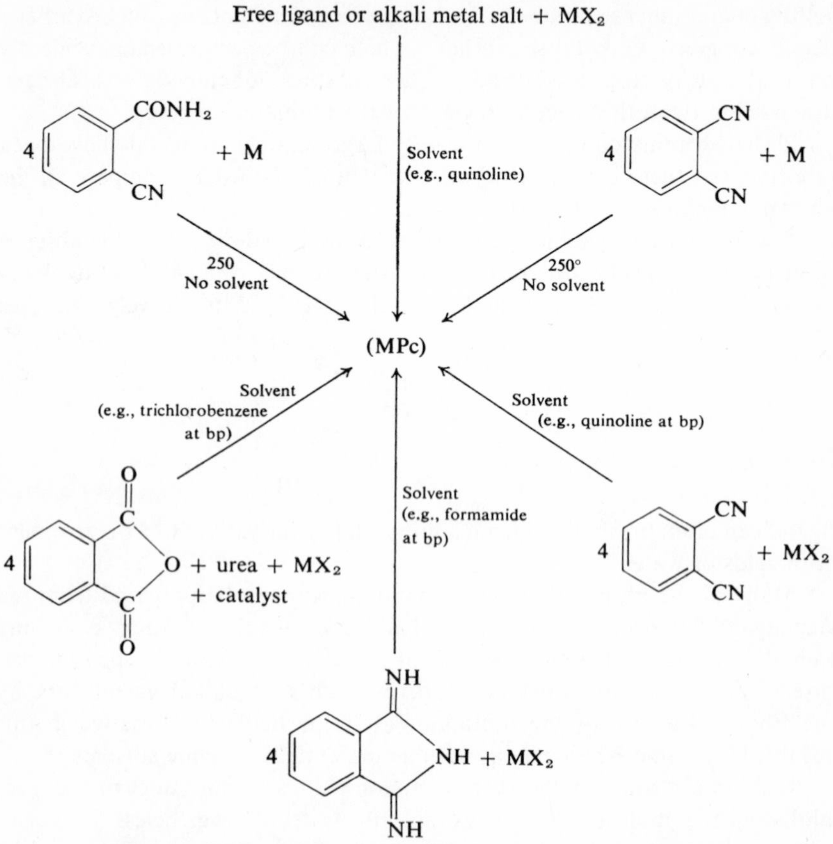

Free ligand or alkali metal salt + MX_2

Solvent (e.g., quinoline)

4 [2-cyanobenzamide: $CONH_2$, CN] + M — 250, No solvent

4 [phthalonitrile: CN, CN] + M — 250°, No solvent

(MPc)

4 [phthalic anhydride] + urea + MX_2 + catalyst — Solvent (e.g., trichlorobenzene at bp)

4 [phthalonitrile: CN, CN] + MX_2 — Solvent (e.g., quinoline at bp)

4 [1,3-diiminoisoindoline: NH, NH, NH] + MX_2 — Solvent (e.g., formamide at bp)

Figure 1. Some methods of preparing phthalocyanine complexes.

Very little work has been done to elucidate the mechanisms of these reactions. The stepwise formation of nickel phthalocyanine from nickel(II) chloride and 1,3-diiminoisoindoline in an appropriate alcohol, however, has been studied.[12] The reaction proceeds by way of an initial complexation of 1,3-diiminoisoindoline (2 moles) with nickel chloride (1 mole). Condensation with 2 additional moles of 1,3-diiminoisoindoline then occurs, followed by reaction with amyl alcohol to yield finally a bisalkoxyiminoisoindolinenine complex (**IV**) that readily converts to nickel phthalocyanine on being heated (Scheme I). Thus it is evident that the ring-closing reaction proceeds by a template process. The negative imino nitrogen displaces the alkoxide ion, which then reduces the ring to give it a dinegative charge:[4]

$$2CH_3CH_2CH_2O^- + \text{ring} \longrightarrow CH_3CH_2CHO + CH_3CH_2CH_2OH + \text{ring}^{2-}$$

Many substitutions on the basic phthalocyanine ring have been effected to yield complexes exhibiting a range of physical and chemical properties.

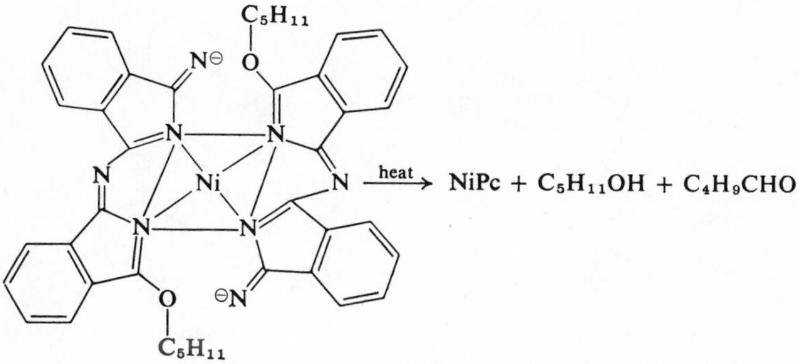

$$\text{IV} \xrightarrow{\text{heat}} \text{NiPc} + C_5H_{11}OH + C_4H_9CHO$$

IV

Scheme I

For example, the metal complexes of the 4,4′,4″,4‴-tetrasulfo derivative are water-soluble, thus enabling their solution chemistry to be investigated.[13–15] The syntheses of these compounds can be carried out in nitrobenzene from the monosodium salt of 4-sulfophthalic acid and the appropriate metal sulphate in the presence of catalysts.[14] For the preparation of other substituted derivatives, however, the substitution reaction is often performed after or during the formation of the unsubstituted phthalocyanine ring.[16]

There are also a number of interesting macrocycles which are closely related to **I** but which contain variations of the basic ring nucleus. 1,3-Diiminoisoindoline reacts readily with 2,6-diaminopyridine in boiling butanol to give a red product that was shown to be the macrocycle **V**.[17] Reaction of **V** with metal salts results in the formation of very stable neutral complexes closely related to those of phthalocyanine.

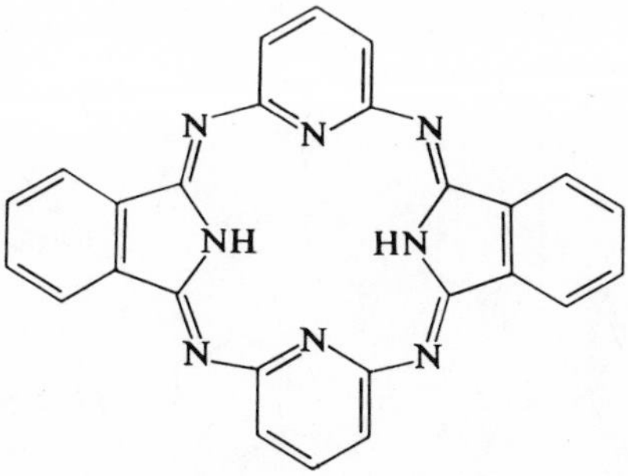

V

The preparation of the free ligand is proposed to proceed by way of a series of successive condensations.[17] Initially a reaction intermediate of the type **VI** is postulated. This intermediate has a syn configuration and favors the cyclization reaction. If its configuration is anti (**VII**), then ultimately linear polymerization will occur. Similar considerations apply to each step of

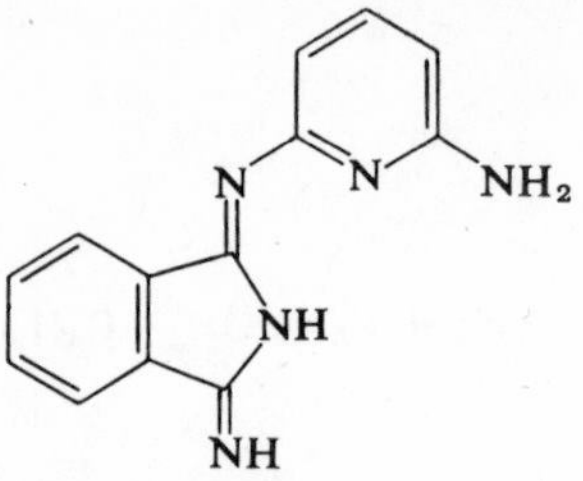

VI 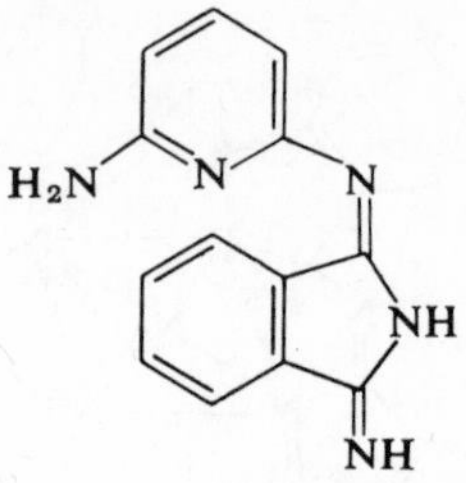 VII

the reaction sequence. Since the overall yield is 40 percent, the syn configuration is apparently largely favored in each successive condensation. This is not unexpected, since molecular models indicate that, in the anti configuration, serious steric clashes occur between the pyridine ring and the benzene backbone of the diiminoisoindoline moiety. Similar arguments can be applied to

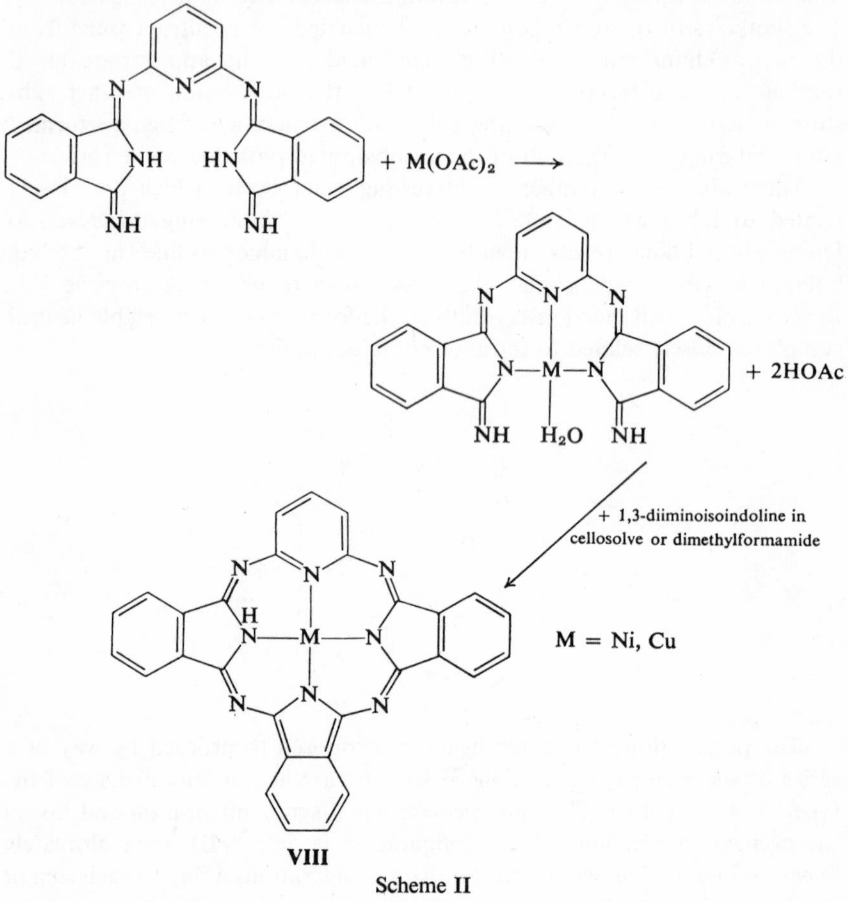

+ $M(OAc)_2$ ⟶

+ 2HOAc

+ 1,3-diiminoisoindoline in cellosolve or dimethylformamide

M = Ni, Cu

VIII

Scheme II

explain the often appreciable yields in the syntheses of phthalocyanine in the absence of metal ions. Thus in these cases the bulkiness of the reactants may direct the steric course of each organic condensation reaction so that cyclization occurs—a process that could be termed the *steric-cyclization effect*.

Recently Bamfield and Mack[18] performed the template synthesis illustrated in Scheme II. It was found that **VIII** tended to be contaminated with the corresponding phthalocyanine complex; this could be reduced to negligible proportions, however, by the inclusion of a small amount of hydroquinone during the ring-closing step. Attempts to prepare **VIII** in the absence of the metal ion were unsuccessful and yielded the macrocycle **V** instead.

Phthalocyanines in which the aza nitrogens linking each isoindoline moiety have been partially or completely replaced by methene groups are known,[8,16,19] and other ligands related to the porphins have been prepared.[8,20–28] Condensation of succinimidine and 1,3-diiminoisoindoline yields tribenzotetraazaporphin (**IX**), which is intermediate in its properties between phthalocyanine and tetraazaporphin.[25] Similarly a number of metal complexes of monobenzotetraazaporphin (**X**) have been prepared by *in situ* condensations.[27]

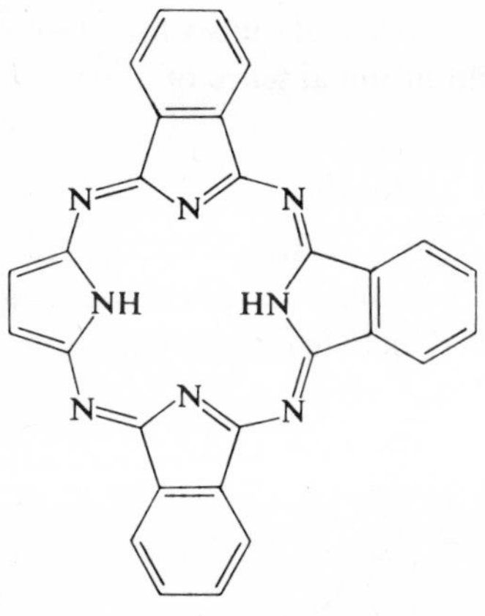

N N N NH HN N N N

IX

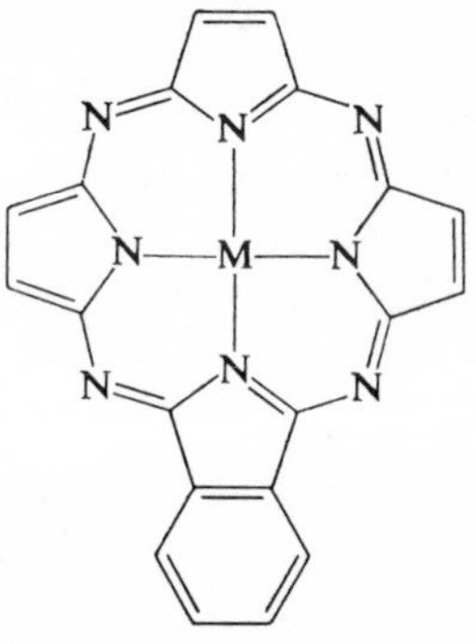

N N N N M N N N N

X

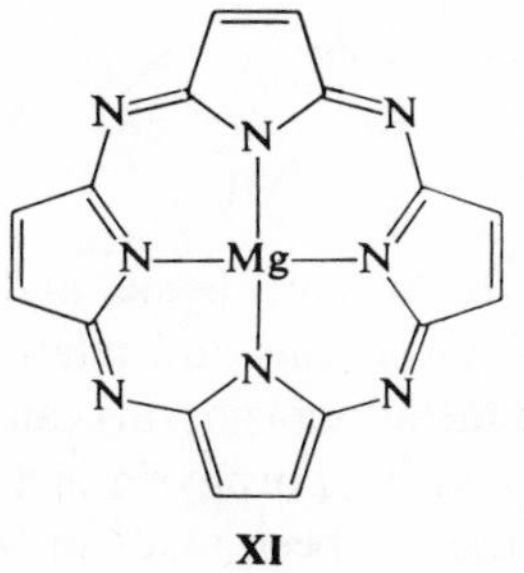

N N N N Mg N N N N

XI

Reaction of maleic dinitrile with magnesium *n*-propoxide in *n*-propyl alcohol yields magnesium tetraazaporphin (**XI**).[22] Various modifications of this reaction have been used to produce substituted derivatives of **XI**. These are listed in Table I. In all cases the free ligand was obtained by treatment of the magnesium complex with glacial acetic acid and then used to prepare other metal derivatives.

TABLE I

Magnesium Tetraazaporphin and Derivatives

Compound	Yield (%)	Reference
Magnesium tetraazaporphin	15	22
Magnesium tetramethyltetraazaporphin	14	26
Magnesium octamethyltetraazaporphin	55	24
Magnesium octaphenyltetraazaporphin	92	20

Although the yield of any reaction depends on a number of factors, the steric-cyclization hypothesis does seem to be important in many of the reactions discussed in this section. This is particularly so for those cases in which the cyclizations occur in high yield in the absence of a metal ion.

B. Porphyrin Syntheses

Porphin refers to the basic ring nucleus **XII**, where R_1, R_2, R_3, etc. = H and X_1, X_2, X_3 X_4 = H, and its substituted derivatives are called porphyrins. The metal complexes of porphyrins include some important biological

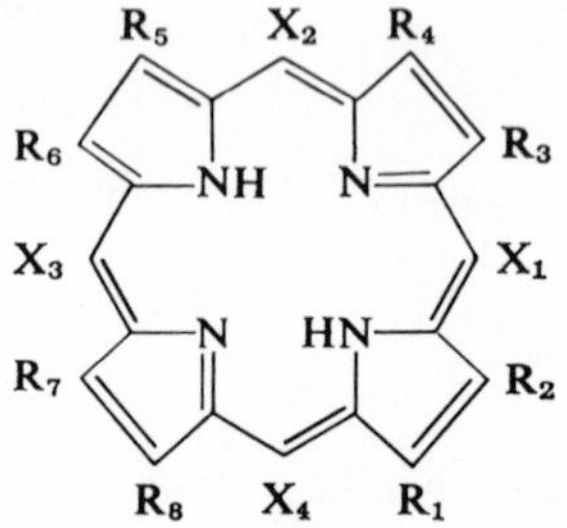

XII

compounds, such as the cytochromes, heme, and chlorophyll. An excellent review of the chemistry of porphyrins and related macrocycles appeared in 1966.[29] A large number of metalloporphyrins can be prepared by heating the appropriate metal acetate and the porphyrin in an organic solvent, although in the case of magnesium use has been made of the Grignard reagent.

Novel ligand-exchange methods for preparing metalloporphyrins from metal carbonyls have also been devised[30]; for example, the first reported chromium(II) porphyrin complex was prepared by heating the ligand with excess chromium hexacarbonyl in *n*-decane for 1.5 hr (under nitrogen). At the end of this time the solvent and unreacted chromium hexacarbonyl were removed under reduced pressure. The product was purified by dissolving in toluene and precipitating with *n*-pentane. It was postulated that the oxidation of the metal was accomplished by the weakly acidic hydrogen atoms attached to the heterocyclic nitrogens of the free porphyrin. Further oxidation to the trivalent state can be accomplished by the treatment of the chromium(II) complex with hydrogen peroxide. Related procedures have been used to prepare porphyrin complexes of cobalt, nickel, iron, and vanadyl, and it has been pointed out that the reaction can probably be applied to the preparation of other metalloporphyrins, such as those of molybdenum and tungsten, which cannot be obtained easily by standard methods.

Other reagents besides metal carbonyls may also be satisfactory for simple ligand-exchange reactions; for example, it has been demonstrated that a similar procedure using σ-bonded diphenyltitanium in place of the metal carbonyl results in the formation of titanium(IV) porphyrin.[30] Undoubtedly other classes of metal-containing reactants could be used for similar syntheses. Such methods should provide facile synthetic pathways to new metalloporphyrins and may well be suitable for extension to the synthesis of specific compounds from the various other classes of macrocyclic complexes.

Metal-free porphyrins have been prepared by a large number of routes using, chiefly, pyrroles, dipyrromethanes, dipyrromethenes, or linear tetrapyrroles as intermediates.[29] In many of these syntheses, however, the reactions are facilitated by the presence of a metal ion, thereby yielding the corresponding metal complex directly.[2] Metal porphyrins containing a variety of β substituents have been prepared. The presence of electron-withdrawing substituents decreases the basicity of the porphyrin nitrogen donors, and as a consequence the corresponding metalloporphyrins are thermodynamically less stable. For a particular porphyrin the general order of thermodynamic stability with different metals has been shown to be Pt(II) > Pd(II) > Ni(II) > Co(II) > Ag(II) > Cu(II) > Fe(II) > Zn(II) > Mg(II) > Cd(II) > Sn(II) > Li_2 > Na_2 > Ba(II) > K_2 > $Ag(I)_2$.[31]

The use of pyrroles for the syntheses of porphyrins has been restricted largely to the preparation of macrocycles containing eight identical β substituents (that is, **XII**, where $R_1 = R_2 = R_3 \cdots$ etc.) or to their derivatives having also four identical meso substituents (**XII**, where $X_1 = X_2 = X_3 = X_4$). When the pyrrole is unsymmetrical, mixtures are obtained.[29]

The reaction, under pressure and at high temperature, of pyrrole with aldehydes gives small yields of the corresponding meso-substituted porphyrins.[32] *meso*-Tetraphenylporphin was originally prepared in this way, but higher yields are obtained when the reaction is performed in the presence of zinc acetate.[33]

This general reaction has been extended to the preparation of a number of related substituted porphyrins,[34] including *meso*-(4-pyridyl)porphin (**XIII**).[35] The free ligand (**XIII**) can be obtained from its zinc complex by displacement with acid. Since it is water-soluble, the metal-complex chemistry of **XIII** has been investigated in some detail. Copper and nickel complexes were obtained by refluxing **XIII** with the metal acetate in acetic acid.[34] Subsequently the kinetics of formation of these and other metalloporphyrins in aqueous solution has been studied.[35] Complexes of the analogous tetraphenyl-substituted compounds have also been prepared.[36]

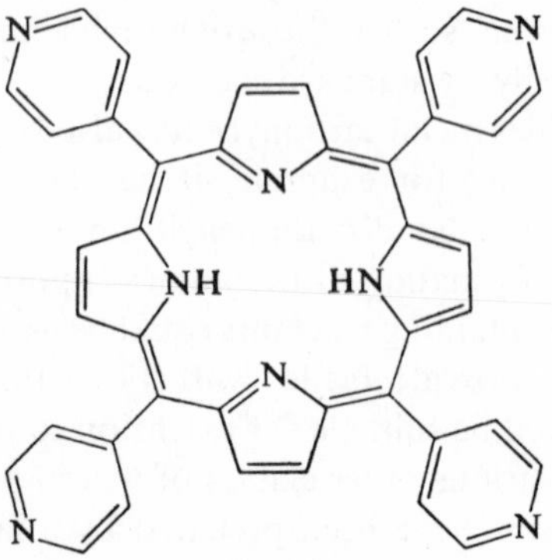

N N N NH HN N N N

XIII

Other porphyrins have been obtained by the self-condensation of pyrroles of the type **XIV**.[37] The reaction is usually performed in the presence of acid or heat and is postulated to proceed by the initial formation of the carbonium ion (**XV**).[29] In many cases the reaction has been carried out in the presence of metallic reagents.

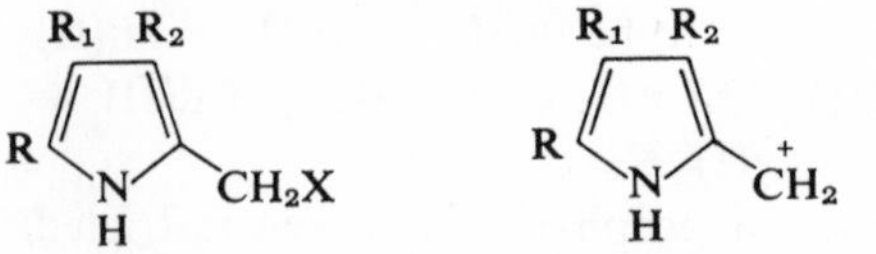

R_1 R_2 R N H CH_2X

XIV

R_1 R_2 R N H $\overset{+}{C}H_2$

XV

X = OH, OMe, OAc, Br, Cl, NH_2, NMe_2
R = H, COOH, $COOCMe_3$

Dipyrromethanes of the type **XVI**, where R_1 and R_6 can be any of a number of functional groups, have been converted into porphyrins by a variety of reactions.[38] The synthesis of chlorophyll-α was achieved by way of a reaction of this class.[39]

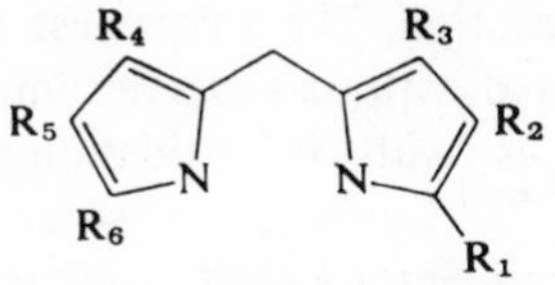

XVI 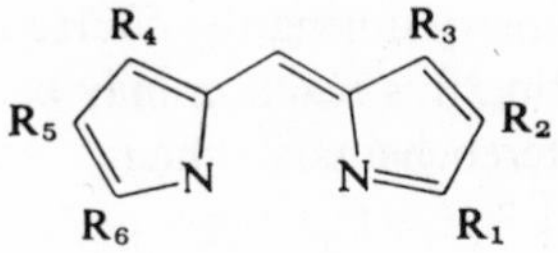 XVII

Fischer[40] has shown that fusion of dipyrromethenes of the type **XVII**, where R_6 = Br and R_1 = CH_2Br or CH_2H, in the presence of acid leads to porphyrin formation (Scheme III). Various modifications of this basic reaction have been used to prepare a large number of porphyrins, and indeed

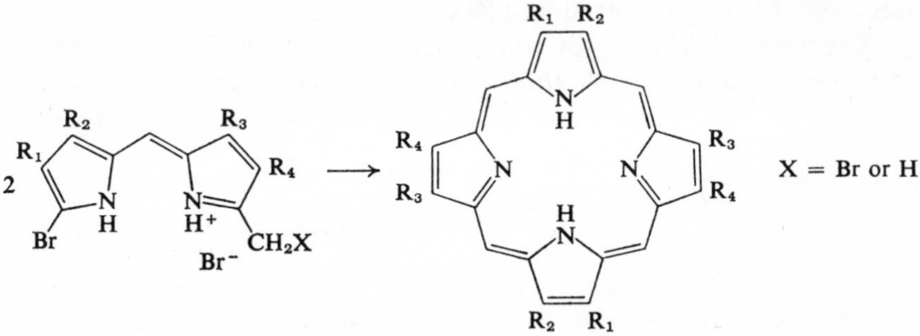

X = Br or H

Scheme III

the synthesis of haemin was achieved as early as 1929 by such a method.[41]

Other syntheses involving dipyrromethenes in the presence of copper salts also yield porphyrins;[42] for example, cyclization of **XVIII** under mild

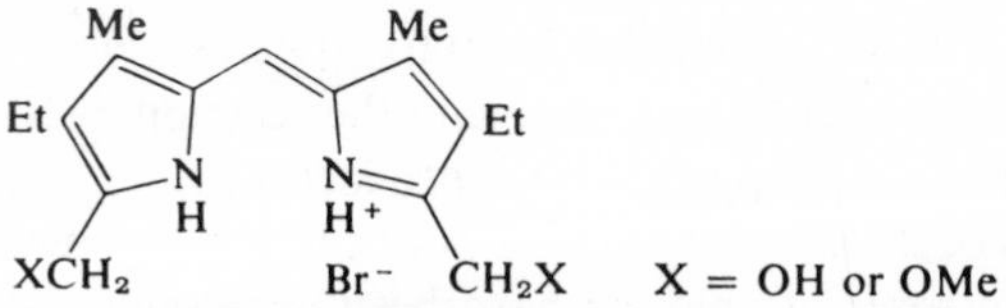

X = OH or OMe

XVIII

conditions in the presence of copper acetate, in methanol, gives up to 21 percent yield of the corresponding porphyrin complex.[43] In the presence of cobalt, zinc, and nickel ions, however, the synthesis is unsuccessful.

The original postulate that bis ligand complexes of 5,5′-di-substituted dipyrromethene cannot be planar, owing to steric hindrance,[44] was later confirmed by a number of investigators.[45] Provided that each ligand is bidentate, the bulky side substituents are expected to force a pseudotetrahedral stereochemistry on the complex, and this stereochemistry has been found in several complexes.[45] If the reaction of **XVIII** with copper acetate proceeds initially by way of a tetrahedral complex, the coordination sphere of the copper ion undergoes an effective transformation from T_d to D_{4h}

microsymmetry during the course of the reaction. The uniqueness of copper in promoting this reaction may be associated with its facile accommodation of this stereochemical change as well as with its oxidation-reduction properties.

Linear tetrapyrroles have been used frequently as intermediates in porphyrin syntheses. In several cases the cyclization of these intermediates proceeds along similar lines to those just discussed for dipyrromethenes except that only one reaction center is involved rather than two.[29] In addition, tetrapyrrole intermediates often provide a route to particular porphyrins that are not obtainable by other procedures. Only one example of this class of reaction is mentioned here.

Treatment of **XIX** or **XX** with copper acetate in boiling methanol for 48 hr leads to oxidative cyclization.[43] After chromatographic purification the

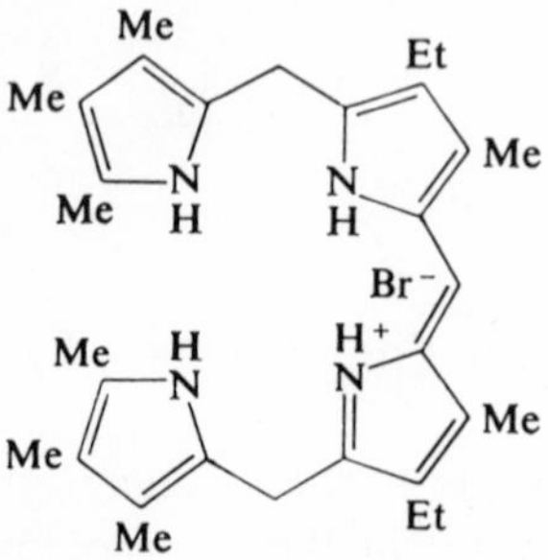

XIX

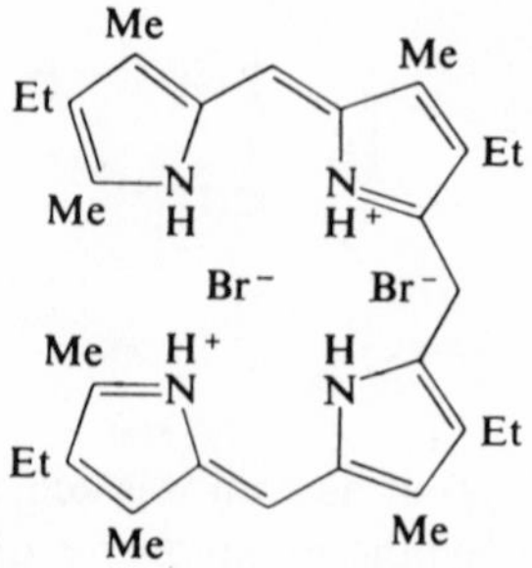

XX

resulting porphyrin can be isolated in 20 to 30 percent yield. Substitution of zinc or cobalt acetate for copper acetate in the reaction of **XIX** yields acyclic products containing eight pyrrole rings per metal ion.

Chlorin (**XXI**), the dihydro form of porphin, is the parent ring nucleus of chlorophyll. Syntheses of chlorins have been achieved by several methods, including reduction of porphins—for example, with sodium in alcohol—and degradation of chlorophyll.[46] Eisner and Linstead obtained **XXI** in low

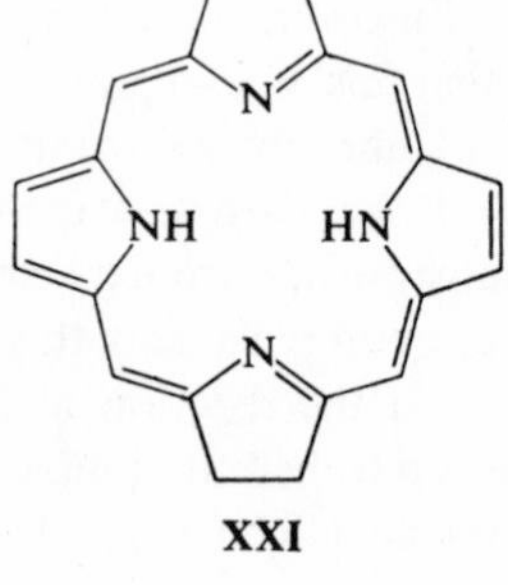

XXI

yield by heating 2-dimethylaminomethylpyrrole with ethyl magnesium bromide in boiling xylene for 7 hr.[47] The magnesium chlorin formed initially was converted to free **XXI** by treatment with dilute hydrochloric acid. Copper chlorin was prepared by heating the free macrocycle, in benzene, with copper acetate in methanol. The purple crystals of product were recrystallized from benzene. By dehydrogenation, the chlorins have provided convenient routes to a number of substituted porphins.

C. Corrinoid Syntheses

Corrin (**XXII**), the reduced form of corrole (**XXIII**), is the ring nucleus of vitamin B_{12}. Once again derivatives of these macrocycles have been obtained

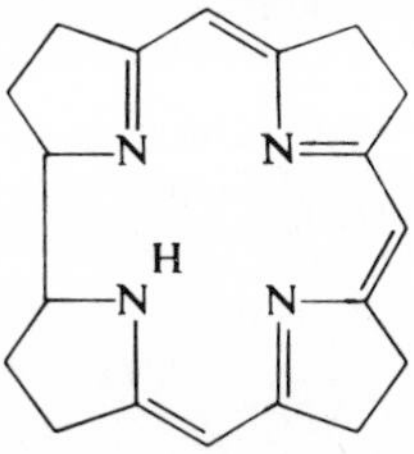

XXII 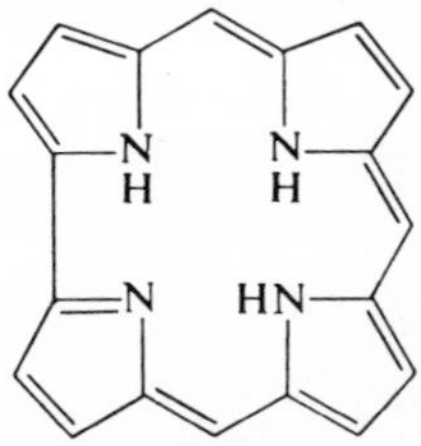 XXIII

by a large variety of organic methods, a number of which yield metal complexes as precursors to the free macrocycles. When the palladium derivative of **XXIV** is heated under reflux in ethanolic hydrochloric acid, a palladium complex of **XXV**, where X = O, is obtained in 45 percent yield.[48]

Treatment of the palladium complex of **XXIV** in pyridine with ammonia or ammonium carbonate at 160° yields a complex of the cyclic amine (**XXV**, where X = NH), and likewise reaction with methylamine or sodium sulphide yields palladium complexes of the macrocycles **XXV**, where X = NMe or S, respectively.[49] An attempt to remove palladium from the amine complex by the action of strong acids was unsuccessful. Reaction of the copper derivative of the dibromide (**XXIV**) with ammonium carbonate in pyridine, however, yields the copper complex of the amine (**XXV**, where X = NH), and treatment of this complex with concentrated sulphuric acid does yield the free ligand.

When the dihydrobromide salt of 1,19-dideoxybiladiene-ac (**XXVI**) is irradiated in dilute basic methanolic solution, cyclization to the corrole (**XXVII**) occurs in 60 percent yield.[50] Nickel, cobalt, copper, and zinc derivatives of **XXVII** were prepared either directly from the free ligand or by the addition of the metal ion, in the case of nickel or copper only, to a solution of **XXVI** before irradiation.

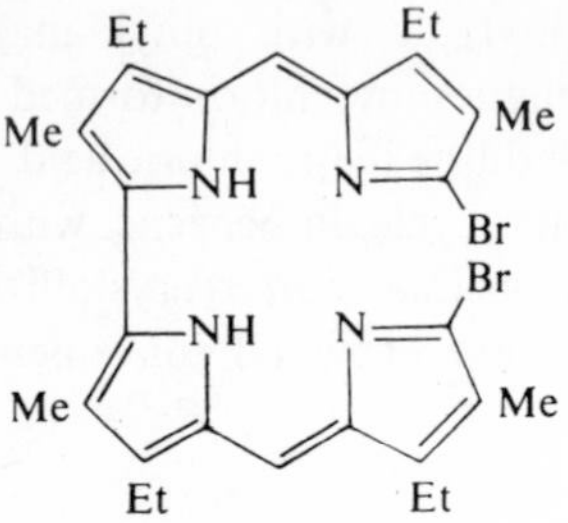

XXIV

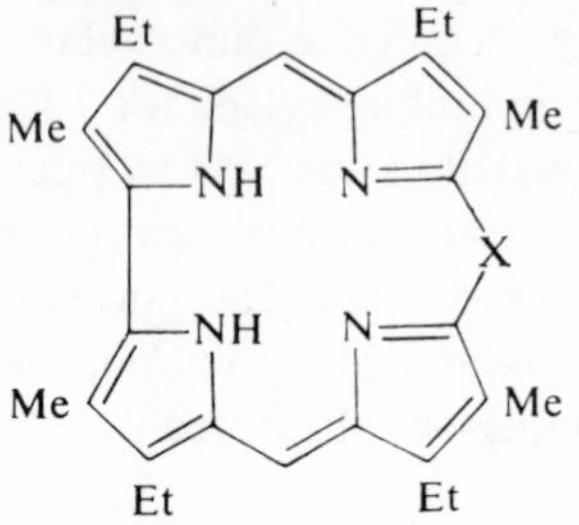

XXV

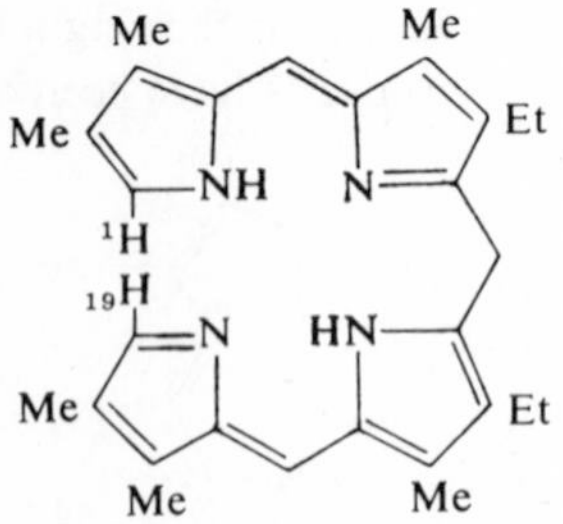

XXVI

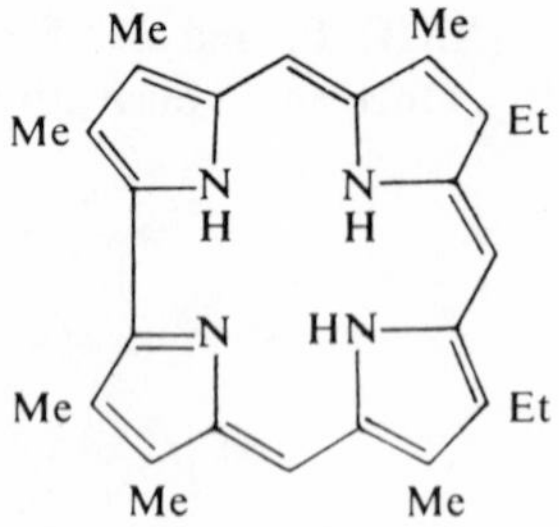

XXVII

Two other kinds of macrocyclic complex can arise from the oxidative base-induced cyclizations of the dihydrobromide salts of 1,19-disubstituted-1,19-dideoxybiladienes-ac (**XXVIII**) in the presence of metal ions. The product

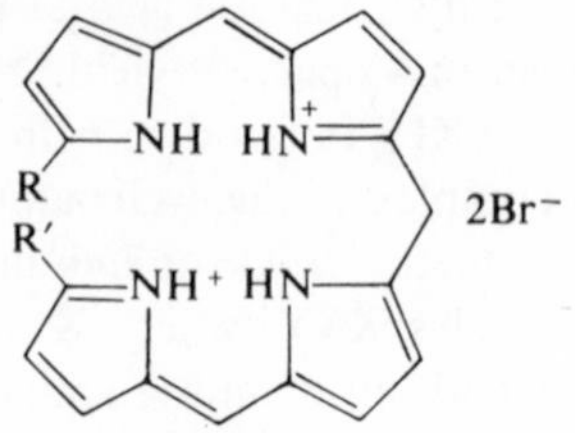

XXVIII

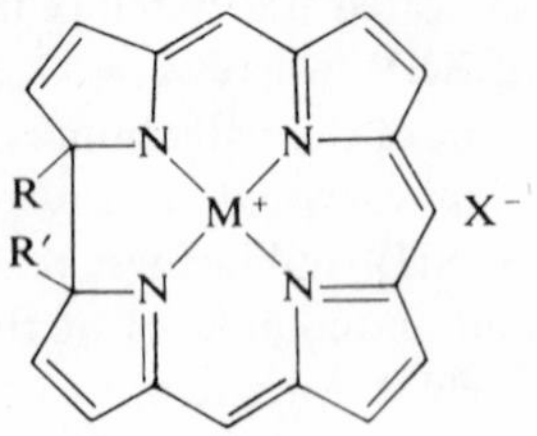

XXIX

depends on the 1 and 19 substituents, namely, R and R′. When R = R′ = Me or CO_2Et and R = CO_2Et, R′ = Me, the products are the 1,19-disubstituted tetradehydrocorrin metal salts (**XXIX**, where M = Ni or Co),[51] and when R = alkyl, R′ = Br, where M = Ni only, nickel 1-alkyltetradehydrocorrins (**XXX**) are obtained in high yield.[52]

Complexes of the type **XXX** react with acids or alky halides, resulting in protonation[52] or alkylation[53] of the C-19 position and concomitant formation of the corresponding cationic complex. Thus for the alkylation reaction, conversion to a product of type **XXIX**, where R = Me, R′ = alkyl, occurs.[53]

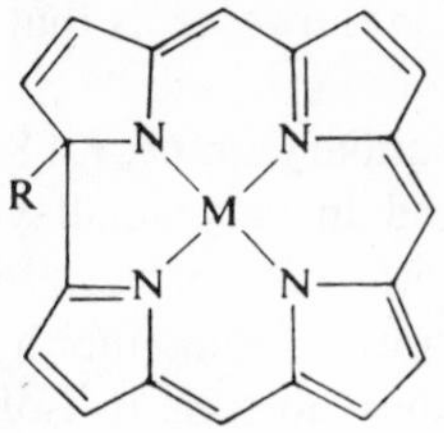

XXX

Eschenmoser et al.[54] were successful in preparing a corrin. The final step of the reaction involves the treatment of the nickel perchlorate complex of **XXXI** with *t*-butoxide in *t*-butyl alcohol to yield **XXXII**, where R = CN. Heating of **XXXII**, where R = CN, in dilute hydrochloric acid at 220° leads

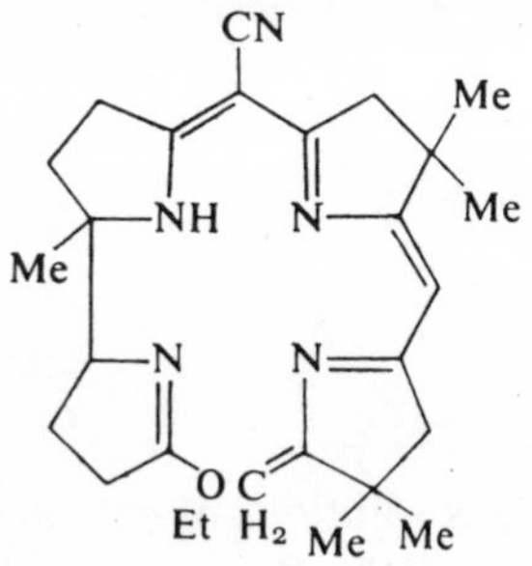

XXXI

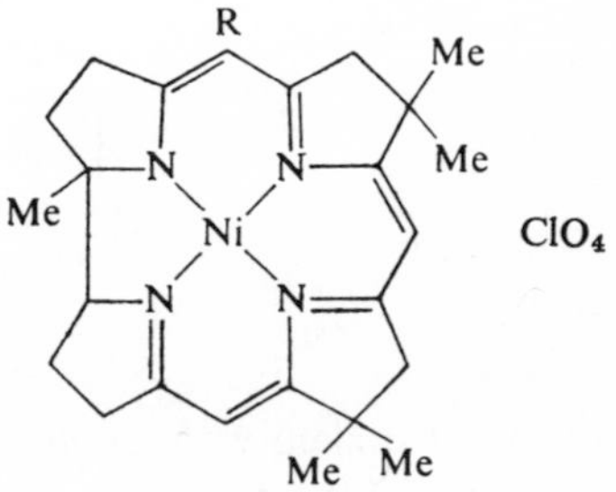

XXXII

to acid hydrolysis of the cyano group with simultaneous decarboxylation to yield the nickel corrin (**XXXII**, where R = H). Other corrin syntheses have subsequently been carried out.[55]

IV. SYNTHETIC MACROCYCLES

A. Direct Syntheses

The isolation of the free organic macrocycle before its use to prepare metal derivatives often has certain advantages compared with *in situ* preparations of the kind discussed in the next section. First, purification of the organic product may be accomplished more readily than purification of its complexes, and, second, the characterization by such physical techniques as gas-liquid chromatography, mass spectrometry, infrared spectroscopy, and nuclear-magnetic-resonance spectroscopy also tends to be less involved for the metal-free macrocycle. Further, the various spectra obtained for the free ligand are usually of great assistance in the interpretation of the corresponding metal-complex spectra. The most undesirable feature of the cyclizations of the kind discussed in the present section, however, is that most of them give only low yields of the desired products.

The cyclic tetradentate secondary amine, 1,4,8,11-tetraazacyclotetradecane (**XXXIII**, cyclam), is prepared in very small yield by refluxing 1,3-bis(2′-aminoethylamino)propane with 1,3-dibromopropane in ethanol for 3 hr, treating the solution with alcoholic potassium hydroxide, and then refluxing for a further 1½ hr.[56] It is then possible to isolate cyclam by codistillation from the reaction mass with additional unreacted 1,3-bis(2′-aminoethyl-amino)propane. It is fortuitous that cyclam is only very slightly soluble in the entraining linear tetraamine and separates from the distillate as a white crystalline solid. The yield varies over the interval of 0 to 3 percent. A second route to cyclam provides a more deliberate but far more tedious synthesis of the compound.[57]

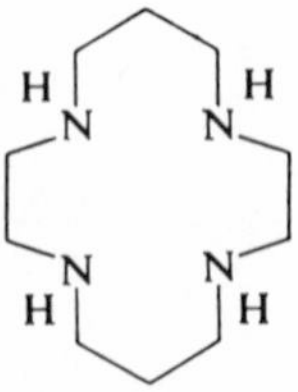

XXXIII

Reaction of cyclam with alcoholic or aqueous solutions of cobalt(II) or nickel(II) salts yields a range of quite stable cobalt(III) (after aerial oxidation of the cobalt(II) solution) and nickel(II) complexes.[56,58,59] From physical measurements, all the diacido nickel-cyclam complexes have been assigned trans configurations in which the four nitrogens of cyclam occupy a single plane. This configuration has been confirmed for $Ni(cyclam)Cl_2$ by an x-ray structure determination.[60]

Kinetic studies in dilute acid reveal that the reaction of cyclam with nickel ions is much slower than the similar reactions using related open-chain tetradentates, such as triethylenetetraamine.[61] It seems that, as a result of the special steric restrictions of cyclam, other metal-amine bond formations apart from the first may also influence the overall rate. The effect of the cyclic ligand structure on ligand dissociation is still more remarkable. The nickel complexes are stable indefinitely in strong acid media.[58]

Both trans and cis diacido cobalt(III) complexes have been prepared.[56] The kinetics of aquation of the dichloro complexes has been investigated over a range of temperatures. At 25°, the cis isomer is approximately 15,000 times more labile than the trans isomer—an effect that reflects the much lower activation energy of the former complex. For both isomers, complete retention of the geometric configuration occurs during substitution reactions.

Collman and Schneider[62] have prepared cobalt(III) and rhodium(III) complexes of 1,4,7,10-tetraazacyclododecane (**XXXIV**, cyclen). The free ligand was synthesized by a modification of the method of Stetter and Mayer.[57] The product was isolated as its tetrahydrochloride, because in this form it is

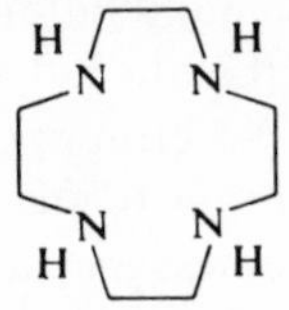

XXXIV

readily purified by recrystallization from hydrochloric acid. The complexes $[Co(cyclen)X_2]^+$, where X = Cl, Br, CO_3, $\frac{1}{2}C_2O_4$, NO_2, and $[Rh(cyclen)Cl_2]^+$ were prepared by standard methods, and the acido groups were shown to occupy the cis positions. This is probably the result of the overall ring size (12 members). This produces a fused chelate ring system composed of four 5-membered rings and is too small to facilitate a planar array of the donors. To be planar the macrocycle must be large enough to encircle the metal ion completely. On the other hand a folded conformation, as occurs in cis diacido complexes, accommodates some shorter distances between donor atoms.

Rosen and Busch[63] have used the reaction sequence given in Scheme IV

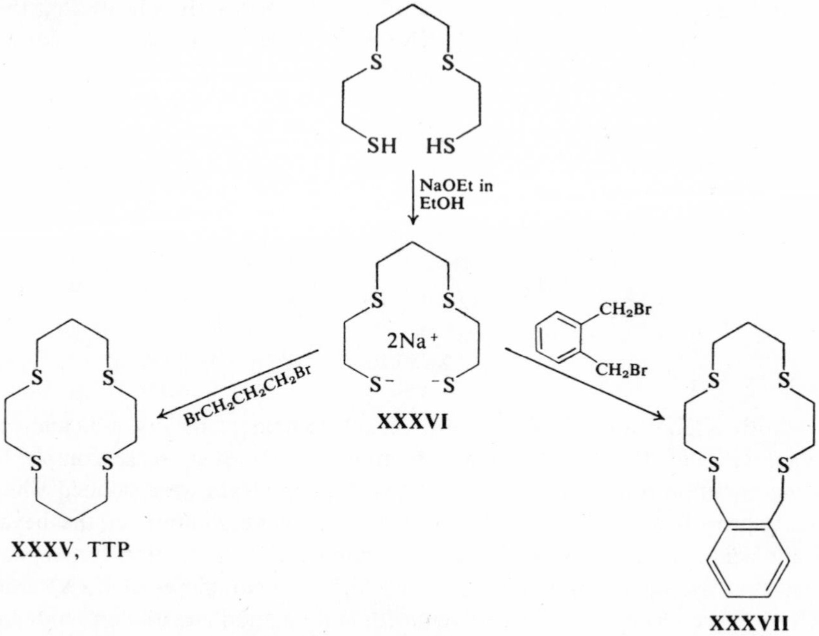

NaOEt in EtOH

$2Na^+$

XXXVI

$BrCH_2CH_2CH_2Br$

XXXV, TTP

XXXVII

Scheme IV

to prepare the quadridentate macrocycles **XXXV** and **XXXVII**. The final ring-closing step gave **XXXVII** in 38 percent yield. Initially the corresponding yield of **XXXV**, the sulfur analog of cyclam, was $7\frac{1}{2}$ percent, although the reaction was performed at moderate dilution.[63] Subsequently this yield was improved by diluting threefold the alcohol solution of **XXXVI** used in the original preparation.[64] The 1,3-dibromopropane was also diluted with alcohol before its dropwise addition to the reaction solution. Under these conditions the yield of the product was increased to about 55 percent. Further dilution failed to result in a significant increase in yield. The comparatively low yields when these condensations are not carried out under conditions of high dilution undoubtedly reflects the competing formation of polymeric species. If one assumes that the reaction of the dihalo compound with the dithiolate anion (**XXXVI**) is a two-step process, high dilution favors cyclization by increasing the probability that the second condensation step will occur at the unreacted mercaptide group of the same molecule of **XXXVI** on which the first condensation occurred.

The crude product from the reaction of 1,3-dibromopropane with **XXXVI** was purified by sublimation at 0.5 mm and 110° to give pure **XXXV** as a white solid with a melting point of 119 to 120°.[63] The unsublimed residue can be extracted with hot ethanol, however, to yield a second white product that has a melting point of 67°.[64] A molecular-weight determination, elemental analyses, nmr, and mass spectral measurements all indicate that this product is the dimeric macrocycle **XXXVIII**. Under the reaction conditions described, **XXXVIII** is formed in about 25 percent yield.

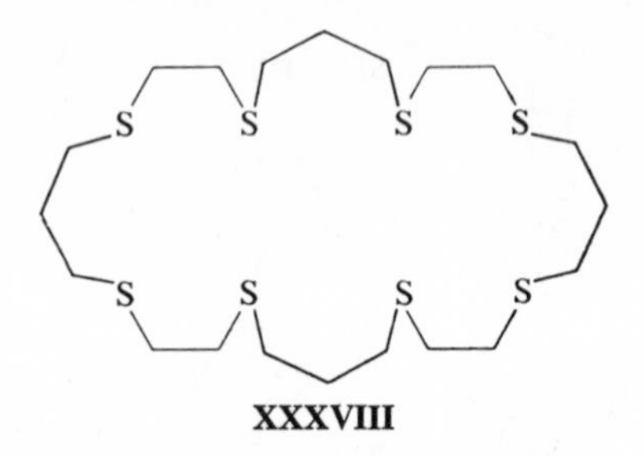

XXXVIII

Both **XXXV** and **XXXVII** form nickel complexes having a ligand-to-metal ratio of 1:1.[63,64] Although attempted syntheses of these complexes using common procedures were unsuccessful, complexes were isolated when the appropriate ligand reacted with a nitromethane solution of the hexaacetic acid complex of nickel(II) tetrafluoroborate. This reaction system was first used by van Leeuwen and Groeneveld.[65] The complexes of **XXXV** and **XXXVII** are unstable in most solvents containing good a-class donors. Such solvents displace the macrocycle, yielding solvated metal ion and free ligand.

Although a number of metal complexes of ligands containing only thioether donors[66] were reported early, no macrocyclic complexes of this kind were reported previously. The quadridentate macrocyclic complexes also provide the first examples of low-spin, square-planar nickel compounds containing four thioether linkages as donors. Metathetical reactions on $Ni(TTP)(BF_4)_2$ in nitromethane using appropriate anions result in the formation of the paramagnetic tetragonal complexes $Ni(TTP)X_2$, where X = NCS, Cl, Br, or I.[63] Complexes of **XXXV** with other metal ions have also been prepared.[64]

Nickel complexes of **XXXVIII** in which the metal-to-ligand ratio is 2:1 have also been isolated. Palladium and platinum form 4:1 complexes.[64]

Three other macrocyclic ligands (**XXXIX** to **XLI**) having only thioether donors have been synthesized by procedures similar to those outlined in Scheme IV.[67] These ligands react with the hexaacetic acid derivative of nickel(II) tetrafluoroborate to yield the complexes $[Ni_2(TTC)_3](BF_4)_4$, $[Ni(TTD)_2](BF_4)_2$, and $[Ni_2(TTE)_3](BF_4)_4$. All the complexes react with

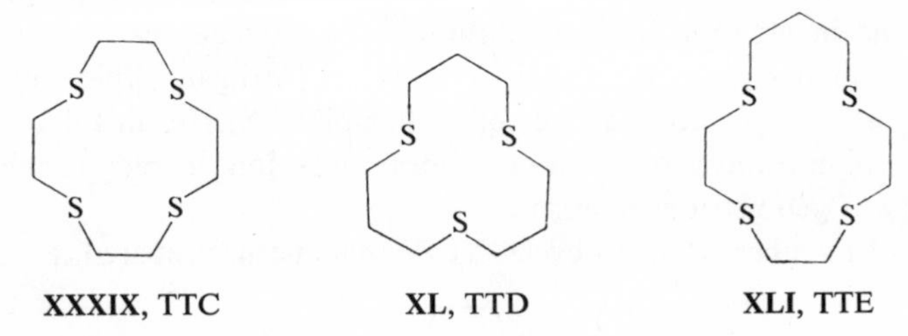

XXXIX, TTC **XL**, TTD **XLI**, TTE

water, which displaces the ligands intact. Physical measurements indicate that for each complex the metal ion is surrounded octahedrally by six sulfur atoms. The dimeric cations are proposed to have the structure **XLII**. It seems that a 14-membered ring, as found in **XXXV**, is the smallest that

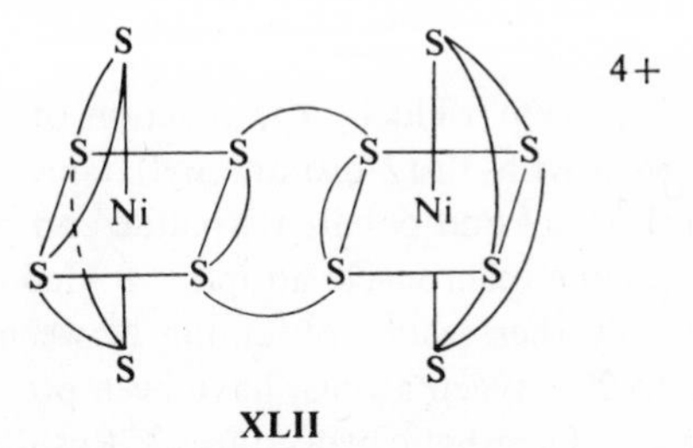

XLII

allows coordination of macrocycles of this general type in a planar configuration.

The hexathioether (**XLIIIa**) has been synthesized in 31 percent yield by reaction of 1,2-dibromoethane with the disodium salt of 3-thiapentane-1,5-dithiol in ethanol in high dilution.[68] A similar procedure using 3-oxapentane-1,5-dithiol gave the corresponding macrocycle containing two oxygen atoms

(**XLIIIb**) in 7 percent yield. No complexes of the latter ligand have yet been isolated, although with **XLIIIa** it has been possible to obtain the compounds

(a) X = S
(b) X = O
(c) X = NH

XLIII

[M(ligand)](picrate)$_2$ in which the ligand seems to be sexadentate.

Complexes containing only thioether donors are most stable when the metal ion has substantial b-class character. Extended studies[69] have shown, however, that chelating ligands containing both thioether and nitrogen donor atoms form relatively stable complexes with a wide range of metals. The potentially sexadentate ligand (**XLIIIc**) has been prepared in 8 percent yield by the reaction of the disodium salt of ethane-1,2-dithiol and di-(2-bromoethyl)amine in ethanol at high dilution.[68] This ligand reacts readily with nickel or cobalt picrate to yield the complexes [M(ligand)](picrate)$_2$, where M = Ni or Co. As a consequence of its flexibility **XLIIIc** might coordinate in two configurations around an octahedral metal ion, thereby providing the possibility of geometric isomerism.

The 12-membered macrocycle **XLIV** has been prepared as a white

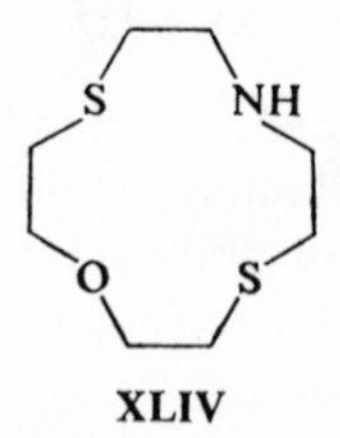

XLIV

crystalline solid in 18 percent yield by the reaction of the disodium salt of 3-oxapentane-1,5-dithiol with di-(2-bromoethyl)amine in ethanol at high dilution.[68] Octahedral nickel and cobalt(II) halide complexes of **XLIV** have been prepared in which the macrocycle adopts a folded configuration.

Forty-two cyclic polyethers each containing between 9 and 60 atoms in the ring, including 3 to 20 oxygen atoms, have been prepared from aromatic vicinal diols by a variety of synthetic procedures.[70] Results indicate that many of the smaller ring compounds can function as ligands, although the donor strength of an ether function toward most metal ions is exceedingly weak. Many of the rings containing 5 to 10 oxygen atoms form complexes with some or all of the following cations: Li, Na, NH_4, RNH_3, K, Rb, Cs, Ag(I), Au(I), Ca, Sr, Ba, Cd, Hg(I), Hg(II), La(III), Tl(I), Ce(III), and Pb(II). The ligand-to-metal ratio in these unusual compounds is most often 1:1. The most remarkable aspect of these ligands is the formation of complexes with large alkali metal ions. The very formation of these species must be attributed to the multiple chelate effect.

The interaction of cobalt(II) chloride with the saturated ligand (**XLV**) in acetic acid has been studied in some detail.[71] Blue crystals of the empirical formula Co_2Cl_4 (ligand) were isolated when the solution above was allowed to stand in the absence of moisture or oxygen. The compound has been

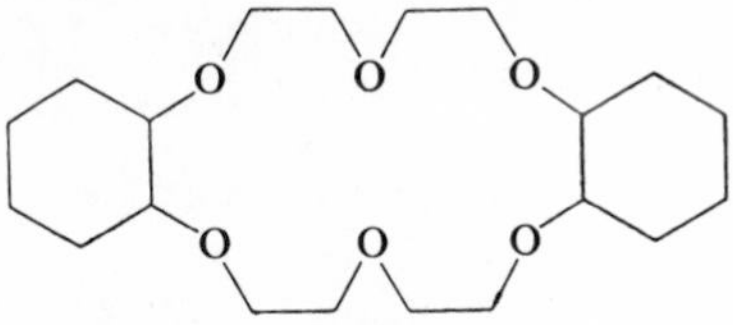

XLV

assigned a dimeric or polymeric salt structure in which the anion is the tetrahedrally coordinated $[CoCl_4]^{2-}$ species and the cation contains the cobalt in an octahedral environment. It has been demonstrated that solvents of high dielectric constant readily dissociate this complex into its components.

A series of interesting cagelike macrocycles containing both nitrogen and oxygen atoms has been reported.[72] Typical of these ligands is **XLVI**, which forms 1:1 complexes with a range of alkali and alkaline earth metal salts.

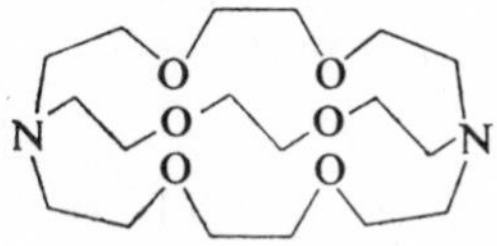

XLVI

Many of the reactions were carried out in chloroform. It was found possible to isolate complexes of the following salts: KSCN, NaI, $Pb(SCN)_2$, $AgNO_3$, $SrBr_2$. The authors suggest that the metal ion in all these complexes is contained in the central cavity of the ligand, and so complete encapsulation of the metal ion by the ligand occurs. Such a phenomenon has been discussed previously, and the name clathro-chelate has been proposed for such complexes.[1] If the structure is as suggested, this presupposes that the ligand must be sterically capable of allowing entry of the particular metal ion into the central cavity. For **XLVI** the combination of 18-membered great rings together with donor atoms of comparatively small radii apparently makes this possible. In the design of other clathro-chelates containing larger donors and (or) smaller ring sizes, however, it may not be possible to effect encapsulation by a similar procedure. In these cases the addition of the metal ion before or during the ligand formation reaction may be necessary to achieve a complex of this kind. There is one example of such an *in situ* reaction,[73] but this is more properly discussed in the next section.

Curtis and Hay[74] have shown that the monohydroperchlorate salt of ethylenediamine reacts with acetone or mesityl oxide to yield **XLVII** (1,7-CT) as its dihydroperchlorate salt. These authors postulated the initial occurrence

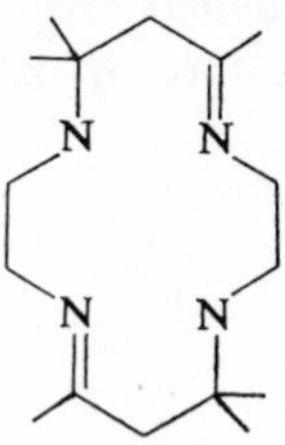

XLVII

of a Michael addition of the nonprotonated amine of ethylenediamine across the double bond of mesityl oxide to yield the β-amino ketone (**XLVIII**). The protonation of the second amine group is assumed to block its participation

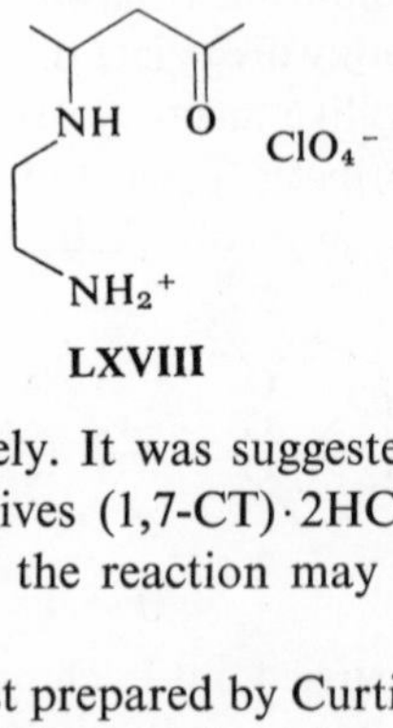

NH O ClO_4^-

NH_2^+

LXVIII

in a similar reaction effectively. It was suggested that condensation of two molecules of **XLVIII** then gives (1,7-CT)·$2HClO_4$. Other mechanisms are also plausible, however, and the reaction may well be more complex than described above.

The ligand **XLVII** was first prepared by Curtis as its nickel complex by an *in situ* reaction.[75] This and similar reactions have been studied at great length and are discussed in the following section.

B. *In Situ* Syntheses

As is also the case for the classical macrocycles, many of the synthetic macrocycles are prepared most successfully in the presence of metal ions to yield the metal complexes directly. For a number of such complexes the macrocycles have not been isolated free of their metal ions and, for these, *in situ* syntheses remain their only mode of preparation. Even for those ligands which can be prepared directly, the addition of a metal ion at some stage of the synthesis is often accompanied by an increase in the yield of macrocycle, as its metal complex. Nevertheless an inherent disadvantage of most *in situ* syntheses is that a small excess of one or another of the organic reactants, which are often chelating agents themselves, may lead to contamination of the required product with acyclic impurities. It is also true that in some cases there remains an element of mystery concerning the actual sequence of reactions and the exact nature of the metal-ion effects. Nevertheless, among other things, a template effect of the kinetic or thermodynamic variety does seem to operate in many of the reactions discussed in this section.

Condensation reactions between carbonyl compounds and primary amines are responsible for a major part of the processes leading to the formation of new macrocyclic ligands. The use of the Schiff base reaction plays a central role in these processes. Such condensations are known to proceed by way of a nucleophilic attack by the amine nitrogen on the carbon of the carbonyl group to yield a carbinol-amine intermediate. Normally the reaction is acid-catalyzed. Coordination of the carbonyl oxygen to a positive center thus favors the reaction by making the carbonyl carbon atom more susceptible to nucleophilic attack. On the other hand, coordination of the amine group decreases its availability as a nucleophile, thereby hindering *in situ* formation of a Schiff base. There is little evidence that such condensations do occur when the amines are coordinated.

In some of the earliest experiments involving *in situ* macrocyclic-ligand synthesis Curry demonstrated that new macrocyclic complexes can be prepared by the reaction of 2,6-diacetylpyridine with certain polyamines in the presence of metal ions.[76] Treatment of 2,6-diacetylpyridine with 3,3′-diaminodipropylamine and nickel chloride in an alcohol-water mixture at 65° for 6 hr leads to isolation of the diamagnetic nickel perchlorate complex, after the addition of perchlorate ion, of the macrocycle **XLIX** (CR).[77,78] As is

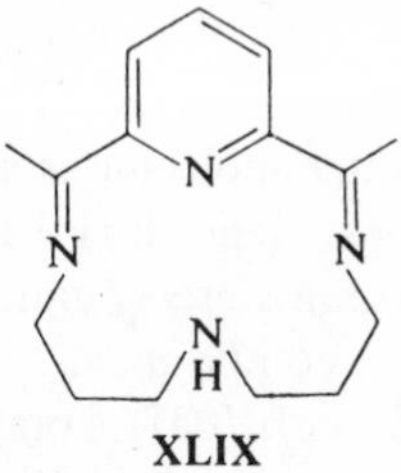

XLIX

often the case with Schiff base condensations of this kind, the addition of a small amount of glacial acetic acid catalyses the reaction. Rich and Stucky[79] also reported the preparation, by a related procedure, of the nickel and copper complexes of **XLIX** as their tetrachlorozincate salts.

The presence of the pyridine group between the two carbonyl functions in 2,6-diacetylpyridine probably assures initial tridentate chelation of this reactant (**L**) with the concomitant activation of the coordinated carbonyls

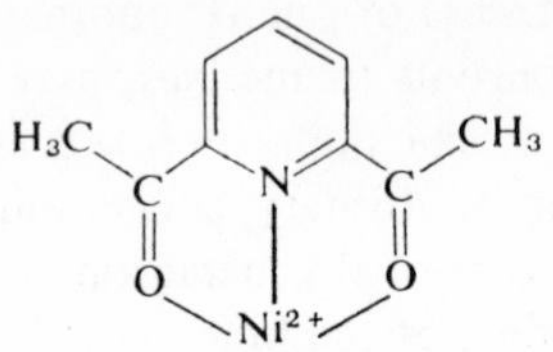

L

toward reaction with a temporarily free amine group from a triamine ligand that is chelated within the same coordination sphere.

The paramagnetic tetragonal complexes $Ni(CR)Cl_2$, $[Ni(CR)(H_2O)_2]$-$(ClO_4)_2$, and $Ni(CR)(NCS)_2$ have been prepared.[77] The chloride and thiocyanate salts are obtained by the addition of lithium chloride or sodium thiocyanate to an acetone solution of the diperchlorate salt, filtering the solid products, and recrystallization from chloroform. The dihydrate of the diperchlorate complex was prepared by allowing the anhydrous salt to stand in a moist atmosphere. A series of diamagnetic compounds, $Ni(CR)Br_2 \cdot H_2O$, $Ni(CR)Br(ClO_4) \cdot H_2O$, and $Ni(CR)Br(BF_4) \cdot H_2O$, each containing the same complex cation, have also been prepared. Addition of a concentrated sodium bromide solution to aqueous $Ni(CR)Cl_2$ results in the precipitation of $Ni(CR)Br_2 \cdot H_2O$. When the sodium bromide solution is added to a solution of $Ni(CR)(ClO_4)_2$ or $Ni(CR)(BF_4)_2$, the mixed salts $Ni(CR)Br(ClO_4) \cdot H_2O$ and $Ni(CR)Br(BF_4) \cdot H_2O$ are obtained. The diamagnetism of these salts was rationalized by postulating that the water molecule was involved in hydrogen bonding with the coordinated bromide and N—H group, and so the donor strength of the bromide ligand was diminished. Structure determination by x-ray diffraction has shown that the macrocyclic molecules in $Ni(CR)Br_2 \cdot H_2O$ are connected by N—H···Br—Ni linkages and that the nickel atom has an approximately square pyramidal stereochemistry with a bromide ligand in the axial position.[80] The water molecule is strongly held between the coordinated and free bromide ions. Hence it is the combination of a number of subtle factors that result in both a stereochemical and a spin-state change in formation of these diamagnetic compounds.

Low-spin pentacoordinate cobalt(II) complexes have also been prepared.[81] The compounds were synthesized by substituting the appropriate metal salt in the general procedure devised by Karn for the preparation of $Ni(CR)Br_2 \cdot H_2O$ or by metathetical reactions on the products obtained by this procedure. Three series of cobalt(II) complexes were obtained. The compounds $Co(CR)X_2 \cdot nH_2O$, where X = Cl, Br, I, NO_3, NCS, and ClO_4 and n = O, ½, or 1, all have magnetic moments of about 1.9 B.M., which are typical of low-spin cobalt(II) ions. Complexes of the type $Co(CR)A(ClO_4)$, A = NH_3 or pyridine, also show normal low-spin magnetic moments. In contrast the compounds $Co(CR)BrX$, where X = ClO_4, PF_6, and $B(C_6H_5)_4$, all show solid-state, room-temperature moments of slightly over 1 B.M., but in solution $Co(CR)Br(ClO_4)$ shows a more normal value of about 2 B.M. In the solid state the low moments of these complexes have been ascribed to antiferromagnetic interactions between adjacent cobalt ions.

The diperchlorate copper salt of **XLIX** has been prepared by an analogous procedure to that used by Karn for the similar nickel complex.[82] A conductometric titration of this compound with tetraethylammonium chloride in methanol gave a 1:1 end point with no evidence of formation of a 2:1 species. Accordingly reaction of methanolic solutions of this copper salt with excess halide ions leads to isolation of the complexes $[Cu(CR)X]ClO_4 \cdot xH_2O$, where X = Cl, Br, I, and NCS and $x = 0$ or 1.

Similarly iron(III) complexes of the pentadentate and sexadentate ligands **LI** and **LII**, respectively, have been prepared by template syntheses from 2,6-diacetylpyridine and the appropriate polyamine.[76] In both cases the

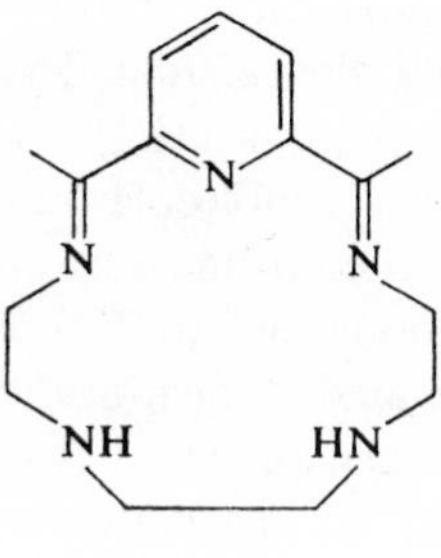

N N N NH HN

LI

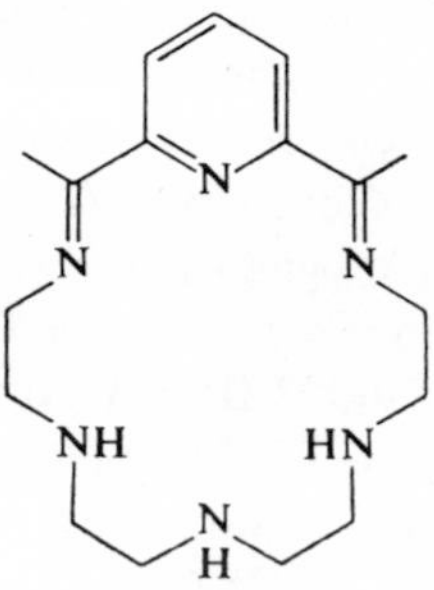

N N N NH HN N H

LII

initial condensation reaction occurred in the presence of iron(II) chloride to yield a purple coloration. As first pointed out by Krumholz, the purple color is typical of iron(II) complexes of ligands containing three conjugated donor nitrogens, as in **LI** and **LII**.[83] Before isolation as their perchlorate salts both complexes were aerially oxidized to iron(III). In order to inhibit the formation of iron oxide, both reactions were performed under slightly acidic conditions.

The product first isolated involves an oxygen-bridged dimer of the type $[(H_2O)(MAC)Fe—O—Fe(MAC)(H_2O)](ClO_4)_4$, where MAC refers to the pentadentate macrocycle.[84] The presence of the bridging unit was first demonstrated by studies on the temperature dependence of the magnetic moment.[84–86]

Stable seven-coordinate mononuclear complexes of the stoichiometry $[Fe(MAC)X_2]Y$, where X = Cl, Br, I, or NCS and Y = ClO_4, BF_4, or NCS, can also be prepared by treating acidified solutions of the oxygen-bridged dimeric complex with the corresponding sodium halide or by substitution of the appropriate salt for sodium perchlorate in the original preparation of the dimeric compound.[86] All these compounds show normal magnetic moments for iron(III) with $S = 5/2$.

An x-ray structure study of the red compound $[Fe(MAC)(NCS)_2](ClO_4)$ shows it to be pentagonal bipyramidal in gross geometry.[87] The metal and the five nitrogen atoms of the ligand are in a single plane with the two thiocyanate groups coordinated through their nitrogen atoms and occupying trans positions along the normal to the pentagonal plane. Treatment of a methanol solution of this compound with ethylenediamine until the red color is discharged leads, on concentration of the solution, to the isolation of brown crystals of the dimer $[(NCS)(MAC)Fe—O—Fe(MAC)(NCS)](ClO_4)_2$, which also shows anomalous magnetic behavior, owing to metal-metal interaction by way of the oxygen bridge.[86] A crystal-structure determination on the complex $[(H_2O)(MAC)Fe—O—Fe(MAC)(H_2O)](ClO_4)_4$ has also established its structure.[87] Each iron atom is surrounded in a plane by the five donor nitrogens of the macrocycle. The oxygen bridge links the two iron atoms, and a water molecule occupies each outer axial position; and so the configuration about each iron is approximately pentagonal bipyramidal. The occurrence of the unusual seven-coordinate structures for these iron(III) complexes can be traced in part to conformational constraints, for molecular models show that the pentadentate ligand is not well suited to coordinate as part of an octahedron. Evidence suggests that the corresponding iron(III) complex of the sexadentate ligand (**LII**) probably has a related oxygen-bridged structure.[82]

A complex of **LI**, $Mn(MAC)Cl_2$, has been prepared from 2,6-diacetylpyridine and 3,3′-diaminodipropylamine by an *in situ* condensation in the presence of $MnCl_2 \cdot 4H_2O$.[88] The reaction was carried out in methanol, and after crystallization from water the product was isolated as orange crystals. Since Mn(II) and Fe(III) are isoelectronic, it seems likely that this manganese complex has a similar structure to that of the monomeric iron(III) complexes discussed above.

Probably the most obvious and most often attempted macrocyclization by way of the Schiff base route involves the condensation of α-diketones with 1,2- or 1,3-diamines (Scheme V). This is true in part because the often

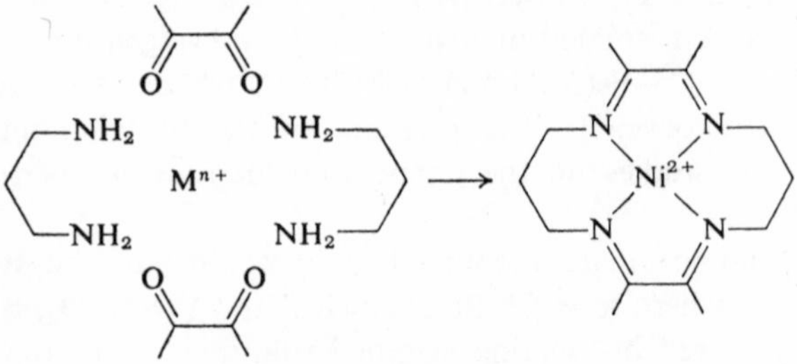

Scheme V

special properties of ligands containing an α-diimine linkage have long been recognized.[89] A range of bidentate and tetradentate ligands containing the *in situ* generated α-diimine group have been known for some time, including the linear quadridentate ligands **LIII**[90] and **LIV**,[91] both of which may be considered precursors of macrocycles (see below).

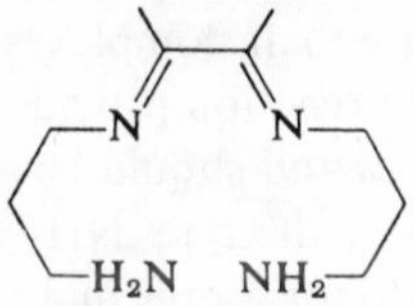

H_2N NH_2

LIII

S S

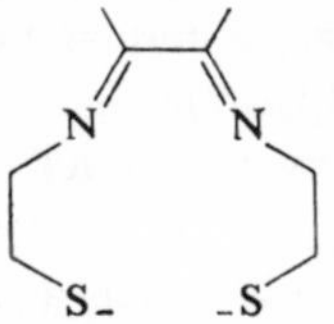

LIV

Baldwin and Rose[92] have reported the preparation of the nickel complex of **LV** (TIM) as its tetrachlorozincate salt. The compound is obtained in

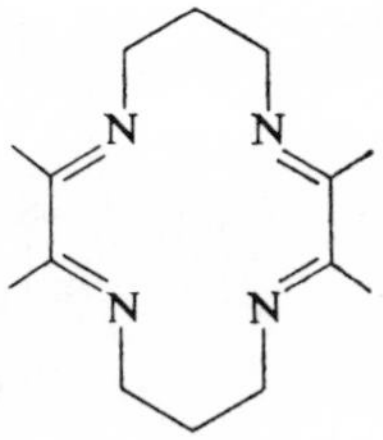

LV

about 20 percent yield by treating 2 moles of 1,3-diaminopropane monohydrochloride in methanol with 2 moles of biacetyl followed by 1 mole of nickel acetate. The solution is allowed to stir, and the complex precipitates after the addition of zinc chloride. Hexafluorophosphate and thiocyanate derivatives were prepared from this compound. The hexafluorophosphate complex contains low-spin, square-planar nickel, and the thiocyanate compound is paramagnetic and six-coordinate. The use of the monohydrochloride amine salt for the cyclization reaction[92] is undoubtedly significant, but its role in the reaction is uncertain. It is possible that it prevents the initial formation of linear polymers by blocking the quarternary nitrogen atom from participating in Schiff base formation; nevertheless other factors, for example, the effect on pH dependent equilibria, could well be important.

The general reaction has been extended to the preparation of the low-spin complexes $[Co(TIM)X]BPh_4$, where X = Cl, Br, I.[93] For these complexes the ligand hydrohalide was first prepared *in situ* by the addition of the appropriate hydrogen halide to a methanol solution of the amine (1:1 mole ratios). After some time, biacetyl was added to the stirred solution, followed by 1 molar equivalent of cobalt acetate and then sodium tetraphenylborate;

whereupon the products crystallized. If these preparations are not performed under an inert atmosphere, cobalt(III) complexes are obtained. A wide variety of cobalt(III) compounds of the type $[Co(TIM)X_2]Y$ have thus been prepared by the aerial oxidation of the corresponding cobalt(II) solutions in the presence of excess halide ion.[94]

By using *N*,*N*′-bis(2-aminopropyl)-1,2-ethanediamine in the general procedure, a range of similar nickel and cobalt complexes of **LVI** has been prepared.[95] It is apparent that this class of reaction provides a facile pathway to a number of new macrocyclic complexes and should be applicable to ring-closing reactions involving other amines of the type $NH_2(CH_2)_nX(CH_2)_mX(CH_2)_nNH_2$, where X equals a range of donor atoms and n and m are either 2 or 3. Preliminary results show that for the case in which X is sulfur n equals 3 and m equals 2, the general reaction does take place to give a nickel complex of **LVII**.[82]

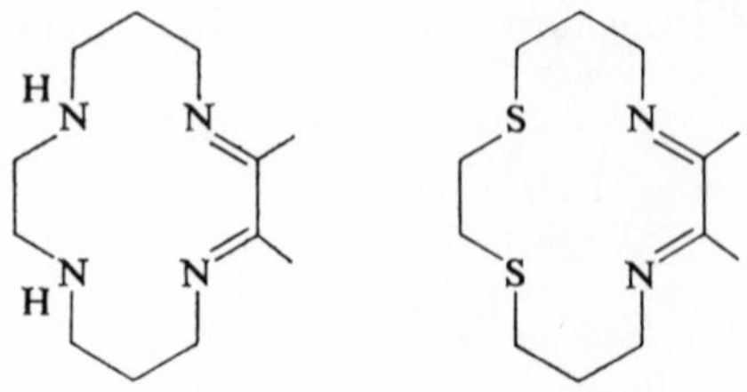

LVI **LVII**

The use of β-dicarbonyl compounds to add new chelate rings to complexes containing cisoid primary amine functions represents a second category of Schiff base condensation yielding macrocycles of distinctive properties. In most cases ionization occurs at the hydrogen of the amine of the resulting unsaturated amino-imine. This is represented by Scheme VI. The product can be represented by either of the alternative structures.

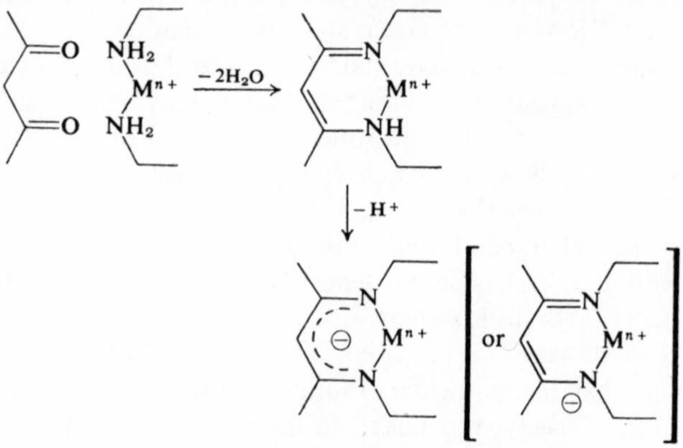

Scheme VI

Studies in this area were pioneered by Jager,[96–99] who used this reaction to prepare series of complexes containing ligands of doubly negative charge. These complexes are of the general type **LVIII**, where X^1 and X^2 are —$(CH_2)_2$— or *o*-phenylene, X^1 is —$(CH_2)_2$— and X^2 is —$(CH_2)_3$—, or X^1 and X^2 are both —$(CH_2)_3$— bridges. The syntheses of these complexes were achieved by the direct reaction of the appropriate diamine with various

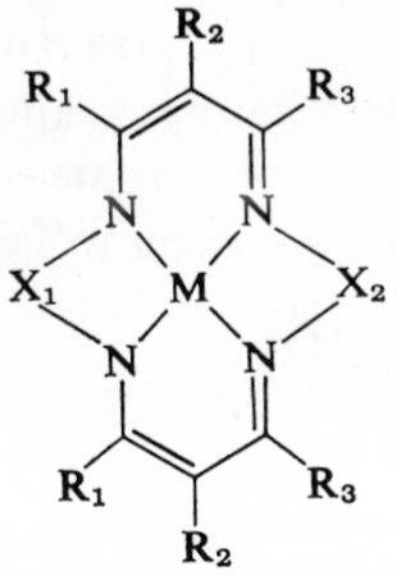

M = Ni or Cu

LVIII

β-ketoiminato and β-diketonato nickel or copper(II) chelates. All have square-planar geometries. The formation of nickel complexes of the type **LVIII**, where $X^1 = X^2 =$ *o*-phenylenediamine, from the β-ketomines **LIX** and **LX** provides an example of this class of reaction.[99]

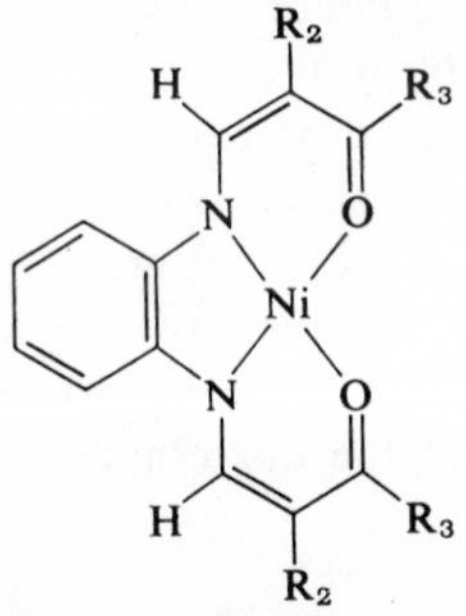

LIX 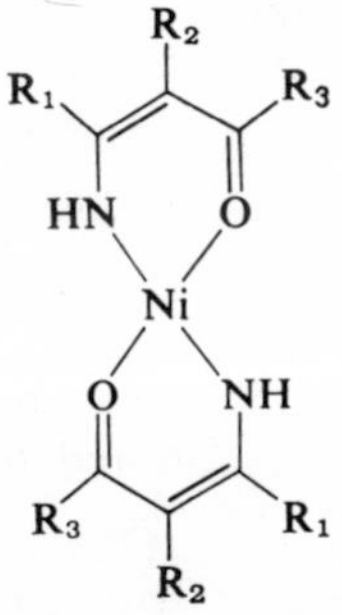 **LX**

R_1	R_2	R_3
H	H	CH_3
H	H	C_6H_5
H	$COCH_3$	CH_3
H	COC_6H_5	CH_3
H	$COOC_2H_5$	C_6H_5
H	H, $COOC_2H_5$	C_6H_5
CH_3	H	CH_3

The cyclizations are performed in the absence of solvent. Powdered **LIX** or **LX** is treated with an excess of molten (about 250°) *o*-phenylenediamine under argon for several hours. The unreacted *o*-phenylenediamine is then distilled off, and the remaining residue extracted with methanol from which the crude product can be isolated. The product can usually be purified by recrystallization from a solvent, such as xylene, or by chromatography of a benzene solution using an alumina column.

It seems that when R_2 contains a carbonyl group, for example, acetyl or benzoyl, electronic effects favor the reaction by making the carbon adjacent to the coordinated oxygen more susceptible to nucleophillic attack by the amine (see **LXI**). The preparative reactions seem to be of a template nature,

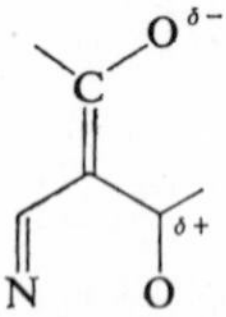

LXI

since attempts to synthesize these macrocycles directly in the absence of metal ions or to displace the macrocycles from the metal complexes have all been unsuccessful.

Many of these complexes containing 14-membered unsaturated rings are chemically very stable—for example, to concentrated acid—and are comparable in this regard with the more stable derivatives of the phthalocyanines or porphyrins. It is noted that **LXII** can be prepared by using either bis(2-aminopenten-(2)-on(4)-ato)nickel(II) or bis(acetylacetato)nickel(II) in the general procedure.[99] This tetramethyl compound is much less stable to acid than the other complexes in the series. It dissolves in dilute acid to give a deep violet solution but then decomposes by cleavage of the azomethine linkages to yield the free nickel ion and the salt of the quaternary cation **LXIII**. The

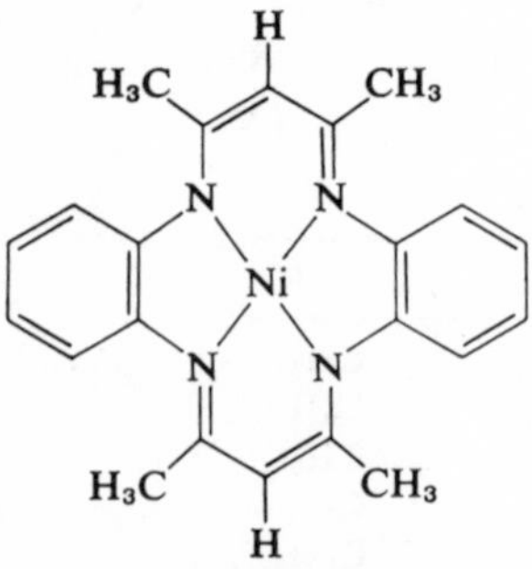

LXII 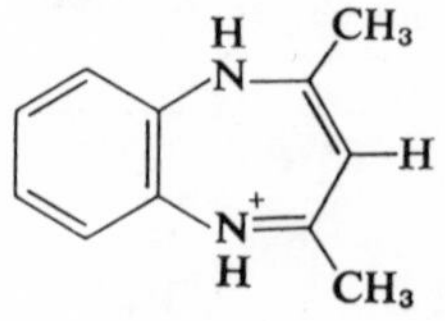 **LXIII**

lower stability of this complex has been attributed to additional steric interactions resulting from the presence of two additional methyl groups, namely, R_1, which interact with the adjacent phenyl rings.

The closely related unsubstituted macrocycle **LXIV** can be prepared directly, in the absence of metal ions, by a different route.[100] Propargylaldehyde and an equimolar amount of *o*-phenylenediamine react exothermically in alcohol or DMF to yield **LXIV**, which, after recrystallization from DMF, is obtained as red-violet flakes (Scheme VII). If the reactants in methanol are mixed at $-20°$, in a 1:1 mole ratio, and the temperature of the solution is allowed to rise slowly to room temperature, the intermediate dialdehyde (**LXV**) can be isolated as almost colorless crystals. Reaction of this product with more *o*-phenylenediamine in DMF at 60 to 80° leads to conversion to **LXIV**.

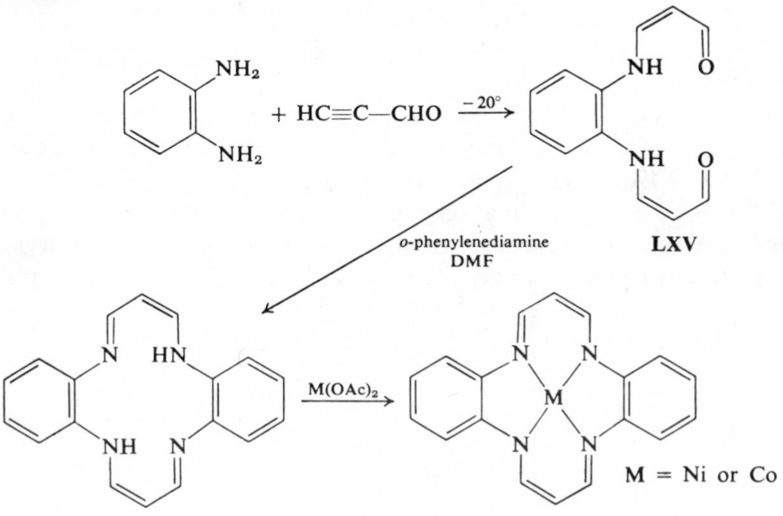

NH_2 NH_2 + HC≡C—CHO $\xrightarrow{-20°}$ NH O NH O **LXV**

o-phenylenediamine DMF

N HN NH N **LXIV** $\xrightarrow{M(OAc)_2}$ N N M N N **LXVI** M = Ni or Co

Scheme VII

Heating of **LXIV** in DMF with nickel, cobalt, or copper acetates yields the respective neutral metal complexes (**LXVI**). Alternatively the macrocycle need not be isolated from its reaction solution before the addition of the metal acetate. Reaction of intermediate **LXV** with the metal salt gives the corresponding stable dialdehyde complex (neutral), and rather surprisingly this does not react in solution with *o*-phenylenediamine to yield a macrocyclic complex of the type **LXVI**. A small yield of the macrocyclic complex, however, can be obtained by reaction in the melt.

No doubt as a partial consequence of the absence of steric hindrances, these complexes exhibit great stability. They can be sublimed unchanged *in vacuo* at 300° and are stable to concentrated sulfuric acid, from which they can be precipitated by the addition of water to the acid solution.

Obviously the procedures described immediately above do not usually involve the simple direct condensation of the β-dicarbonyl compound with the amine groups. Examples of such processes have recently been reported. Olson and Vasilevskis[101] have shown that when a mixture of aqueous nickel acetate, *N*-methylethylenediamine (2 moles), acetic acid (2 moles), and acetylacetone (2 moles) is refluxed overnight and then made basic, a nickel complex of the negatively charged ligand (**LXVII**) can be isolated from the reaction solution.

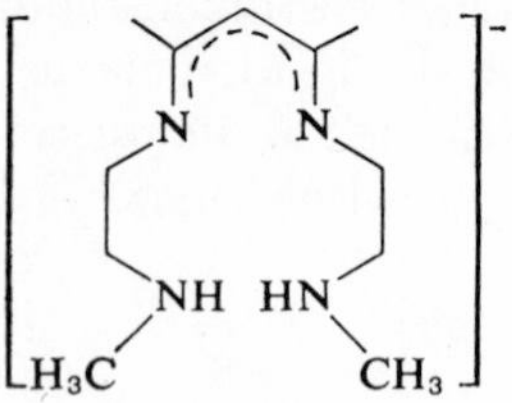

LXVII

Cummings and Sievers[102] have demonstrated that by a similar procedure using 1 mole of β-diketone and 1 mole of triethylenetetramine, complexes of the type **LXVIII** can be prepared. Previous studies indicate that Schiff base condensations usually occur at only one of the oxygens of a β-diketone (enol form). The ligands are particularly interesting, since they carry a uninegative charge that is delocalized over the six-membered chelate ring. Attempts to

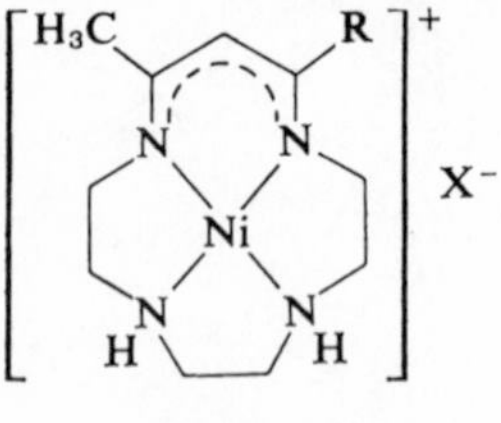

$R = CF_3$ or CH_3
$X = Br^-, I^-, NCS^-, NO_3^-, BF_4^-, PF_6^-$

LXVIII

obtain these quadridentate macrocycles free of their metal ions were unsuccessful. These workers have also prepared the potentially sexadentate ligand (**LXIX**) by condensation of two equivalents of trifluoroacetylacetone with one equivalent of triethylenetetramine in alcohol. This ligand has been

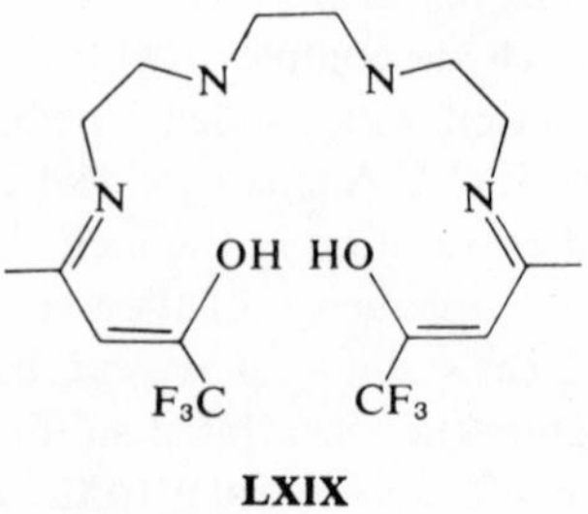

LXIX

used in alternative syntheses of complexes of the type **LXVIII**; for example, refluxing the octahedral nickel complex of this ligand for 6 hr results in a color change from yellow-brown to red. Treatment of the red solution with excess sodium iodide and adjusting to pH 10 with sodium hydroxide leads to isolation of **LXVIII**, where $R = CF_3$ and $X = I$.

Honeyburne and Webb[103] showed that a new series of conjugated 1,3-bidentate Schiff bases can be prepared by heating 1,1′,3,3′-tetramethoxypropane (1 mole) under reflux with an aromatic amine (2 moles) in aqueous ethanol containing one equivalent of hydrochloric acid. The product is the corresponding 3-iminopropeneamine monohydrochloride (**LXX**). The reaction appararently proceeds by the initial acid hydrolysis of the methoxy

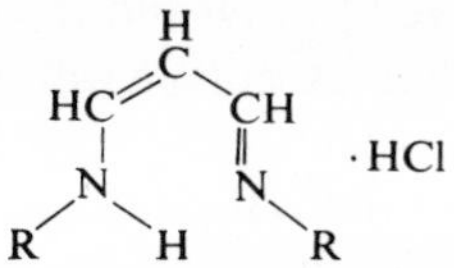

H
C
HC CH
N N ·HCl
R H R

LXX

compound to produce a carbonyl species that is then involved in Schiff base formation.

The reaction has been adapted to the synthesis of a copper chloride complex of the uncharged ligand (**LXIV**).[104] 1,1,3,3-Tetramethyoxypropane and sufficient HCl to cause its hydrolysis were heated in ethanol. The $CuCl_2 \cdot 2H_2O$ was then added, and the solution was refluxed for 8 hr. The insoluble yellow-green product was isolated and then suspended in an ethanol solution of *o*-phenylenediamine. The suspension was heated at the reflux for 4 hr more. On cooling, the intensely colored solution yielded the yellow-brown copper chloride salt of **LXIV**, which contained 1 mole of HCl. The HCl can be removed by heating the complex in alcohol, with potassium *t*-butoxide or tetrabutyl ammonium hydroxide. It has not been possible to convert this product into a neutral complex of type **LXVI** by deprotonation with base; it is claimed that the complex is stable to 10 *N* aqueous sodium hydroxide at 60° for 15 hr.

Green and Tasker[105] have prepared the dialdehyde (**LXXI**). Reaction of **LXXI** (1 mole) with the appropriate diamine (1 mole) in the presence of the hydrated metal acetate, all in methanol, leads to the formation of complexes of the type **LXXII**. In the absence of metal ions, high-melting-point polymers are obtained. The basicity of the reaction solution resulting from the use of acetate salts is sufficient to ionize both secondary nitrogen protons of the ligand and yield neutral complexes. The dianion ligand is, of course, a close analog of those ligands obtained from condensations involving β-diketones (above).

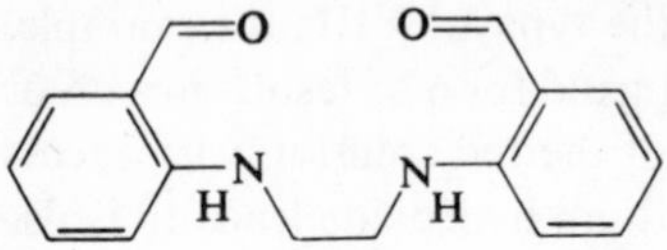

LXXI

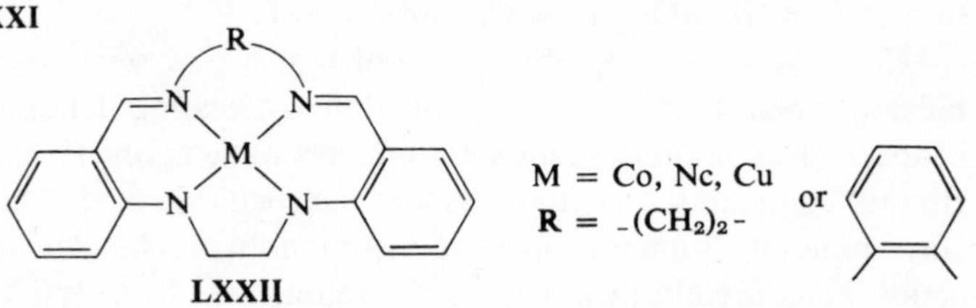

M = Co, Nc, Cu
R = $-(CH_2)_2-$ or

LXXII

The complex **LXXII**, where M = Ni and R = $—(CH_2)_2—$, has also been prepared by reaction of 1,2-dibromoethane with a chloroform solution of the nickel complex of the open-chain quadridentate formed from *o*-aminobenzenealdehyde (2 moles) and ethylenediamine (1 mole).[106] Such *N* alkylation reactions of two cis-coordinated —NH groups provide another kind of template reaction by which ring closure may be effected.

The synthesis of **LXXIII** has provided a route to a number of new macrocyclic complexes.[107] The *in situ* condensation of this dialdehyde with rhe appropriate diamine in an alcohol solution of nickel perchlorate leads to isolation of square-planar nickel complexes of the ligand (**LXXIV**).[107] In the

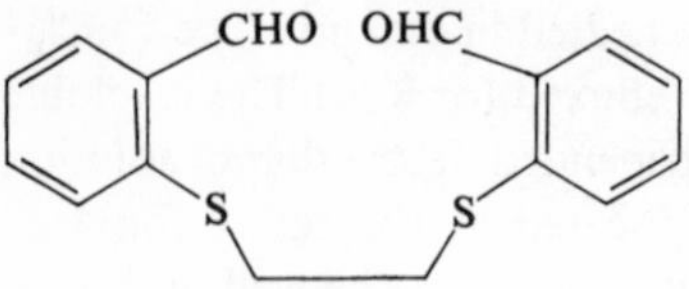

LXXIII

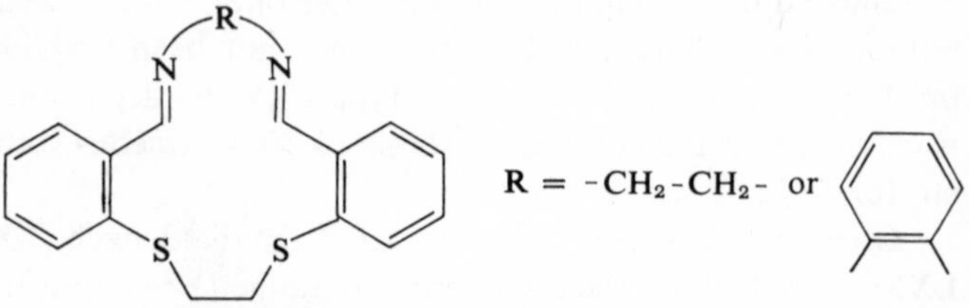

R = $-CH_2-CH_2-$ or

LXXIV

presence of coordinating anions the condensation produces a series of paramagnetic complexes that seem to be five-coordinate. It should also be possible to prepare these and similar compounds by reaction of suitable difunctional alkylating agents with the recently reported dithiolo complexes of the type **LXXV**.[108]

Condensation of **LXXIII** and **LXXVI** in alcohol in the presence of the appropriate metal salts leads to isolation of complexes of the type **LXXVII.**[107]

The nickel complexes are both high spin whereas the cobalt(II) complex is low spin with a moment of 1.9 BM. To aid in the characterization of these

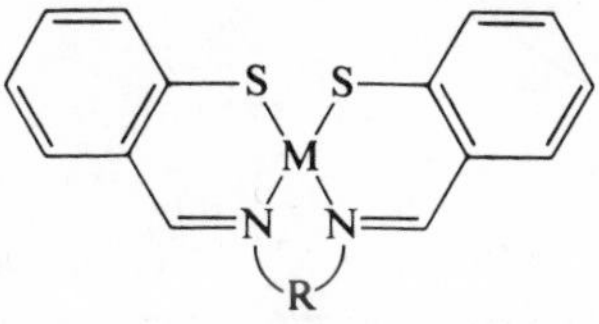

LXXV

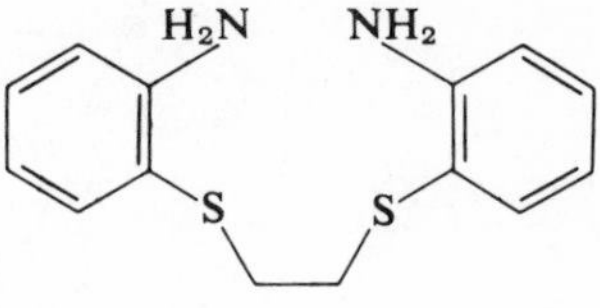

LXXVI

complexes, the linear sexadentate ligand (**LXXVIII** LS) was prepared by a direct condensation of the dialdehyde (**LXXIII**) in alcohol, with 2 moles of

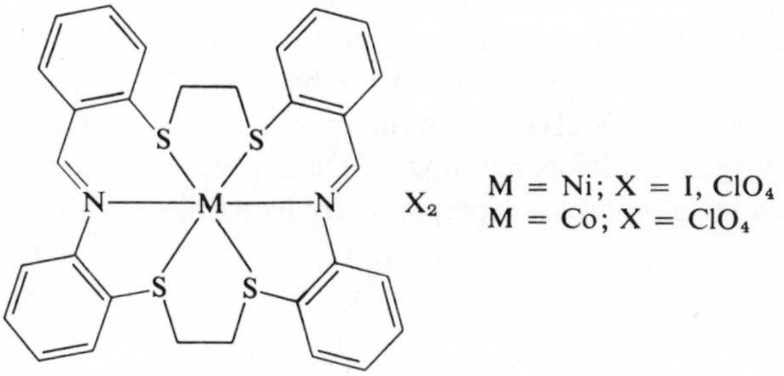

X_2 M = Ni; X = I, ClO_4
M = Co; X = ClO_4

LXXVII

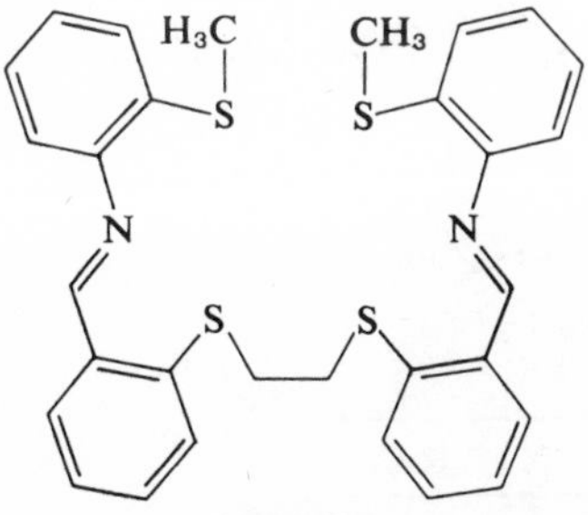

LXXVIII

o-methylthioaniline. The yellow crystalline solid can then be used to prepare the complex $Ni(LS)(ClO_4)_2$, which has very similar properties to the related nickel complex **LXXVII**, where X = ClO_4. The visible spectra of both nickel complexes indicate that they are essentially octahedral. Because of the planarity of each S—N—S unit, both these sexadentate ligands can coordinate only around an octahedron in the manner shown in **LXXIX** (or its optical isomer).

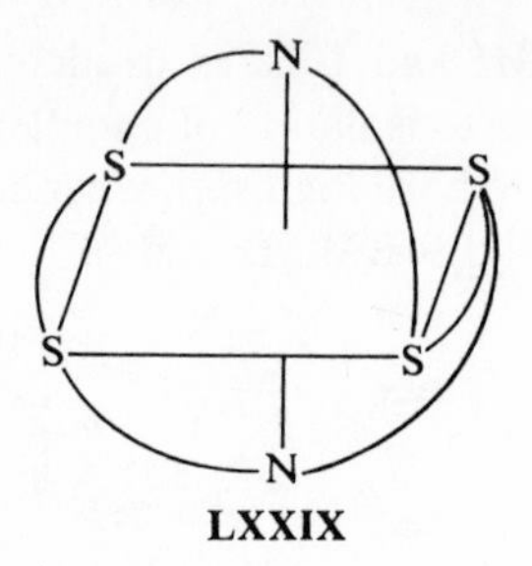

LXXIX

Treatment of the nickel complex (**LXXVII**, where X = ClO_4) with hot DMF leads to ligand displacement, and the macrocycle separates from the solution as yellow crystals. The free ligand was characterized by infrared, mass spectrometry, and elemental analyses and has been used to reprepare the nickel complex by reaction in alcohol with nickel perchlorate.

Subsequently, related compounds have been prepared by similar procedures, using methanol as a solvent, from the dialdehyde (**LXXI**), which contains secondary nitrogen atoms in the backbone in contrast to the thioether sulfur atoms of **LXXIII**.[109] A number of diamines related to **LXXVI** but containing oxygen or nitrogen donors in the backbone were also used for the reaction. The metal complexes prepared are listed below (**LXXX**).

NH HN
N—M—N $(ClO_4)_2 \cdot nCH_3OH$
X X
$(CH_2)_n$

LXXX

	M	X	R	*n*
(a)	Ni	S	2	1
(b)	Ni	O	2	1
(c)	Co	S	2	1
(d)	Fe	S	2	1
(e)	Fe	NH	2	1
(f)	Fe	NH	3	0
(g)	Zn	S	2	1
(h)	Zn	NH	2	1
(i)	Zn	NH	3	0

A crystal-structure determination on **LXXXd**[109] confirms that its structure is similar to that previously predicted[106] for the complexes (**LXXVII**), namely, that the metal ion is essentially octahedral with the ligand adopting the geometrical configuration **LXXIX**.

Treatment of acetone solutions of the complexes with pyridine leads to ligand-exchange reactions, and the metal-free macrocycles can be isolated directly. Attempts[106,109] to prepare these macrocycles by Schiff base condensation, in the absence of a metal ion, were unsuccessful, thus demonstrating the template nature of the reactions. The successful preparation and isolation of the free macrocycles can thus be traced to both the template effect associated with the particular metal used and the lability of the resultant metal macrocyclic complexes. Consideration of both these factors is obviously important in the selection of a suitable metal ion when designing similar reactions.

The self-condensation of *o*-aminobenzaldehyde has been shown by McGeachin to yield a number of polycyclic products, including **LXXXIa** and **LXXXIb**.[110] In the presence of metal ions the condensation yields metal complexes of either the cyclic trimer (**LXXXII**) or tetramer (**LXXXIII**).[111–113] The metal-free trimer (**LXXXIa**) rearranges in the presence of excess nickel nitrate, in boiling alcohol, to yield the complex $Ni(TRI)(NO_3)_2 \cdot H_2O$.[114] An

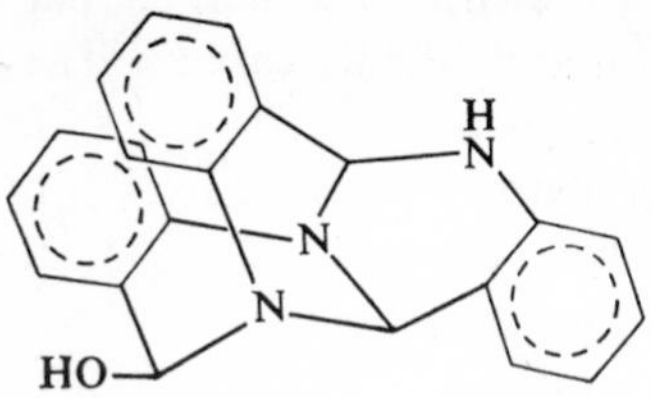

LXXXIa

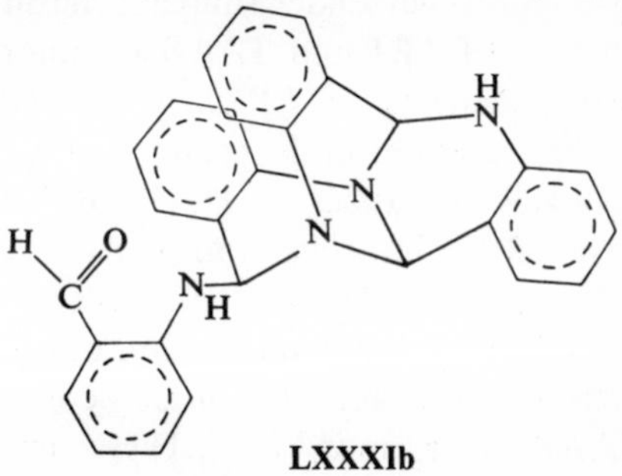

LXXXIb

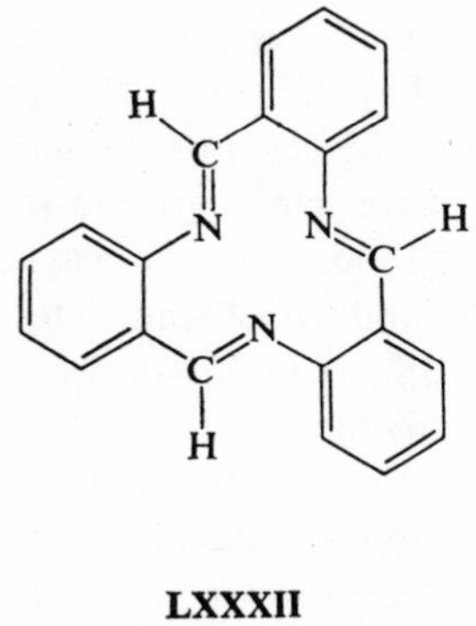

LXXXII

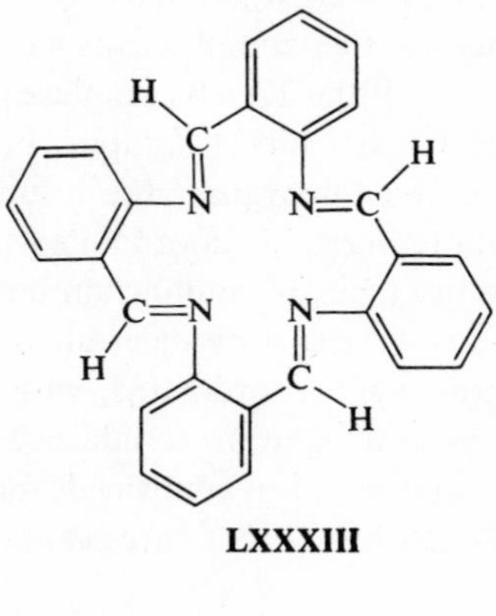

LXXXIII

x-ray study confirmed the earlier conclusion that the similar complex containing three water molecules is pseudooctahedral with the TRI ligand coordinated on one face of the octahedron as in **LXXXIV**.[115] Since the complex

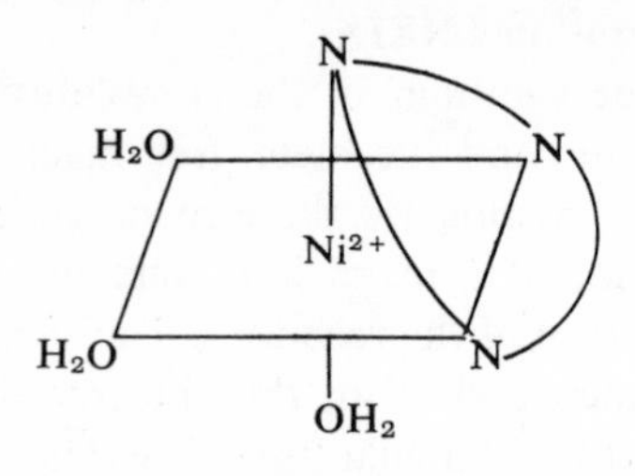

LXXXIV

has no plane of symmetry, center of inversion, or alternating rotation-reflection axis, it exists as optically active isomers. Resolution of this complex was achieved using a column packed with either potato starch or microcrystalline cellulose.[116] If the trimer (**LXXXIa**) and nickel nitrate are permitted to react similarly in a 2:1 mole ratio, in methanol, the bisligand complex $Ni(TRI)_2(NO_3)_2 \cdot H_2O$ is obtained.[114,117] In the latter complex each tridentate ligand occupies an opposite face of the coordination octahedron, and so a novel sandwich structure results. The trisanhydrotetramer (**LXXXIb**) also rearranges under similar conditions in the presence of nickel ion, but a mixture of TRI and TAAB complexes is produced in this case.[117] These rearrangements of **LXXXIa** and **LXXXIb** are metal-ion–dependent, for similar reactions in the presence of copper(II) ion yield only complexes of the quadridentate macrocycle TAAB. This result has been rationalized on the basis of the stereochemical preferences of the copper(II) ion. Nevertheless in both cases template mechanisms seem to be operating.

When the self-condensation of *o*-aminobenzaldehyde is performed in the presence of nickel salts, in an alcohol solution, complexes of both TRI and TAAB are formed.[112,113] These products can be separated, as their perchlorate salts, by fractional crystallization from aqueous sodium perchlorate. A range of other complex salts has been obtained by metathesis. The magnetic properties of the TAAB complexes are interesting.[118] With ClO_4^-, BF_4^-, and $B(C_6H_5)_4^-$ the nickel complex is low-spin, and in the cases of the iodide, nitrate, and thiocyanate the resultant complexes have triplet ground states and are tetragonally coordinated. For the anions Cl^- and Br^- the magnetic properties indicate equilibrium between the singlet and triplet spin states.

A crystal-structure determination on $Ni(TAAB)I_2 \cdot H_2O$ has shown it to be tetragonally coordinated, with the TAAB occupying a meridianal plane and the water and an iodide occupying the axial positions.[119] A structure determination also confirmed that the coordination sphere of Ni(II) in $Ni(TAAB)(BF_4)_2$ is square-planar. In both cases the benzene rings of the ligand lie alternatively above and below the plane containing the four nitrogen atoms and the metal atom.[119]

The condensation of *o*-aminobenzaldehyde in the presence of cobalt(II) ions has yielded a variety of products. Refluxing an alcohol solution, containing a few drops of concentrated nitric acid, of freshly prepared *o*-aminobenzaldehyde (4 moles) and cobalt nitrate (1 mole) for several hours gives a gray-black precipitate.[120] This product is digested with hot methanol until only a maroon solid remains, and this is isolated and dissolved in a large amount of methanol containing nitric acid, to ensure complete oxidation to the Co^{3+} state. From this solution it is possible to obtain two series of compounds of the type $Co(TRI)_2X_3 \cdot nH_2O$, where X is NO_3, Cl, Br, I, NCS, BF_4, PF_6, and ClO_4 and n is 1 to 5. The two series represent diastereoisomers, and their existence derives from the disymmetry of chelated TRI. A meso form occurs when the two ligand molecules have opposite chiralities; when the chiralities are the same, a racemic isomer results. The racemic isomer *dl*-$Co(TRI)_2{}^{3+}$ has been resolved into its optical antipodes,[120] and Wing and Eiss[121] have used x-ray diffraction to solve the structure and determine the absolute configuration of *d*-$Co(TRI)_2I_3 \cdot 3H_2O$.

Apparently the analogous nickel complex is formed in a stereospecific process, for only one form is known and it has been assigned the meso configuration.[117]

From the reaction, under nitrogen, of *o*-aminobenzaldehyde and cobalt(II) salts in alcohol, with no nitric acid added, it has been possible to isolate complexes of the type $Co(TAAB)X_2$.[93] Similar reactions with copper(II)[112] and iron(II)[122] seem to be product-specific, yielding only TAAB complexes.

Anhydrous zinc chloride and *o*-aminobenzaldehyde react in ether to give a 50 percent yield of a yellow complex that has been formulated as $[ZnTAAB][ZnCl_4]$.[123] This product is an excellent synthetic intermediate for preparing other TAAB complexes by metal exchange;[124] for example, treatment of a slurry of the zinc complex in ethanol, under nitrogen, with cobalt chloride for 1 hr leads to the formation of green $CoTAABCl_2$ in high yield.[93,124] By related procedures, nickel, palladium, and iron complexes have been prepared from this zinc compound.[124]

All the TAAB complexes show remarkable stability to concentrated acids but are reactive toward nucleophiles. $Ni(TAAB)X_2$ reacts with ethoxide ion to form the neutral complex **LXXXV**, which is converted back to the parent compound by acid.[125] Examples of the addition of water[126] or methanol[127] across a coordinated imine group have been reported previously.

The bisalkoxide ions $CH_3N(CH_2CH_2O^-)_2$ and $S(CH_2CH_2O^-)_2$ readily react with two trans-azomethine positions of TAAB in the ions $Ni(TAAB)^{2+}$ and $Cu(TAAB)^{2+}$.[128] The product in each case is a basket complex (**LXXXVI**). Since $[O(CH_2)_5O]^{2-}$ does not form a similar basket complex, it is assumed

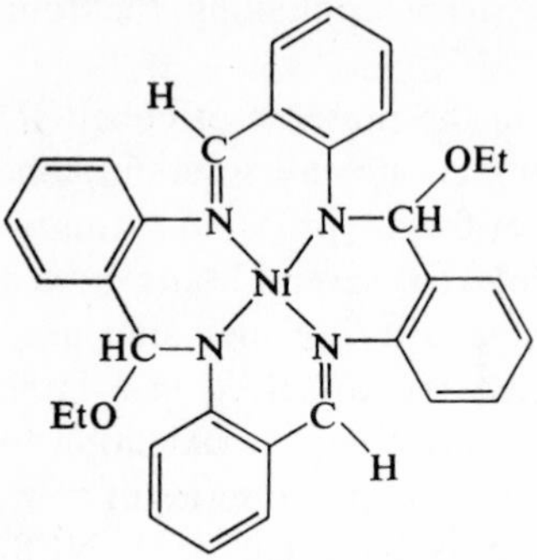

LXXXV

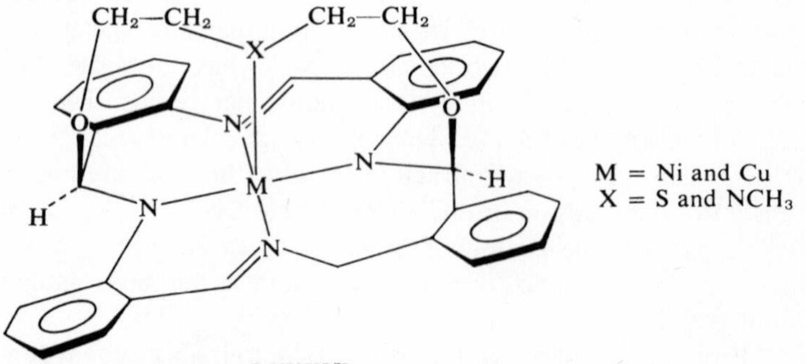

M = Ni and Cu
X = S and NCH_3

LXXXVI

that the reaction proceeds by a template mechanism involving coordination of the central donor atom of the dialkoxide to an axial position of the metal ion. It follows that these three-dimensional macrocyclic ligands are produced from *o*-aminobenzaldehyde by two major processes, each of which involves a template mechanism.

One of the most widely studied groups of synthetic macrocycles was discovered by Curtis,[75] who showed that when trisethylenediaminenickel(II) perchlorate is allowed to stand for several days in anhydrous acetone solution, the color slowly changes from blue-violet to yellow-brown. The major products of the reaction are the chemically stable positional isomers 6,7,7,12,12,14-hexamethyl-1,4,8,11-tetraazacyclotetradeca-1(14), 4-dienenickel (II) (**LXXXVII**, $Ni(1,4\text{-}CT)^{2+}$) and 5,5,7,12,12,14-hexamethyl-1,4,8,11-tetra-azacyclotetradeca-1(14),7-dienenickel(II) (**XLVII**, $Ni(1,7\text{-}CT)^{2+}$).[129,130] Since the perchlorate complex of the 1,4-CT isomer is the least soluble, pure samples of both isomers can be obtained by fractional crystallization using acetone, alcohol, water, or their mixtures as solvents.[129] It is possible to monitor the process with infrared spectra.

Substitution of trisethylenediaminenickel(II) fluoroborate for the perchlorate salt in the general reaction gives the 1,7-CT isomer almost ex-

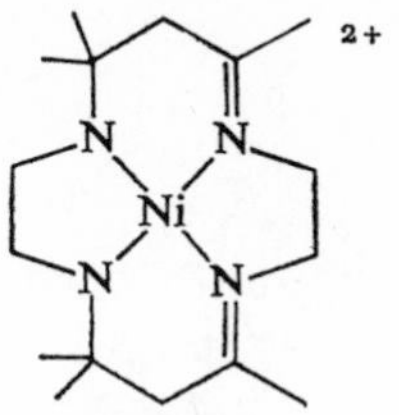

LXXXVII

clusively, with the formation of less than 10 percent of the 1,4-CT product.[131] Owing to the restricted inversion about the asymmetric secondary amines, each of the coordinated positional isomers can exist in both meso and racemic forms. Two such forms have been isolated only in the case of the 1,7 isomer, and they can be interconverted in basic solution.[129,131,132]

Metathetical reactions on both the square-planar perchlorate and the fluoroborate salts have produced a range of tetragonally distorted nickel compounds exhibiting triplet ground states.[133] These compounds have provided excellent models for the study of the electronic spectra of tetragonally distorted complexes containing constant in-plane ligand fields but varying axial fields.[4]

Curtis has discussed the preparation of complexes of this general class in a comprehensive review.[5] The cyclization process entails the addition of two new chelate rings by way of the formation of two 3-carbon bridges. Each bridge arises from the condensation of two molecules of acetone with two cis primary amine groups. The mechanism of the reaction is far from understood. It has been suggested that Schiff base condensation may occur to yield a species, such as **LXXXVIII**, together with the liberation of ethylenediamine.[5] Further reaction with acetone might then occur to produce a complex containing the coordinated β-aminoketone (**LXXXIX**), for this fragment has been found in such solutions. Reaction of the coordinated β-aminoketone with ethylenediamine and acetone has been postulated to

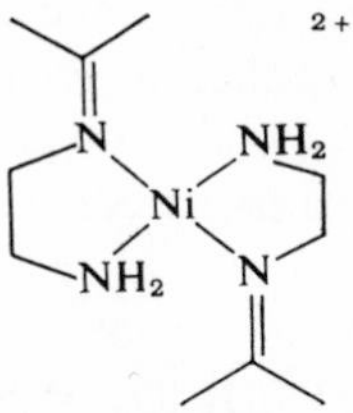

LXXXVIII 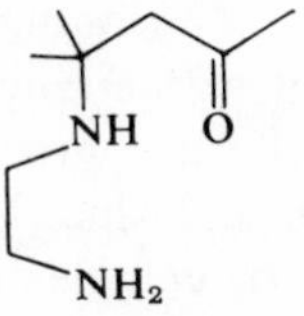 LXXXIX

occur to yield the complex **XC**, which has been isolated from the reaction solution during the course of the reaction. A further aldol reaction of **XC** with acetone followed by a Schiff base condensation would ultimately lead to cyclization.

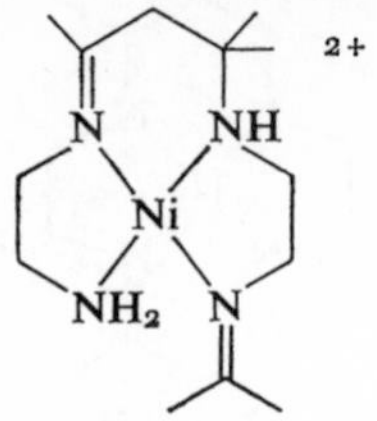

XC

Bis(ethylenediamine)copper(II) perchlorate also reacts with acetone, and a compound of the structure **XCI** has been isolated. It readily converts to the

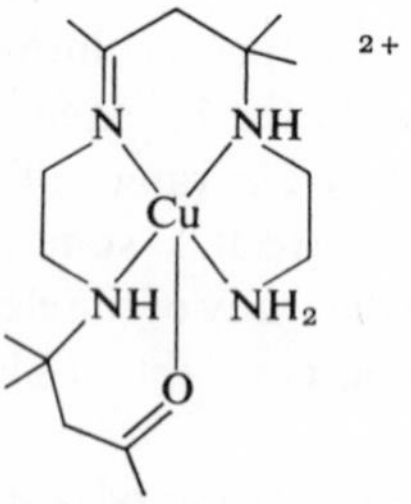

XCI

1,7-CT complex under basic conditions.[5] If excess ethylenediamine is added to the reaction solution above, a mixture of the 1,4-CT and 1,7-CT complexes is obtained. The pure isomers can be isolated by fractional crystallization.[134] The copper complexes have also been obtained by a number of modifications of the basic method above.[5,134–136]

It was mentioned in the last section that (1,7-CT)·$2HClO_4$ can be prepared in the absence of metal ions.[74] Sadasivan and Endicott[136] also isolated this dihydroperchlorate salt from the reaction mixture produced by trisethylenediamineiron(II) perchlorate and acetone. It is difficult to say whether a kinetic template effect is operating during this condensation. The (1,7-CT)·$2HClO_4$ reacts directly with a range of metal carbonates[136] or acetates[137] in methanol or methanol-water mixtures to form metal complexes. Cobalt(II), nickel(II), copper(II), and zinc(II) derivatives have been obtained in this manner. The cobalt(III) complex can be obtained from the corresponding reaction with $Na_3Co(CO_3)_3$ or by oxidation of a solution of the appropriate cobalt(II) complex. Reaction of (1,7-CT)·$2HClO_4$ with H_2PdCl_4 has been reported to yield [Pd(1,7-CT)]$(ClO_4)_2$.[5]

The free ligand 1,4-CT can be obtained in nonaqueous solution by removal of the metal ion with potassium cyanide from a solution of its nickel complex, the nickel being precipitated as $K_2Ni(CN)_4$.[138] From the ligand solution the complexes [Cu(1,4-CT)]$(ClO_4)_2$, [Zn(1,4-CT)][$ZnCl_4$)]·CH_3OH, and [Fe(1,4-CT)][$FeCl_4$]·H_2O have been prepared, as well as a series of cobalt(II) compounds.[138]

Treatment of the ligand solution above with dilute perchloric acid leads to isolation of the diperchlorate salt of the β-aminoketone (**XCII**).[139] Copper and nickel complexes of the structures **XCIII** and **XCIV** can be obtained in

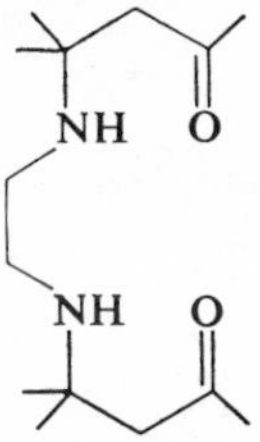

XCII 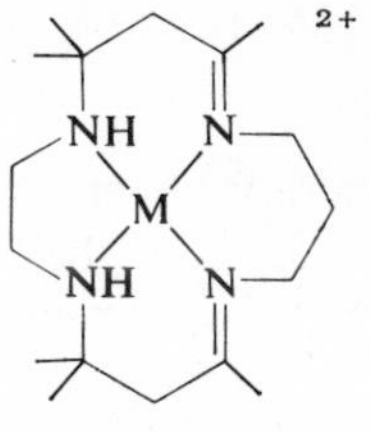 XCIII 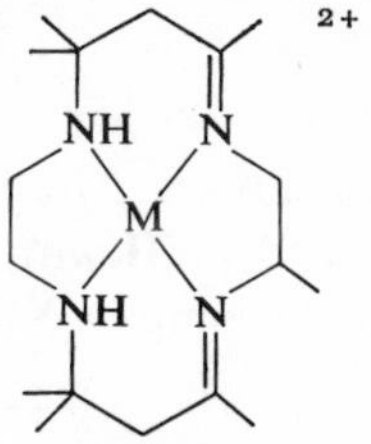 XCIV

20 to 50 percent yield by condensation of the appropriate diamine with **XCII** in the presence of a methanol or aqueous solution of the metal ion.

Condensation of several nickel and copper diamine complexes with a range of aliphatic aldehydes and ketones has led to the syntheses of related macrocycles containing varying peripheral substituents or varying macrocyclic ring

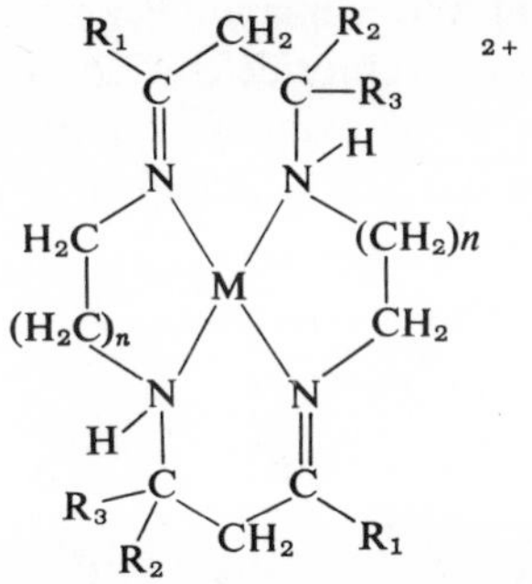

M = Ni or Cu
R_1, R_2 = Me or Et
R_3 = Me
R_4 = H or Me
n = 1 or 2

XCV

sizes (**XCV**).[134,140] It has been shown[141] that chelate rings of the kind formed in these cyclization reactions can be prepared by the reaction of the trisethylenediaminenickel ion with appropriate β-hydroxyketones or β-hydroxyaldehydes. In several instances this method gives more readily controlled reactions than the previously reported condensations using acetone or other simple carbonyl compounds.

The general reaction has been extended to cyclizations using linear quadridentate ligands as their nickel or copper complexes. These reactants contain cis amine donors. For example, triethylenetetramine nickel(II) and copper(II) salts react with acetone at 100° to yield the cyclic complexes having the structure **XCVI**. In the case of nickel other carbonyl compounds besides

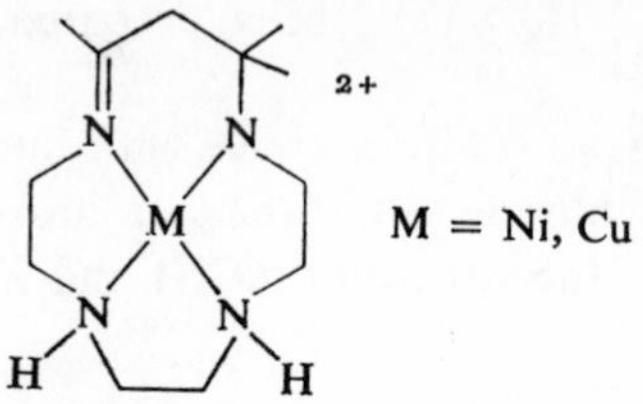

$M = Ni, Cu$

XCVI

acetone can be used to prepare a range of substituted complexes.[142] Similarly reaction of **XCVII** with acetone leads to the formation of the heterodonor macrocyclic complex **XCVIII.**[143]

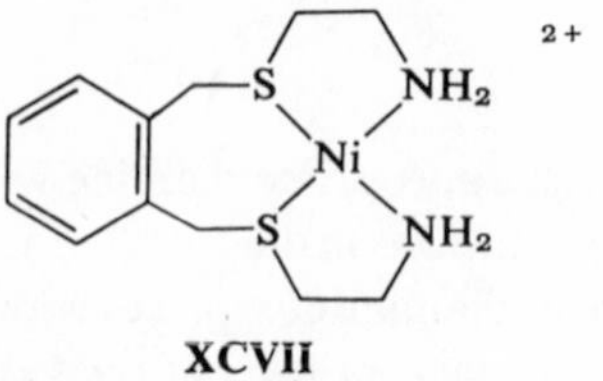

XCVII 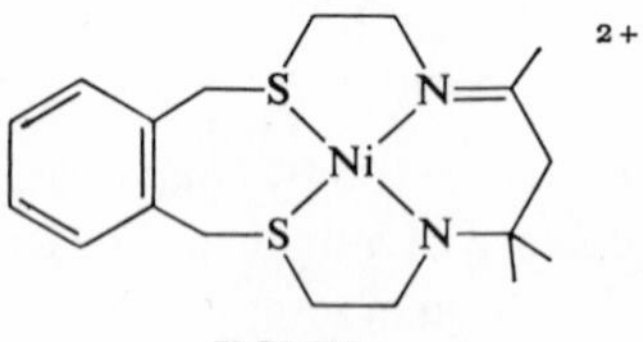 **XCVIII**

The macrocycles derived from dimethylgloxime, which were prepared by Schrauzer,[144] as well as by Thierig and Umland,[145] are related to the α-diimine macrocycle. Reaction of boron trifluoride etherate with bis(dimethylgloxime)nickel(II) leads to reaction at the oxime functions to yield the very stable complex **XCIX**, which can be purified by sublimation. Use of BCl_3 or

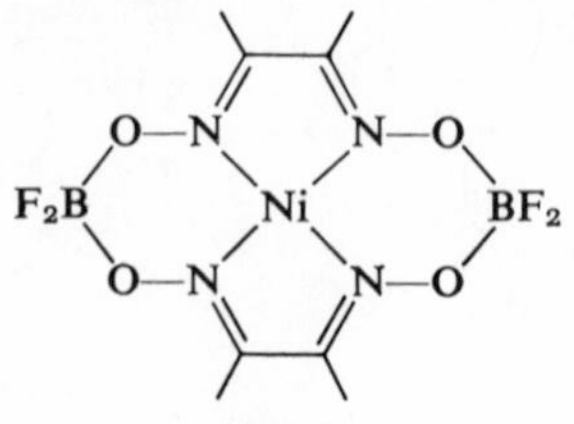

XCIX

$AlCl_3$ in this reaction leads to similar macrocycles, which are, however, considerably less stable than **XCIX**; for example, the chlorides are sensitive to atmospheric moisture, although **XCIX** is not decomposed by water. Compounds analogous to **XCIX** but with the fluorine atoms replaced by alkyl or phenyl groups have also been prepared.

The uninegative macrocycle **C** has been isolated by a similar reaction from the corresponding dioxime tetradentate.[146]

Boston and Rose[147] used a single reaction to prepare the first reported clathro-chelate complex. After some difficulty they were able to isolate highly hygroscopic potassium tris(dimethylglyoximato)cobalt(III) by bubbling

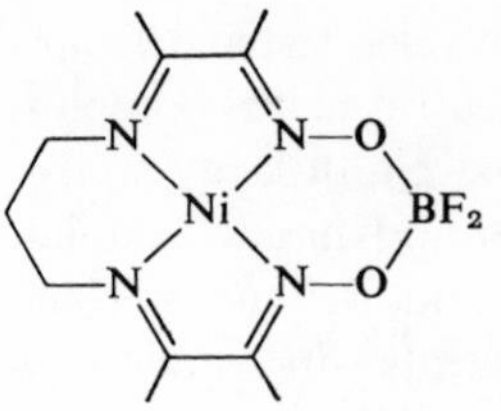

C

air through an aqueous solution of dimethylglyoxime, potassium hydroxide, and cobalt sulphate present in the respective mole ratios of 3:5:1. If this product is refluxed with boron trifluoride etherate in ether, a solution of **CI** as its BF_4^- salt is formed. Excess potassium tetrafluoroborate can be pre-

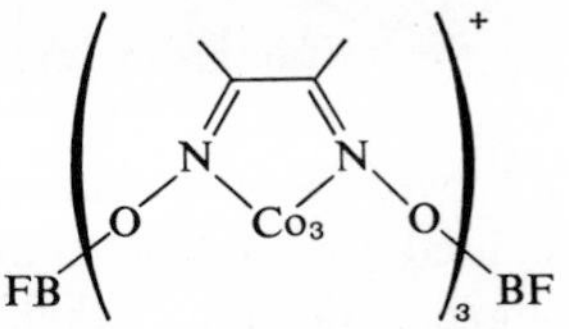

CI

cipitated from the solution by the addition of CH_2Cl_2, and the orange-red product can be isolated from the remaining solution. The PF_6 salt can be made by metathesis. Physical measurements are in accordance with the proposed structure, in which the cobalt ion occupies the central cavity of the ligand. Because of the tetrahedral distribution of bonds around the capping boron atoms, it might be expected that the cage, once formed, would be fairly rigid; hence it would not be possible to remove the metal ion from such a structure without prior ligand-bond rupture.

A range of sulfur-nitrogen macrocyclic ligands has been prepared by template reactions. In fact, only for this class of derivative has the kinetic coordination template effect been adequately demonstrated. Condensation, in alcohol, of an α-diketone with 2-aminoethanethiol, in a 1:2 molar ratio, leads to isolation of the bisthiazolidinyl (**CII**).[148] In solution, however, this compound probably exists in equilibrium with small amounts of the tautomeric Schiff base (**CIII**), which is more ideally suited to coordination than **CII**. Thus, in the presence of nickel ion, the Schiff base form preferentially

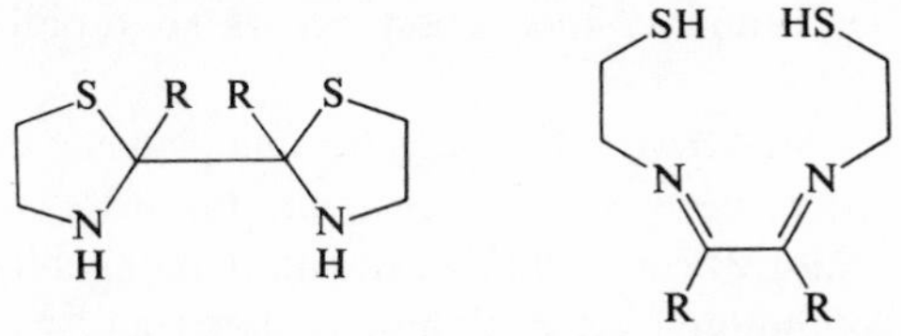

CII **CIII**

coordinates and is effectively sequestered from the equilibrium and precipitated in high yield as its neutral nickel complex.[1,91,148] The structure of the product has been confirmed by an x-ray study.[149] This reaction provides a classic example of the thermodynamic template effect. Such equilibria are usually pH-dependent,[150] and the use of basic conditions—for example, metal acetates or basic solvents—in reactions of this kind usually enhances the formation of the Schiff base tautomer.

Extensive studies have demonstrated that the coordinated mercaptide ion is very often susceptible to electrophilic attack by an alkyl halide, and so a coordinated thioether is formed.[151] The following is a typical reaction:

$$\underset{\text{Square-planar–diamagnetic}}{Ni(SCH_2CH_2NH_2)_2} + 2RX \longrightarrow \underset{\text{Octahedral-paramagnetic}}{Ni(RSCH_2CH_2NH_2)_2X_2}$$

Kinetic studies revealed that the sulfur atom remained coordinated during the alkylation. This reaction and the previous rearrangement reaction of sulfur chelates have been reviewed in more detail elsewhere.[152]

Both reactions can be used to produce a new series of macrocyclic complexes. If the nickel complex **CIV**, in dimethylformamide, is treated with a suitable difunctional alky halide—for example, α,α′-dibromo-*o*-xylene—the metal fixes the positions of the mercaptide ions in an array that is favorable for ring closure to occur.[153] The reaction is illustrated in Scheme VIII. The

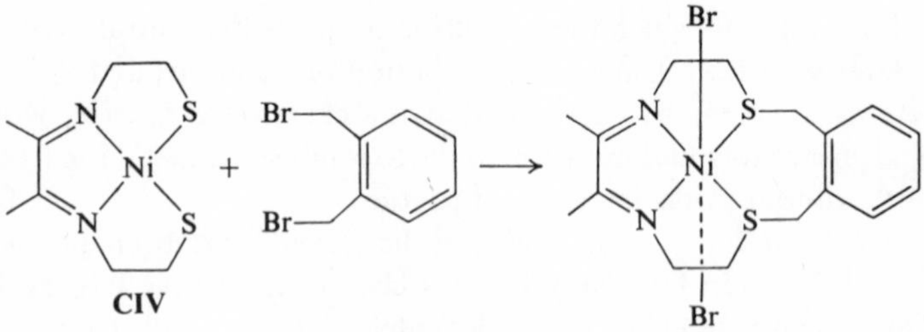

CIV

Scheme VIII

reaction proceeds by a sterically controlled mechanism[154] (kinetic template effect). Reaction of **CIV** with benzyl bromide indicates that the alkylations occur consecutively with a second-order rate constant for each step. In contrast the reaction with α,α′-dibromo-*o*-xylene also proceeds by an initial slow step, but the second condensation occurs so rapidly that it is not observed.

Similar macrocycles derived from 1,3-dibromopropane and 1,4-dibromobutane have also been prepared.[4,153] An interesting extension of this work is the reaction in chloroform of **CIV** with sulfur monochloride to produce the complex **CV** containing a tetrasulphide chelate ring.[155]

Alkylation reactions, in 1,2-dichloroethane, on the aromatic analog of

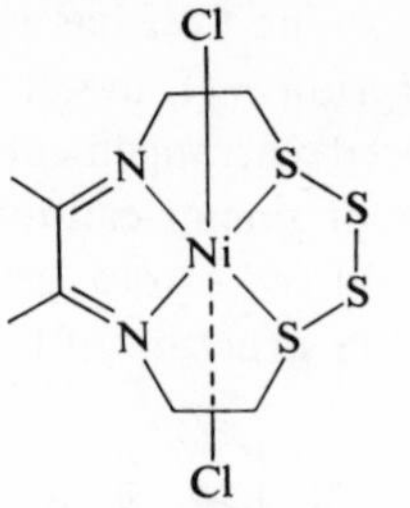

Cl
N S S S S Ni N
Cl
CV

CIV have been used to produce a new series of related octahedral macrocyclic complexes.[156] The preparative sequence for one such complex is illustrated in Scheme IX.

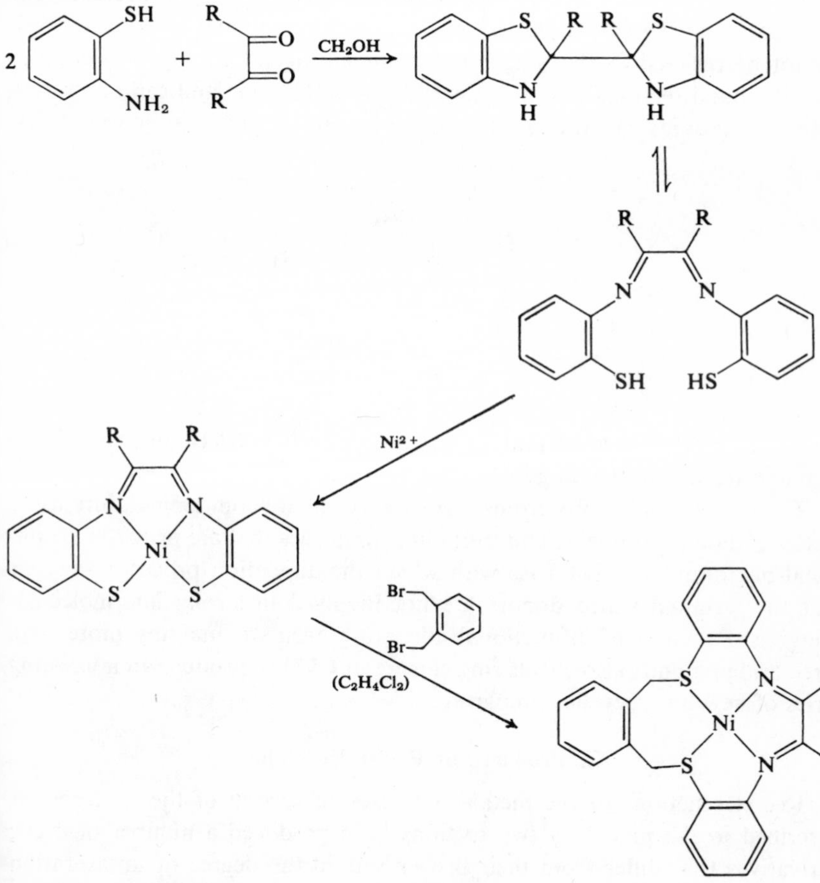

Scheme IX

Recently the bisbenzothiazoline **CVI** has been prepared by the reaction, in alcohol, of 2,6-diacetylpryridine (1 mole) with 2-aminobenzenethiol (2 moles).[82] The metal-ion–induced rearrangement of the yellow bisbenzothiazole occurs readily in the presence of zinc or cadmium acetate (solvent: DMF or acetone) to yield unusual bright red or orange complexes that seem to be of type **CVII**. Confirmation of the structure of these complexes must await the

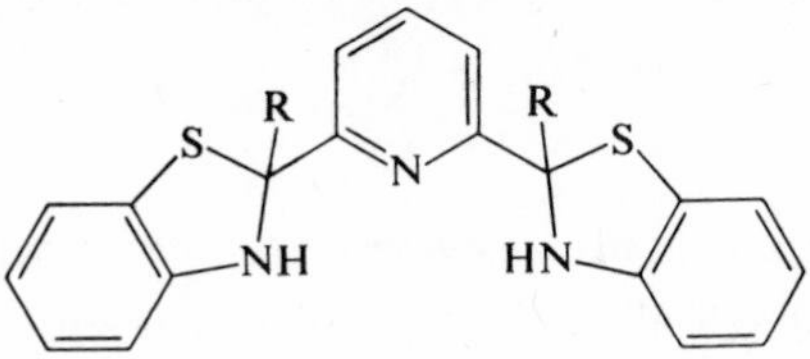

CVI

results of a crystal-structure determination in progress at present. The possibility that these compounds are essentially pentagonal-planar, however, cannot be rejected at this stage. The complete ligand conjugation coupled with the nondirectional electron distribution in the zinc and cadmium ions, both d^{10}, provides a rationale for the adoption of such a stereochemistry.

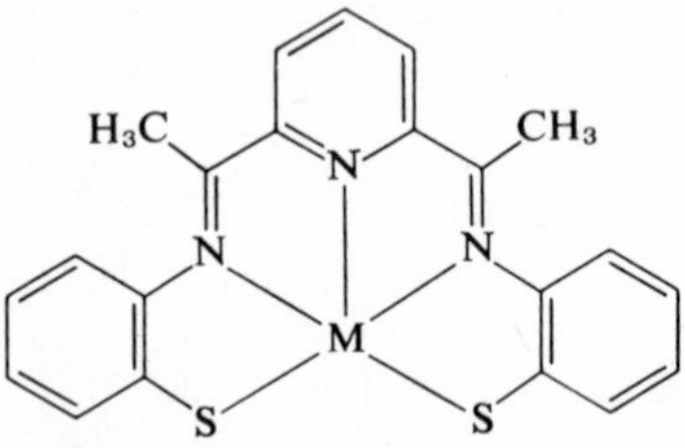

M = Zn, Cd

CVII

Molecular models indicate that, in a structure such as **CVII**, all donor-metal contacts are within bonding distance.

The cis terminal sulfur atoms readily alkylate at room temperature using methyl iodide to form zinc and cadmium complexes that are probably pentagonal-bipyramidal.[82] The ease with which the alkylation proceeds suggests that the terminal sulfur donors are not involved in strong intermolecular bridging. Reaction of difunctional alkylating agents containing more than three bridging units also effects ring closure of **CVII** to produce an interesting series of new macrocyclic complexes.

C. Produced by Redox Reactions

Redox reactions of the metal complexes of several of the macrocycles described in the preceding two sections have produced a number of cyclic derivatives that differ from their progenitors in the degree of unsaturation present. Much of the recent increase in activity in this area is directed toward the production, by selective redox reactions, of macrocycles whose ring size and unsaturation pattern are identical or closely related to those of specific natural rings. The existence of a range of possible redox reagents and the recent ready availability of macrocyclic complexes suitable as reactants have both contributed to this intensification of interest. Nevertheless this aspect of the chemistry of macrocyclic complexes is still in its infancy. In a number of cases it has been found that the coordination of a metal ion with an organic substrate profoundly influences the latter's redox properties. Little work has been carried out on the effect of variation of the central metal ion on a particular redox reaction. For a particular metal ion the oxidation state, spin state, coordination number, stereochemistry, and nature of other ligands may all influence the reaction. In addition the presence of specific ring substituents on the macrocyclic ligand, together with the possibility of its adoption of a number of conformations or configurations, increases the number of factors that may affect the reaction.

The results of electrochemical studies—polarography, cyclic voltammetry, and coulometry—can be very useful in the design of synthetic redox reactions. Nevertheless it must be remembered that, because of possible mechanistic differences, the duplication of an electrochemical result may not be possible using a particular chemical redox reagent, although such data as redox potentials suggest that the reaction should occur. Controlled potential electrolysis may often provide a more reliable synthetic procedure for obtaining the required product.

Hydrogenation[157] in a Parr bomb at 50 psi (platinum oxide catalyst) of an aqueous solution of the nickel perchlorate complex of the quadridentate ligand having the structure **XLIX** (CR) results in reduction of the two imine linkages to produce the nickel complex of the structure **CVIII**, $Ni(CRH)^{2+}$.

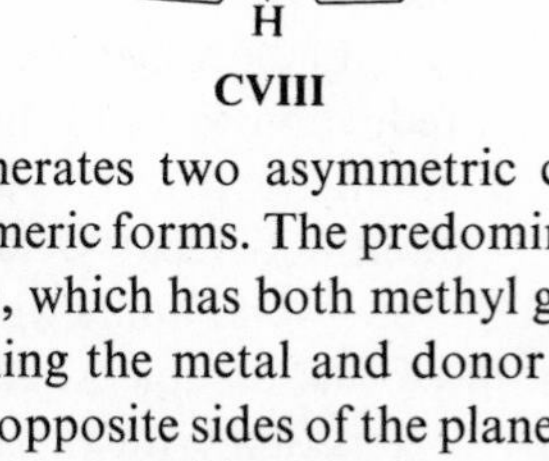

CVIII

Because this process generates two asymmetric carbon atoms, this ligand exists in two diastereoisomeric forms. The predominate isomer obtained in the reaction is the meso form, which has both methyl groups located on the same side of the plane containing the metal and donor atoms. The racemic form has the methyl groups on opposite sides of the plane. The diastereoisomers can be separated by fractional crystallization from water; the racemic form is more soluble in this solvent. The structure of the racemic isomer has been confirmed by an x-ray study.[158] From Ni(*meso*-CRH)$(ClO_4)_2$ the following compounds have been prepared directly: Ni(CRH)$X_2 \cdot xH_2O$, where X is Cl, Br, I, ClO_4, BF_4, NO_2, NCS, N_3, or 0.5 C_2O_4.[157] The formation of the violet oxalate complex reflects the greater flexibility of the reduced ligand, which is able to fold and allow the donors of bidentate ligands to coordinate to adjacent octahedral sites. This cis complex is thought to be dimeric, with the four oxygen atoms of the oxalate ion coordinated simultaneously to two nickel ions.[157]

Each of the three tetrahedral secondary nitrogen atoms of the ligand can have either of two configurations. Consequently when *meso*-CRH coordinates in a plane, there is the possibility of as many as six configurational isomers. Steric considerations, however, indicate that only two of these isomers are probable, and Ochiai and Busch[159] have isolated them. As prepared directly, Ni(*meso*-CRH)$(ClO_4)_2$ is red and was designated the α isomer. The dimeric oxalato complex described above is prepared from the α isomer by treatment with boiling aqueous sodium oxalate for 20 min. Treatment of this oxalate complex with perchloric acid gives yellow β-Ni(*meso*-CRH)$(ClO_4)_2$. The regeneration of the oxalate complex from this isomer is more facile than from the α isomer. The β isomer is converted to the α isomer by a base. The reactions are illustrated in Scheme X, which also shows the structural assignments for the α and β isomers.

Treatment of an aqueous solution of $Ni(CRH)^{2+}$ with sodium cyanide leads to liberation of the ligand, which can be isolated as a white solid.[157] Reaction of the appropriate metal salt directly with a solution of the free meso isomer has led to the isolation of complexes of the type M(*meso*-CRH)X_2, where M is iron,[160] cobalt,[161] or copper.[162] Similarly the cobalt(III) complexes [Co(*meso*-CRH)XY]$(ClO_4)n$, where XY = $(Cl^-)_2$, $(Br^-)_2$, (I^-Cl^-), $(H_2O)_2$, $(OH^-)_2$, $(N_3^-Cl^-)$, $(N_3^-)_2$, $(NCS^-)_2$, $(CN^-)_2$, $(NO_2^-)_2$, CO_3^{2-}, $C_2O_4^{2-}$, acac$^-$, and en, have been prepared by standard procedures.[163] As with the nickel complexes, both the α and β trans configurational isomers can be isolated. In general the secondary nitrogen atoms in these complexes are configurationally stable in acid solution, but under basic conditions acidic ionization and inversion may occur.

Reaction of Ni(CR)$(ClO_4)_2$ with borohydride, in the respective molar ratio of 4:1, results in partial reduction to yield the corresponding complex of the monoimine (**CIX**) in about 30 percent yield.[164] Alternatively this product can be obtained in about 7 percent yield by heating an aqueous solution of α-Ni(*meso*-CRH)$^{2+}$ for several hours while air is bubbled through the solution.[164,165] Reduction of this complex back to the CRH complex can be accomplished readily by reaction with excess borohydride ion or by catalytic hydrogenation over platinum oxide.[164]

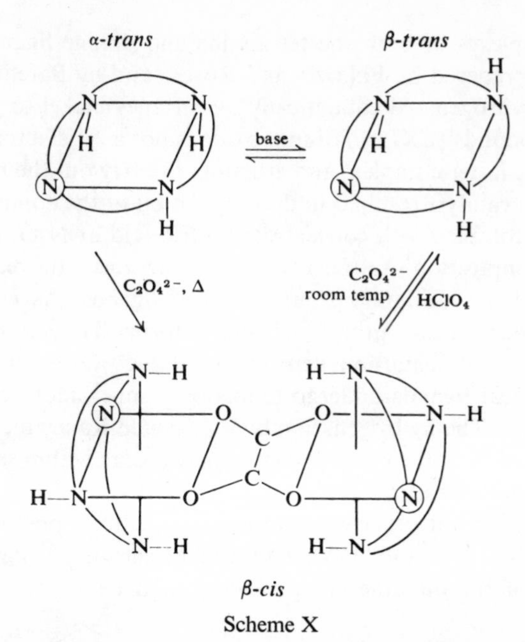

Scheme X

Oxidation of the ligand in the complex $Ni(CR)(ClO_4)_2$ has been accomplished.[164] If the dark-colored solution that results from the dissolution of this complex in concentrated nitric acid is warmed slightly, the color changes to orange, and a brown gas is liberated. A complex of the triimine ligand can be isolated from the solution (structure **CX**).

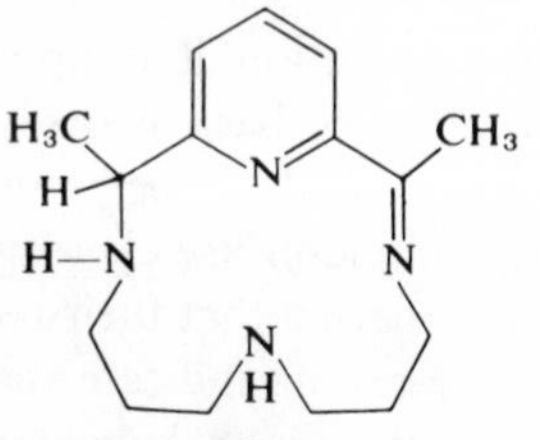

CIX

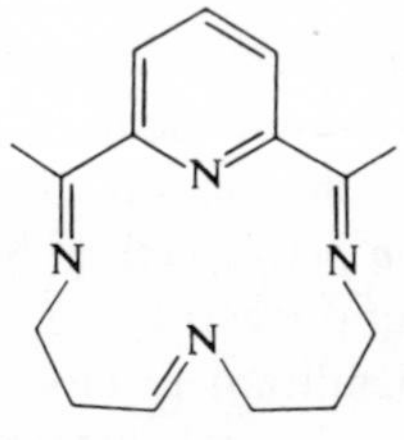

CX

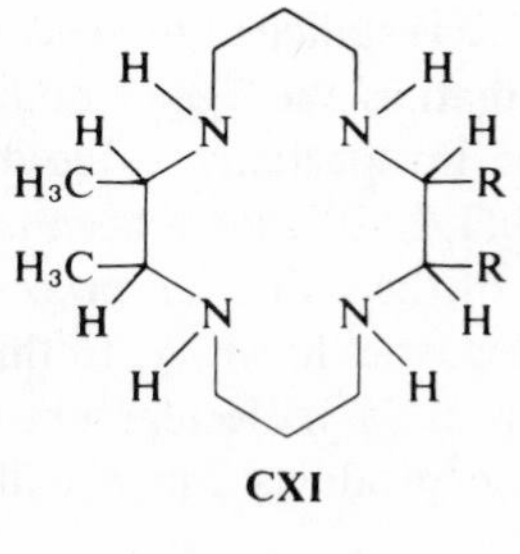

R = CH_3 or H

CXI

Nickel complexes of both the tetraimine and dimine ligands (structures **LV** and **LVI**) prepared by Baldwin and Rose[92] and by Barefield,[164] respectively, can be hydrogenated catalytically over Raney nickel to yield the fully saturated compounds (**CXI**). Platinum oxide is not a satisfactory catalyst for reduction of the latter complex, and attempts to carry out the reaction in the presence of this catalyst resulted in decomposition of the complex. Attempts to oxidize this complex with concentrated nitric acid or N_2O_4 in acetonitrile also led to decomposition. A difficulty in such reactions is the ready hydrolysis or solvolyses of the α-diimine groups in these ligands. As expected, complexes of the type **CXI** exist in several isomeric forms. The ease with which the reduction occurs is metal-ion–dependent; the corresponding cobalt(III) complexes of these ligands undergo reduction[93] much more easily than the nickel derivatives. The hydrogenation over activated Raney nickel catalyst of a methanol solution of the cobalt(III) complex occurs within several minutes even at a pressure of 1 atm.

Facile oxidations of other complexes are sometimes possible using mild oxidizing agents; for example, Vassian and Murmann[166] have shown that the oxidation of the dioxime complex illustrated below occurs readily. In

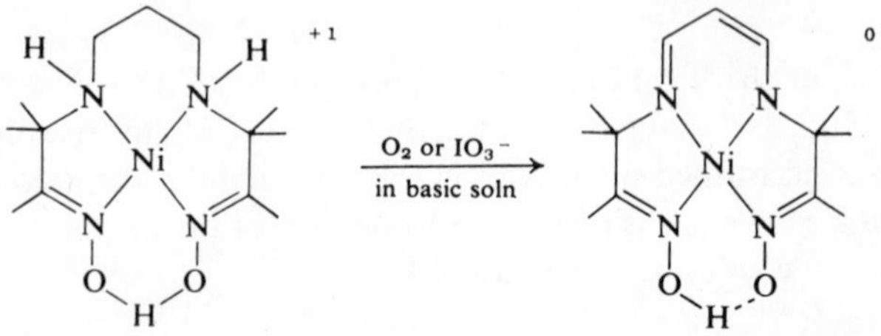

O_2 or IO_3^- in basic soln

view of the experimental conditions during oxidation it is apparent that neither the reactant nor the product is susceptible to basic hydrolysis.

Polarographic studies on the cobalt, nickel, and copper complexes of TAAB have been performed.[167] Since these complexes are closely related to some of the natural macrocycles, it might be expected that their redox properties may be significant to biochemistry. The results indicate that there is considerable ligand-metal interdependence in the redox behavior of these complexes. Formally the reduction of the central metal occurs stepwise until the number of electrons added is sufficient to yield a d^{10} electronic configuration. At more cathodic potentials the ligand double bonds are reduced in protonic solvents. Three of the partially reduced products, $M(TAAB)^{+1}$, where M = Cu, Ni, and $Ni(TAAB)^0$, have been isolated and characterized. Formally the end products of the reduction process are complexes of Cu^+, Ni^0, and Co^{-1}. It is more realistic, however, to think of these complexes as containing the added electrons in molecular orbitals that have significant ligand character. In fact the products can equally well be formulated as

Cu^{3+}, Ni^{2+}, and Co^{+1} derivatives of the doubly reduced anion $TAAB^{2-}$, a porphine analog.

Catalytic hydrogenation over platinum oxide of $Ni(TAAB)^{2+}$ in methanol leads to the isolation of a product whose infrared spectrum shows the absence of imine functions and the presence of secondary nitrogens.[168] This product is the fully reduced complex **CXII**, in which all azomethine linkages have been hydrogenated. A similar reaction using $Cu(TAAB)^{2+}$ does not result in ligand reduction, but instead $Cu(TAAB)^{+}$ can be isolated from the solution. This copper(I) complex is also obtained by stirring a methanol solution of the copper(II) complex over mercury. Catalytic reduction of the imine functions in the positional isomers $Ni(1,7\text{-CT})^{2+}$ and $Ni(1,4\text{-CT})^{2+}$

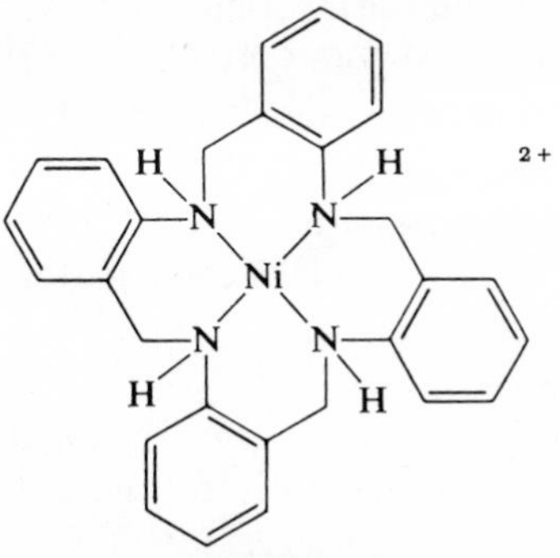

CXII

leads to the formation of a pair of saturated compounds of structures **CXIII** and **CXIV** respectively.[169] The reductions have also been performed success-

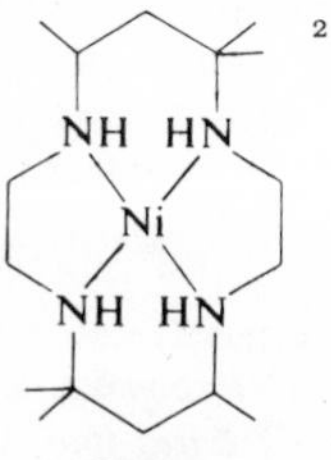

CXIII 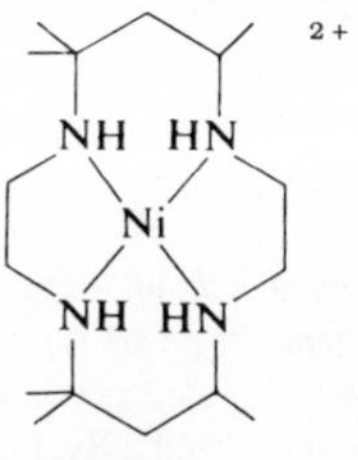 **CXIV**

fully using sodium borohydride, nickel-aluminum alloy in basic solution or electrolysis.[5,170] The very large number of isomers of **CXIII** and **CXIV** that result from the presence of both two asymmetric carbon atoms and four asymmetric nitrogens has been discussed in detail elsewhere.[171] Several of these isomers have been separated, and their structures elucidated.

Recently an investigation of the polarographic behavior in acetonitrile of square-planar nickel complexes of **CXIII** and **CXIV**, as well as of the unreduced $Ni(1,7\text{-CT})^{2+}$ and $Ni(1,4\text{-CT})^{2+}$, has been carried out.[101] All these complexes undergo one-electron reductions and oxidations to form products having nickel ions with formal oxidation states of +1 and +3, respectively. The complexes of both oxidation states are stable enough to be isolated. These were prepared by both controlled potential electrolysis and, for the Ni(I) complexes, chemical reduction using sodium amalgam in acetonitrile. The polarographic evidence indicates that the redox reactions primarily involve the metal ion and not the ligand. It is claimed that the use of the aprotic solvent acetonitrile for these electrochemical redox reactions partly determines the course of the reaction. Numerous previous results show that when chemical reagents are used in aqueous solution, almost invariably reduction or oxidation occurs at the ligand, and such reactions have been used to produce a whole family of nickel(II) compounds containing varying degrees of unsaturation.[5] Derivatives of both **CXIII** and **CXIV** containing one to four imine functions are known. Nitric acid was found to be a useful oxidant for the synthesis of the tetraimine complex,[172] and it has been suggested that initially nickel(III) intermediates are formed in this and other oxidations of this sort.[5,164] Such an intramolecular redox pathway may explain the observation that the oxidation almost always occurs at a secondary amine donor atom to produce an imine function.

A polarographic study of $Cu(1,7\text{-}CT)^{2+}$ in water shows that both metal and ligand reductions occur. The electrochemical behavior is somewhat complicated; for example, the half-wave potentials are pH-dependent.[173]

References

1. D. H. Busch, *Record Chem. Progr.*, **25**, 107 (1964).
2. D. St. C. Black and E. Markham, *Rev. Pure Appl. Chem.*, **15**, 109 (1965).
3. A. B. P. Lever, *Advan. Inorg. Chem. Radiochem.*, **7**, 27 (1965).
4. D. H. Busch, *Helv. Chim. Acta*, Fasciculus extraordinarius Alfred Werner, 174 (1967).
5. N. F. Curtis, *Coord. Chem. Rev.*, **3**, 3 (1968).
6. B. N. Figgis, *Introduction to Ligand Fields*, Interscience, New York, 1967.
7. R. Lemberg and J. W. Legge, *Haematin Compounds and Bile Pigments*, Interscience, New York, 1949; R. Lemberg, *Rev. Pure Appl. Chem.*, **6**, 1 (1956); E. Margoliash, *Ann. Rev. Biochem.*, **30**, 549 (1960); J. E. Falk, J. N. Phillips, R. Hill, and C. H. Gray in *Comprehensive Biochemistry*, Elsevier, Amsterdam, 1963, Vol. IX, pp. 1–98.
8. F. H. Moser and A. L. Thomas, *Phthalocyanine Compounds*, Monograph no. 157, A.C.S., Washington, D.C., 1963.
9. G. T. Byrne, R. P. Linstead, and A. R. Lowe, *J. Chem. Soc.*, 1017 (1934).
10. P. A. Barrett, C. E. Dent, and R. P. Linstead, *J. Chem. Soc.*, 1719 (1936).
11. P. A. Barrett, D. A. Frye, and R. P. Linstead, *J. Chem. Soc.*, 1157 (1938).
12. T. J. Hurley, M. A. Robinson, and S. I. Trotz, *Inorg. Chem.*, **6**, 389 (1967).
13. D. H. Busch, J. H. Weber, D. H. Williams, and N. J. Rose, *J. Am. Chem. Soc.*, **86**, 5161 (1964).

14. J. H. Weber and D. H. Busch, *Inorg. Chem.*, **4**, 469, 472 (1965).
15. K. Fenkart and C. H. Brubaker, *Inorg. Nucl. Chem. Letters*, **4**, 335 (1968).
16. R. P. Linstead and F. T. Weiss, *J. Chem. Soc.*, 2975 (1950).
17. J. A. Elridge and R. P. Linstead, *J. Chem. Soc.*, 5008 (1952).
18. P. Bamfield and P. A. Mack, *J. Chem. Soc.* (*C*), 1961 (1968).
19. J. H. Helberger and A. von Rebay, *Ann.*, **531**, 279 (1937); J. H. Helberger, *Ann.*, **529**, 205 (1937); C. E. Dent, *J. Chem. Soc.*, 1 (1938); P. A. Barrett, R. P. Linstead, and G. A. P. Tuey, *J. Chem. Soc.*, 1809 (1939); P. A. Barrett, R. P. Linstead, F. G. Rundall, and G. A. P. Tuey, *J. Chem. Soc.*, 1079 (1940).
20. A. H. Cook and R. P. Linstead, *J. Chem. Soc.*, 929 (1937).
21. H. Fischer, H. Haberland, and A. Muller, *Ann.*, **521**, 122 (1935); H. Fischer and W. Friedich, *Ann.*, **523**, 154 (1936); H. Fischer and A. Muller, *Ann.*, **528**, 1 (1937); H. Fischer and F. Endermann, *Ann.*, **531**, 245 (1937); F. Endermann and H. Fischer, *Ann.*, **538**, 172 (1939); H. Fischer, H. Guggemos, and A. Schafer, *Ann.*, **540**, 30 (1939).
22. R. P. Linstead and M. Whalley, *J. Chem. Soc.*, 4839 (1952).
23. R. P. Linstead, *J. Chem. Soc.*, 2873 (1953).
24. M. E. Baguley, H. France, M. Whalley, and R. P. Linstead, *J. Chem. Soc.*, 3521 (1955).
25. J. A. Elvidge and R. P. Linstead, *J. Chem. Soc.*, 3536 (1955).
26. P. M. Brown, D. B. Spiers, and M. Whalley, *J. Chem. Soc.*, 2882 (1957).
27. A. Parkinson, British patents 762,778 and 763,084.
28. A. W. Johnson and I. T. Kay, *J. Chem. Soc.*, 1620 (1965); R. L. N. Harris, A. W. Johnson, and I. T. Kay, *J. Chem. Soc.* (*C*), 22 (1966).
29. R. L. N. Harris, A. W. Johnson, and I. T. Kay, *Quart. Rev.*, **10**, 211 (1966).
30. M. Tsutsui, R. A. Velapoldi, K. Suzuki, F. Vohwinkel, M. Ichikawa, and T. Koyano, *J. Am. Chem. Soc.*, **91**, 6262 (1969).
31. J. E. Falk, *Porphyrins and Metalloporphyrins*, Elsevier, Amsterdam, 1964; J. E. Falk and J. N. Phillips, in *Chelating Agents and Metal Chelates*, F. P. Dwyer and D. P. Mellor (eds.), Academic, New York, 1964, chap. 10.
32. P. Rothemund, *J. Am. Chem. Soc.*, **57**, 2010 (1935).
33. P. Rothemund and A. R. Menotti, *J. Am. Chem. Soc.*, **63**, 267 (1941); R. H. Ball, G. D. Dorough, and M. Calvin, *J. Am. Chem. Soc.*, **68**, 2278 (1946); J. H. Priesthoff and C. V. Banks, *J. Am. Chem. Soc.*, **76**, 937 (1954).
34. G. M. Badger, R. A. Jones, and R. L. Laslett, *Australian J. Chem.*, **17**, 1028 (1964); D. W. Thomas and A. E. Martell, *J. Am. Chem. Soc.*, **78**, 1335 (1956).
35. E. B. Fleischer, *Inorg. Chem.*, **1**, 493 (1962); E. I. Choi and E. B. Fleischer, *Inorg. Chem.*, **2**, 94 (1963); E. B. Fleischer, E. I. Choi, P. Hambright, and A. Stone, *Inorg. Chem.*, **3**, 1284 (1964), and references therein.
36. E. B. Fleischer and A. Laszlo, *Inorg. Nucl. Chem. Letters*, **5**, 373 (1969), and references therein.
37. W. Seidel and F. Winkler, *Ann.*, **554**, 162 (1943); E. Bullock, A. W. Johnson, E. Markham, and K. B. Shaw, *J. Chem. Soc.*, 1430 (1958); A. W. Johnson, I. T. Kay, E. Markham, R. Price, and K. B. Shaw, *J. Chem. Soc.*, 3416 (1959); A. H. Jackson, P. Johnston, and G. W. Kenner, *J. Chem. Soc.*, 2262 (1964); M. Friedmann, *J. Org. Chem.*, **30**, 859 (1965).
38. Ref. 29, p. 218.
39. R. B. Woodward, W. A. Ayer, J. M. Beaton, F. Bickelhaupt, R. Bonnett, P. Buchschacher, G. L. Closs, H. Dutler, J. Hannah, F. P. Hauck, S. Ito, A. Langemann, E. Le Goff, W. Leimgruber, W. Lwowski, J. Sauer, Z. Valenta, and H. Volz, *J. Am. Chem. Soc.*, **82**, 3800 (1960).

40. H. Fischer and J. Klarer, *Ann.*, **448**, 178 (1926); H. Fischer, H. Friedrich, W. Lamatsch, and K. Morgenroth, *Ann.*, **466**, 147 (1928).
41. H. Fischer, *Naturwiss*, **17**, 611 (1929).
42. A. H. Corwin and V. L. Sydow, *J. Am. Chem. Soc.*, **75**, 4484 (1953); I. T. Kay, *Proc. Acad. Nat. Sci.*, **48**, 901 (1962).
43. A. W. Johnson and I. T. Kay, *J. Chem. Soc.*, 2418 (1961).
44. C. R. Porter, *J. Chem. Soc.*, 368 (1938).
45. D. P. Mellor and W. H. Lockwood, *Proc. Roy. Soc., New South Wales*, **74**, 141 (1940); J. E. Fergusson and C. A. Ramsay, *J. Chem. Soc.*, 5222 (1965); J. Ferguson and B. O. West, *J. Chem. Soc.*, 1565, 1569 (1966); M. Elder and B. R. Penfold, *J. Chem. Soc. (A)*, 2556 (1969).
46. H. Fischer and A. Stern, *Die Chemie des Pyrrols*, Liepzig, 1940, vol. llii, p. 144.
47. V. Eisner and R. P. Linstead, *J. Chem. Soc.*, 3742 (1955).
48. A. W. Johnson and I. T. Kay, *Proc. Chem. Soc.*, 168 (1961).
49. A. W. Johnson, I. T. Kay, and R. Rodrigo, *J. Chem. Soc.*, 2336 (1963).
50. A. W. Johnson and I. T. Kay, *Proc. Chem. Soc.*, 89 (1964); *J. Chem. Soc.*, 1620 (1965).
51. D. Dolphin, R. L. N. Harris, A. W. Johnson, and I. T. Kay, *Proc. Chem. Soc.*, 359 (1964); D. Dolphin, R. L. N. Harris, J. L. Huppatz, A. W. Johnson, and I. T. Kay, *J. Chem. Soc. (C)*, 30 (1966).
52. R. L. N. Harris, A. W. Johnson, and I. T. Kay, *Chem. Commun.*, 355 (1965); D. A. Clark, R. Grigg, R. L. N. Harris, A. W. Johnson, I. T. Kay, and K. W. Shelton, *J. Chem. Soc. (C)*, 1648 (1967).
53. R. Grigg, A. W. Johnson, and K. W. Shelton, *J. Chem. Soc. (C)*, 1291 (1968).
54. A. Eschenmoser, *Pure Appl. Chem.*, **7**, 297 (1963); E. Bertele, H. Boos, J. D. Dunitz, F. Elsinger, A. Eschenmoser, I. Felner, H. P. Gribi, H. Gschwend, E. F. Meyer, M. Pesaro, and R. Scheffold, *Angew. Chem. Intern. Ed.*, **3**, 490 (1964).
55. Y. Yamada, D. Miljkovic, P. Wehvli, B. Golding, P. Loliger, R. Keese, K. Muller, and A. Eschenmoser, *Angew. Chem. Intern. Ed.*, **8**, 343 (1969), and references therein.
56. B. Bosnich, C. K. Poon, and M. L. Tobe, *Inorg. Chem.*, **4**, 1102 (1965); C. K. Poon and M. L. Tobe, *J. Chem. Soc. (A)*, 2069 (1967); 1549 (1968).
57. H. Stetter and K. H. Mayer, *Chem. Ber.*, **94**, 1410 (1961).
58. B. Bosnich, M. L. Tobe, and G. A. Webb, *Inorg. Chem.*, **4**, 1109 (1965).
59. B. Bosnich, C. K. Poon, and M. L. Tobe, *Inorg. Chem.*, **5**, 1514 (1966).
60. B. Bosnich, R. Mason, P. Pauling, G. B. Robertson, and M. L. Tobe, *Chem. Commun.*, 97 (1965).
61. T. Kaden, *Prog. Coord. Chem.*, M. Cais (ed.), Elsevier, Amsterdam, 1968, p. 114.
62. J. P. Collman and P. W. Schneider, *Inorg. Chem.*, **5**, 1380 (1966).
63. W. Rosen and D. H. Busch, *Chem. Commun.*, 148 (1969); *J. Amer. Chem. Soc.*, **91**, 4694 (1969).
64. K. Travis and D. H. Busch, to be published.
65. P. W. N. M. van Leeuwen and W. L. Groeneveld, *Rec. Trav. Chim.*, **87**, 86 (1968).
66. S. E. Livingstone, *Quart. Rev.*, **19**, 386 (1965); R. Backhouse, M. E. Foss, and R. S. Nyholm, *J. Chem. Soc.*, 1714 (1957); R. L. Carlin and E. Weissberger, *Inorg. Chem.*, **3**, 611 (1964); F. A. Cotton and D. L. Weaver, *J. Am. Chem. Soc.*, **87**, 89 (1965); C. D. Flint and M. Goodgame, *J. Chem. Soc. (A)*, 2178 (1968).
67. W. Rosen and D. H. Busch, *Inorg. Chem.*, **9**, 262 (1970).
68. D. St. C. Black and I. A. McLean, *Chem. Commun.*, 1004 (1968); *Tetrahedron Letters*, 3961 (1969).

69. L. F. Lindoy, S. E. Livingstone, and T. N. Lockyer, *Australian J. Chem.*, **19**, 1391 (1966); P. S. K. Chia, S. E. Livingstone, and T. N. Lockyer, *Australian J. Chem.*, **19**, 1835 (1966); P. S. K. Chia, S. E. Livingstone, and T. N. Lockyer, *Australian J. Chem.*, **20**, 239 (1967); L. F. Lindoy, S. E. Livingstone, and T. N. Lockyer, *Australian J. Chem.*, **20**, 471 (1967).
70. C. J. Pedersen, *J. Am. Chem. Soc.*, **89**, 7017 (1967); *Fed. Proc.*, **27**, 1305 (1968); *J. Am. Chem. Soc.*, **92**, 386, 391 (1970).
71. A. C. L. Su and J. F. Weiher, *Inorg. Chem.*, **7**, 176 (1968).
72. B. Dietrich, T. M. Lehn, and J. P. Sauvage, *Tetrahedron Letters*, 2885, 2889 (1969).
73. D. R. Boston and N. J. Rose, *J. Am. Chem. Soc.*, **90**, 6859 (1968).
74. N. F. Curtis and R. W. Hay, *Chem. Commun.*, 524 (1966).
75. N. F. Curtis, *J. Chem. Soc.*, 4409 (1960); D. A. House and N. F. Curtis, *Chem. Ind. (London)*, 1708 (1961).
76. J. D. Curry and D. H. Busch, *J. Am. Chem. Soc.*, **86**, 592 (1964).
77. J. L. Karn and D. H. Busch, *Nature*, **211**, 160 (1966).
78. J. L. Karn and D. H. Busch, *Inorg. Chem.*, **8**, 1149 (1969).
79. R. L. Rich and G. L. Stucky, *Inorg. Nucl. Chem. Letters*, **1**, 61 (1965).
80. E. B. Fleischer and S. W. Hawkinson, *Inorg. Chem.*, **7**, 2312 (1968).
81. K. M. Long and D. H. Busch, *Inorg. Chem.*, **9**, 505 (1970).
82. L. F. Lindoy and D. H. Busch, unpublished work.
83. P. Krumholz, *Inorg. Chem.*, **4**, 612 (1965).
84. S. M. Nelson, P. Bryan, and D. H. Busch, *Chem. Commun.*, 641 (1966).
85. W. M. Reiff, C. J. Long, and W. A. Baker, *J. Am. Chem. Soc.*, **90**, 6347 (1968).
86. S. M. Nelson and D. H. Busch, *Inorg. Chem.*, **8**, 1859 (1969).
87. E. Fleischer and S. Hawkinson, *J. Am. Chem. Soc.*, **89**, 720 (1967).
88. A. V. Heuvelen, M. D. Lundeen, H. G. Hamilton, and M. D. Alexander, *J. Chem. Phys.*, **50**, 489 (1969).
89. L. F. Lindoy and S. E. Livingstone, *Coord. Chem. Rev.*, **2**, 173 (1967).
90. L. T. Taylor, N. J. Rose, and D. H. Busch, *Inorg. Chem.*, **7**, 785 (1968).
91. M. C. Thompson and D. H. Busch, *J. Am. Chem. Soc.*, **86**, 213 (1964).
92. D. A. Baldwin and N. J. Rose, *Abstr. 157th Natl. Meeting A.C.S.*, Minneapolis, Minn., 1969.
93. K. Farmery and D. H. Busch, to be published.
94. K. Farmery, S. C. Doherty, N. J. Rose, and D. H. Busch, to be published.
95. E. K. Barefield, K. Farmery, and D. H. Busch, to be published.
96. E. G. Jäger, *Z. Chem.*, **4**, 437 (1964).
97. E. G. Jäger, *Z. Chem.*, **8**, 30 (1968).
98. E. G. Jäger, *Z. Chem.*, **8**, 392 (1968).
99. E. G. Jäger, *Z. Anorg. Allg. Chem.*, **364**, 178 (1969).
100. H. Hiller, P. Dimroth, and H. Pfitzner, *Leibigs Ann. Chem.*, **717**, 137 (1968).
101. D. C. Olson and J. Vasilevskis, *Inorg. Chem.*, **8**, 1611 (1969).
102. S. C. Cummings and R. B. Sievers, *J. Am. Chem. Soc.*, **92**, 215 (1969); *Inorg. Chem.*, in press.
103. C. L. Honeybourne and G. A. Webb, *Chem. Commun.*, 739 (1968).
104. P. Chave and C. L. Honeybourne, *Chem. Commun.*, 279 (1969).
105. M. Green and P. A. Tasker, *Chem. Commun.*, 518 (1968).
106. E. Uhlemann and M. Plath, *Z. Chem.*, **9**, 30 (1969).
107. L. F. Lindoy and D. H. Busch, *Chem. Commun.*, 1589 (1968); *J. Am. Chem. Soc.*, **91**, 4690 (1969); L. F. Lindoy and D. H. Busch, *Inorg. Nucl. Chem. Letters*, **5**, 525 (1969).

108. E. Hoyer and B. Lorenz, *Z. Chem.*, **8**, 28 (1968).
109. E. B. Fleischer and P. A. Tasker, *Abstr. 158th Natl. Meeting A.C.S.*, New York, September, 1969; E. B. Fleischer, personal communication.
110. S. G. McGeachin, *Canadian J. Chem.*, **44**, 2323 (1966).
111. G. A. Melson and D. H. Busch, *Proc. Chem. Soc.*, 233 (1963).
112. G. A. Melson and D. H. Busch, *J. Am. Chem. Soc.*, **86**, 4834 (1964).
113. G. A. Melson and D. H. Busch, *J. Am. Chem. Soc.*, **87**, 1706 (1965).
114. L. T. Taylor, S. C. Vergez, and D. H. Busch, *J. Am. Chem. Soc.*, **88**, 3170 (1966).
115. E. B. Fleischer and E. Klem, *Inorg. Chem.*, **4**, 637 (1965).
116. L. T. Taylor and D. H. Busch, *J. Am. Chem. Soc.*, **89**, 5372 (1967).
117. L. T. Taylor and D. H. Busch, unpublished work.
118. G. A. Melson and D. H. Busch, *J. Am. Chem. Soc.*, **86**, 4830 (1964).
119. S. W. Hawkinson and E. B. Fleischer, Inorg Chem., *8*, 2402 (1969).
120. S. C. Cummings and D. H. Busch, *J. Am. Chem. Soc.*, **92**, 1924 (1970).
121. R. W. Wing and R. Eiss, *J. Am. Chem. Soc.*, **92**, 1929 (1970).
122. S. C. Vergez, V. Katovic, and D. H. Busch, unpublished work.
123. F. Seidel, *Ber.*, **59B**, 1894 (1926).
124. V. Katovic and D. H. Busch, unpublished work.
125. L. T. Taylor, F. L. Urbach, and D. H. Busch, *J. Am. Chem Soc.*, **91**, 1072 (1969).
126. D. H. Busch and J. C. Bailar, Jr., *J. Am. Chem. Soc.*, **78**, 1139 (1956).
127. C. M. Harris and E. D. McKenzie, *Nature*, **196**, 670 (1962).
128. V. Katovic, L. T. Taylor, and D. H. Busch, *J. Am. Chem. Soc.*, **91**, 2122 (1969).
129. N. F. Curtis, Y. M. Curtis, and H. K. J. Powell, *J. Chem. Soc.* (*A*), 1015 (1966).
130. R. R. Ryan, B. T. Kilbourn, and J. D. Dunitz, *Chem. Commun.*, 910 (1966); M. F. Bailey and I. E. Maxwell, *Chem. Commun.*, 908 (1966); B. T. Kilbourn, R. R. Ryan, and J. D. Dunitz, *J. Chem. Soc.* (*A*), 2407 (1969).
131. L. G. Warner, N. J. Rose, and D. H. Busch, *J. Am. Chem. Soc.*, **90**, 6938 (1968); **89**, 703 (1967).
132. N. F. Curtis and Y. M. Curtis, *J. Chem. Soc.* (*A*), 1653 (1966).
133. J. Karn, thesis, The Ohio State University (1966).
134. M. M. Blight and N. F. Curtis, *J. Chem. Soc.*, 3016 (1962).
135. N. F. Curtis, D. A. Swann, T. N. Waters, and I. E. Maxwell, *J. Am. Chem. Soc.*, **91**, 4588 (1969).
136. N. Sadasivan and J. F. Endicott, *J. Am. Chem. Soc.*, **88**, 5468 (1966).
137. L. G. Warner, N. J. Rose, and D. H. Busch, *Abstr. 152nd Natl. Meeting A.C.S.*, New York, 1966.
138. J. L. Love and H. K. J. Powell, *Inorg. Nucl. Chem. Letters*, **3**, 113 (1967).
139. J. L. Love and H. K. J. Powell, *Chem. Commun.*, 39 (1968).
140. M. M. Blight and N. F. Curtis, *J. Chem. Soc.*, 1204 (1962); D. A. House and N. F. Curtis, *J. Am. Chem. Soc.*, **86**, 223 (1964).
141. T. E. MacDermott and D. H. Busch, *J. Am. Chem. Soc.*, **89**, 5780 (1967).
142. D. A. House and N. F. Curtis, *J. Am. Chem. Soc.*, **84**, 3248 (1962); *J. Am. Chem. Soc.*, **86**, 1331 (1964).
143. F. L. Urbach and D. H. Busch, unpublished results.
144. G. N. Schrauzer, *Chem. Ber.*, **95**, 1438 (1962).
145. D. Thierig and F. Umland, *Angew. Chem.*, **74**, 1438 (1962).
146. N. J. Rose, V. Litvinov, and D. H. Busch, unpublished results.
147. D. R. Boston and N. J. Rose, *J. Am. Chem. Soc.*, **90**, 6859 (1968).
148. H. Jadamus, Q. Fernando, and H. Freiser, *Inorg. Chem.*, **3**, 928 (1964).

149. Q. Fernando and P. Wheatly, *Inorg. Chem.*, **4**, 1726 (1965).
150. E. Bayer, *Angew. Chem.*, **73**, 659 (1961).
151. D. H. Busch, J. A. Burke, D. C. Jicha, M. C. Thompson, and M. L. Morris, *Advan. Chem. Ser.*, **37**, 125 (1963); D. H. Busch, D. C. Jicha, M. C. Thompson, J. W. Wrathall, and E. Blinn, *J. Am. Chem. Soc.*, **86**, 3642 (1964); E. L. Blinn and D. H. Busch, *J. Am. Chem. Soc.*, **90**, 4280 (1968); N. J. Rose, C. A. Root, and D. H. Busch, *Inorg. Chem.*, **6**, 1431 (1967).
152. L. F. Lindoy, *Coord. Chem. Rev.*, **4**, 41 (1969).
153. M. C. Thompson and D. H. Busch, *J. Am. Chem. Soc.*, **86**, 3651 (1964).
154. E. L. Blinn and D. H. Busch, *Inorg. Chem.*, **7**, 820 (1968).
155. N. B. Egen and R. A. Krause, *J. Inorg. Nucl. Chem.*, **31**, 127 (1969).
156. M. S. Elder, G. M. Prinz, P. Thornton, and D. H. Busch, *Inorg. Chem.*, **7**, 2426 (1968).
157. J. L. Karn and D. H. Busch, *Inorg. Chem.*, **8**, 1149 (1969).
158. E. B. Fleischer and R. Dewar, *Nature*, **222**, 372 (1969).
159. E. Ochiai and D. H. Busch, *Inorg. Chem.*, **8**, 1798 (1969).
160. P. Merrell and D. H. Busch, unpublished work.
161. K. Long and D. H. Busch, unpublished work.
162. L. F. Lindoy, N. Tokel, and D. H. Busch, unpublished work.
163. E. Ochiai and D. H. Busch, *Inorg. Chem.*, **8**, 1474 (1969).
164. E. K. Barefield, Ph.D. thesis, The Ohio State University (1969).
165. E. Ochiai and D. H. Busch, unpublished work.
166. E. G. Vassian and R. K. Murmann, *Inorg. Chem.*, **6**, 2043 (1967).
167. N. E. Tokel, V. Katovic, K. Farmery, L. B. Anderson, and D. H. Busch, *J. Am. Chem. Soc.*, **92**, 400 (1970).
168. G. A. Melson, F. L. Urbach, L. T. Taylor, V. Katovic, and D. H. Busch, unpublished work.
169. N. F. Curtis, *J. Chem. Soc.*, 2644 (1964).
170. N. F. Curtis, *J. Chem. Soc.* (*C*), 1979 (1967); 924 (1965).
171. L. G. Warner and D. H. Busch, *Coordination Chemistry*, Plenum, New York, 1969, pp. 1–17.
172. N. F. Curtis, *Chem. Commun.*, 882 (1966).
173. J. M. Palmer, E. Papaconstantinou, and J. F. Endicott, *Inorg. Chem.*, **8**, 1516 (1969).

Sulfanuric Compounds

THERALD MOELLER and RONALD L. DIECK
Department of Chemistry
Arizona State University, Tempe

I. INTRODUCTION

In principle a number of aquo-ammono sulfuric acids can be related to one another and to aquo monosulfuric acid $O_2S(OH)_2$ in terms of a hydration-dehydration, ammonation-deammonation scheme.[1,2] Nearly all the compounds that have been characterized have linear molecular structures. Early experiments did suggest, however, and later studies confirm, the existence of heterocyclic polysulfimides $(O_2SNH)_n$ or their derivatives $(O_2SNX)_n$. Thus the thermal decomposition of the diamide $O_2S(NH_2)_2$ allowed Traube to isolate, not free sulfimide, but presumably its ammonium salt, from which he obtained the silver(I), sodium, potassium, and barium salts.[3] Hantzsch and Holl established the presence of a trimeric sulfimide group in solution,[4] but the white solid they believed to be free trimeric sulfimide proved to be impure imidodisulfamide $HN(SO_2NH_2)_2$.[5] Reaction of the silver salt with methyl iodide yielded a trimeric methyl derivative $(O_2SNCH_3)_3$.[4] Other salts have been obtained,[6] but the isolation of free sulfimide has not been completely established. The trimeric dinegative ion $S_3O_6N_3H^{2-}$ does exist in aqueous solution, and the dipyridinium salt is isolable in large yields from the reaction of pyridine with the amido acid chloride H_2NSO_2Cl.[6] All this information suggests formulation **I** for trisulfimide. Hantzsch recognized the direct formula analogy between trisulfimide and cyanuric acid and suggested the existence of the tautomeric isomer, sulfanuric acid (**II**).*

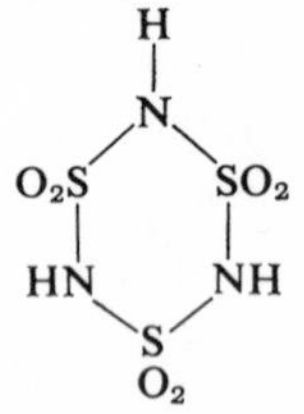

I

Trisulfimide

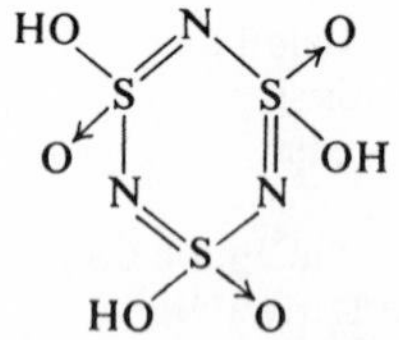

II

Trimeric sulfanuric acid

The formation of the tetramer $(O_2SNCH_3)_4$ from the reaction products of sulfur(VI) oxide with ammonia in nitromethane and the presence of high-molecular-weight products of the type $HOSO_2(—NH—SO_2—)_nOH$ in this system suggest the formation of tetra and polysulfimides and thus allow for the possible existence of tetrameric and more highly polymeric sulfanuric acids.[7]

There is no experimental evidence for the direct isolation of any sulfanuric acid. Trimeric sulfanuric chloride (**III**), however, can be prepared quite readily, and from it by displacement reactions a number of derivatives. It is these compounds which are discussed in this summary. Since only trimeric species are known, the unqualified term "sulfanuric" implies a trimeric composition. Trimeric sulfanuric and phosphonitrilic chlorides (**IV**)

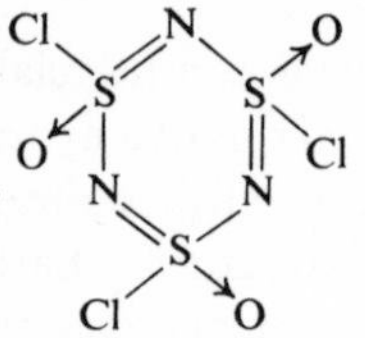

III

Trimeric sulfanuric chloride

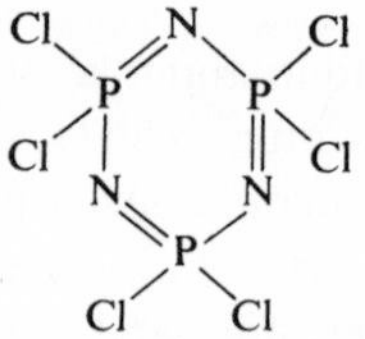

IV

Trimeric phosphonitrilic chloride

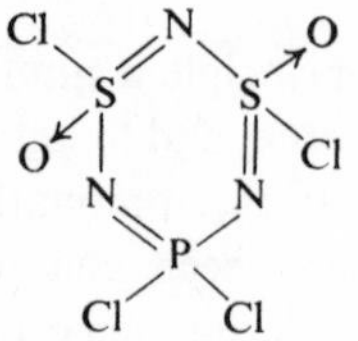

V

Mixed trimeric sulfanuric-phosphonitrilic chloride

* In this formula and those which follow, localized double and single bonds are written for convenience in comparison. Electron density in these cyclic systems is actually delocalized by way of $p_\pi - d_\pi$ interactions.[18]

are isoelectronic. At least one isoelectronic mixed species (V) has been described also.

Previously published summaries on sulfanuric compounds are limited in scope.[2,8–10]

II. α- AND β-SULFANURIC CHLORIDES

A. Methods of Synthesis

Sulfanuric chloride was first prepared by Kirsanov[11] by the sequence of reactions summarized by the equations

$$H_2NSO_2OH(s) + 2PCl_5(s) \xrightarrow{\text{heat}} Cl_3N{=}PSO_2Cl(s) + POCl_3(g) + 3HCl(g)$$

$$3Cl_3N{=}PSO_2Cl(s) \xrightarrow{\text{heat}} (NSOCl)_3(s) + 3POCl_3(g)$$

The thermal decomposition reaction gave a mixture of isomers, from which the so-called α and β isomers were isolated by extraction with *n*-heptane, followed by fractional crystallization or fractional sublimation after removal of the solvent. A third isomer was detected but disappeared during the purification process. This procedure, with some modification, is the best available in terms of yield and purity of products.

Other preparative procedures include the reaction of a mixed solution of thionyl chloride ($SOCl_2$) and sulfuryl chloride (SO_2Cl_2) in petroleum ether with dry ammonia gas at $-80°C$ under an atmosphere of dry nitrogen.[12] Another procedure is based on the reaction of trithiazyl chloride ($N_3S_3Cl_3$) with sulfur(VI) oxide under nitrogen at 30 to 40°C to form ultimately a product $(NSCl)_3 \cdot 2.8SO_3$, which, when heated at 140 to 160°C and 20 atm for 30 min, gives sulfanuric chloride.[13] Neither of these methods gives a high yield of product.

B. Preparative Procedure

Directions for synthesis in terms of a modification of the Kirsanov procedure are as follows[14]:

Trichlorophosphazosulfonyl chloride* $Cl_3P{=}NSO_2Cl$ (362 g, 1.40 moles) is placed in a 500-ml, pear-shaped distillation flask fitted with a 30-cm Vigreux column and an inlet tube for admitting dry air or nitrogen. A Liebig distillation head is fitted to the top of the Vigreux column, a condenser is attached, and the entire system is connected to a vacuum pump through a

* Prepared as described by Kirsanov.[11]

receiving flask maintained at −70°C. Either air or nitrogen, previously dried with calcium chloride, is passed through the contents of the flask at such a rate that the pressure in the system is maintained at 2.5 to 5 mm Hg. The flask is slowly heated to 118 to 120°C, in which temperature range thermal decomposition, with the evolution of phosphorus(V) oxotrichloride, proceeds rapidly. After 1 hr at this temperature, the reaction flask is warmed to 130°C and maintained at this temperature for an additional hour. It is then warmed at 140°C for a short time to ensure complete removal of the oxochloride. The brown residue in the flask is extracted 10 times with 100- to 150-ml quantities of hot *n*-heptane. When cooled to 0°C, the solution obtained deposits large crystals of primarily α-sulfanuric chloride. The solvent is removed *in vacuo* from the filtrate to give a second crop of crystals. The two crops are combined, and the two isomers separated by fractional crystallization from *n*-heptane or by vacuum sublimation. At 0.005 mm Hg, the β form, together with small quantities of a third and less stable isomer, sublimes readily at room temperature; the α form requires a temperature of 80°C for sublimation.

The pure β isomer is obtained as long needlelike crystals by extracting the material sublimed at room temperature with as little *n*-heptane as possible and then cooling to 0°C. The α isomer crystallizes as rhombic prisms. The yields are α, 27.60 g or 20 percent; β, 20.0 g or 14.3 percent.

All directions must be followed very closely. Repetition of the procedure several times may be required to make it produce the indicated yields of products consistently.

C. Properties

α-Sulfanuric chloride is a colorless, crystalline compound that melts at 144 to 145°C.[14] It is soluble in the common organic solvents (Table I). β-Sulfanuric is also a colorless, crystalline compound, but its melting point is 46 to 47°C. It is generally more soluble than the α isomer, but quantitative

TABLE I

Solubility Data for α-Sulfanuric Chloride at 25°C[14]

Solvent	Solubility (g/100 g solvent)
Benzene	22.50
Acetonitrile	13.15
Carbon disulfide	4.10
Carbon tetrachloride	2.95
Petroleum ether (bp 90–110°C)	2.32
Cyclohexane	1.63
n-Heptane	1.56

data are lacking. The α compound is stable with respect to conversion to any other isomer in the solid state or in solution in any solvent from which it is recoverable. The β compound is stable with respect to isomerization in the solid state and in solution in nonpolar solvents, such as cyclohexane or benzene. Isomerization to the α compound occurs rapidly in a more polar medium, such as diethyl ether or acetonitrile. An increase in temperature accelerates this isomerization process.

TABLE II

Infrared Spectra of Cyclohexane Solutions of Sulfanuric Chlorides[14]

α Isomer (cm^{-1})	β Isomer (cm^{-1})
665 s	665 s
700 vs	700 w
816 m	822 s
1110 vs	1110 vs
1344 w	1340 vs

The more significant absorption bands in the infrared region are compared in Table II.[14] Although the bands are similarly located, certain of them vary widely in intensity. No absolute assignments of vibrational modes have been made. The spectrum of an acetonitrile solution of the β isomer changes within an hour to that of the α isomer; but that of a cyclohexane solution remains unaltered for at least 4 days.

The planar cis-trans molecular structures assigned by Kirsanov[11] are incorrect.[15–17] X-ray diffraction, single-crystal analyses[15–17] indicate that orthorhombic crystals of the α isomer contain four molecules to the unit cell of dimensions $a = 7.552$, $b = 11.540$, and $c = 10.078$ Å. The space group is *Pnma*. The molecular structure involves a six-membered sulfur-nitrogen ring in a chair conformation, with the chlorine atoms in axial positions and the oxygen atoms in equatorial positions (Figure 1). The arrangement of bonds around each sulfur atom is roughly tetrahedral. Bond distances and angles are S—N = 1.571(4) Å, S—Cl = 2.003(3) Å, S—O = 1.407(7) Å, N—S—N = 112.8(4)°, S—N—S = 122.0(4)°, N—S—O = 111.9(3)°, N—S—Cl = 106.3 (3)°, and O—S—Cl = 107.9(3)°. Since all S—N bond distances are the same and between calculated single- and double-bond distances, $d_\pi - p_\pi$ delocalization of ring electron density is indicated.[18] The molecular symmetry is C_{3v}. The ^{35}Cl nuclear quadrupole resonance spectrum is consistent with this molecular structure.[33]

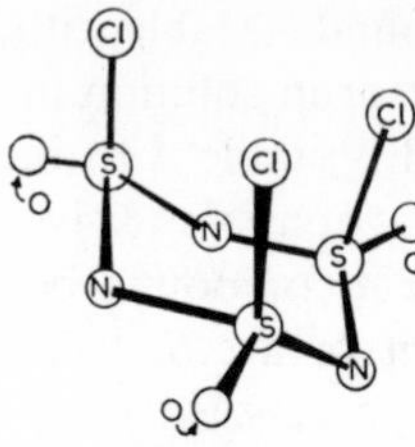

Figure 1.

No comparable investigation has yet been reported for β-sulfanuric chloride. The dipole moment of the α isomer (3.88 D at 25°C) is consistent with the axial orientation of the chlorine atoms; the small moment of the β isomer (1.91 D at 25°C) is indicative of a more symmetrical structure.[14] An analysis of dipole-moment data rules out boat-form conformations and suggests that the molecular structure of the β isomer is that of another chair conformation with at least one of the chlorine atoms in an equatorial position.[14] Infrared data are not inconsistent with this conclusion. The ^{35}Cl nuclear quadrupole resonance spectrum of the β isomer also supports this conclusion.[33] Furthermore the lack of ready conversion of the β isomer to the α in nonpolar solvents is reasonable in the light of the large energy of activation needed to reorient bonds; the ready conversion in polar media can result from lowered activation energy resulting from nucleophilic interaction between solvent molecules and ring sulfur atoms.[14]

α-Sulfanuric chloride sublimes slowly at room temperature below 100°C, but above the melting point some thermal decomposition is noted.[19] Rapid heating results in explosive decomposition, probably as a consequence of the rapid interaction of sulfur(VI) with adjacent nitrogen(-III).[19] Mixtures with powdered metals, decaborane, or lithium aluminum hydride explode violently when heated. Pure α-sulfanuric chloride is not particularly shock-sensitive, however. Explosive decomposition yields as major products elemental molecular nitrogen and sulfur(IV) oxide and at least the compounds $SOCl_2$, SCl_2, SO_3, and HCl, with sulfur nitrides, sulfur oxonitrides, and nitrogen oxides as other probable products. Mass-spectral analysis shows $N_3S_3O_3Cl_2^+$, SN^+, SO^+, $HOSN^+$, SO_2^+, and HCl^+ as major fragmentation products.[19]

Unlike trimeric phosphonitrilic chloride, α-sulfanuric chloride commonly reacts with nucleophilic reagents by ring cleavage rather than substitution.[14] Water forms imidodisulfamide, sulfuric acid, and hydrogen chloride, rather than the parent acid, and ammonia effects solvolysis, followed by polymerization to analogs of melam and melem.[21] Early attempts at substitution with morpholine yielded only *N,N'*-dimorpholinosulfamide.[14] More recent studies, however, have shown that morpholine and other weakly basic nucleophiles can give substitution products under controlled conditions.[19,22,23]

D. Derivatives of α-Sulfanuric Chloride

1. Morpholino Derivatives

With benzene as a solvent and sufficient excess morpholine present to function as an acceptor for released hydrogen chloride, a trimorpholino derivative

$$N_3S_3O_3\left[N\begin{matrix} CH_2CH_2 \\ \\ CH_2CH_2 \end{matrix}O\right]_3$$

(isomer I), with a melting point of 171 to 172°C, was recovered in 40 to 50 percent initial yield after recrystallization from acetonitrile and absolute ethanol.[22] With benzene as a solvent and triethylamine present as a preferential hydrogen chloride acceptor, a mixed product (mp 160 to 162°C) was recovered in an initial yield of 45 to 54 percent. Recrystallization from absolute ethanol yielded a second trimorpholido derivative (isomer II) of mp 196 to 197°C in 34 to 49 percent yield.[22] With acetonitrile as a medium and excess morpholine present, isomer II was obtained in about 56 percent yield. Each reaction is effected by dropwise addition of a solution of α-sulfanuric chloride to a cooled, well-stirred solution of morpholine, with full protection from moisture. Oily residues remaining after removal of precipitated amine hydrochloride and evaporation of solvent are rendered crystalline with ethanol. Final recrystallization is effected from absolute ethanol. Molecular-weight data support the trimeric formulations.

Both isomers dissolve in absolute ethanol and acetonitrile, but neither one is particularly soluble in acetone, dioxane, diethyl ether, tetrahydrofurane, nitrobenzene, or pyridine. Isomer I is soluble in water but not in anhydrous benzene; isomer II is soluble in anhydrous benzene but not in water. The infrared spectra of the two compounds are closely similar and resemble those of the α- and β-chlorides. Absorptions around 1075 cm^{-1} indicate ring S—N vibrations, and those in the 942 to 948 cm^{-1} region exocyclic S—N vibrations. The former are at lower frequencies than those for the chlorides as a consequence of weakening of S—N bonds by decreasing polarization of sulfur atoms. The latter are absent in the spectra of the chlorides. X-ray powder data indicate the two isomers to be different crystallographically but provide no quantitative indication of molecular structural differences. It is reasonable to conclude that isomerism is again a consequence of change in orientation of substituents rather than in ring conformation.[22]

2. *Phenyl Derivatives*

A diphenyl derivative can be prepared by the reaction of α-sulfanuric chloride with diphenyl mercury (1:2 mole ratio) in anhydrous benzene[19]:

$$N_3S_3O_3Cl_3 + 2(C_6H_5)_2Hg \xrightarrow{C_6H_6} N_3S_3O_3Cl(C_6H_5)_2 + 2(C_6H_5)HgCl(s)$$

The reaction occurs at 30 to 33°C over a period of stirring of about 67 hr. The crude product is recovered by filtering out the phenylmercury chloride and volatilizing the solvent. A pure product (mp 120°C) is recovered in 58.5 percent yield by extracting with *n*-heptane and purifying chromatographically on silica with benzene as solvent. Reaction in a 1:3 mole ratio yields the same product and not the triphenyl derivative.

Diphenylsulfanuric chloride is soluble in benzene, toluene, acetonitrile, hot *n*-heptane, cyclohexane, and anhydrous ethanol. It is not moisture-sensitive. It does not react with phenyl lithium. The compound is obtained in only one isomeric form by the above-mentioned procedure. The compound sublimes *in vacuo* and undergoes thermal decomposition at 200 to 300°C. Thermal decomposition products are similar to those from α-sulfanuric chloride, plus chlorobenzene and benzene. The mass spectral fragmentation pattern is complex.[19]

3. *Miscellaneous Derivatives*

Preparations of a variety of other products, including sulfur(VI) oxide adducts,[24] have been attempted, but no absolute characterizations have been reported.

III. CIS AND TRANS SULFANURIC FLUORIDES

A. Methods of Synthesis

Although it might be assumed that the preparation of sulfanuric fluorides could be effected by a pyrolysis reaction comparable with that used for the chlorides, the preparation of the fluorine analog $F_3P{=}NSO_2F$ has not been effected.[25] The thermal stabilities of the P—F and S—F bonds might well preclude such a reaction even if the compound were available.

To date, the only successful syntheses of the sulfanuric fluorides have involved fluorination of α-sulfanuric chloride.[23,26] Seel and Simon[26] treated the chloride with potassium fluoride under pressure at 145°C, using carbon tetrachloride as a reaction medium:

$$N_3S_3O_3Cl_3 + 3KF \longrightarrow N_3S_3O_3F_3 + 3KCl(s)$$

Removal of solid potassium salts and the carbon tetrachloride yielded a colorless liquid, from which by gas chromatography two products, cis (mp 17.4°C, bp 138.4°C) and trans (mp −12.5°C, bp 130.3°C) sulfanuric fluorides were isolated in very small quantities.* A modification of this procedure gives a larger yield of the more readily separable cis isomer.[23]

B. Preparative Procedure

Directions for synthesis at ambient pressure and lower temperature are as follows[23]:

To a finely powdered mixture of α-sulfanuric chloride (30 g, 0.10 mole) and potassium fluoride (35 g, 0.60 mole) are added 100 ml of acetonitrile and 0.2 to 0.3 ml of water. The mixture is maintained under reflux at 90 to 95°C (oil bath temperature) for 15 hr while being stirred vigorously. The solid is then removed by filtration and washed with acetonitrile. The combined filtrate and washings are distilled *in vacuo* at 100°C, leaving a sticky, resinous residue that sets to a yellowish-white powder on cooling. This residue is a highly polymeric sulfanuric fluoride[27] and is discarded. The distillate is fractionally distilled at 760 mm Hg, using a 50-cm column, to a product of sp gr 1.1 to 1.2 to remove most of the more volatile acetonitrile. This product is re-distilled at 1 atm, giving a later fraction of sp gr 1.80. From this mixture of sulfanuric fluorides, the pure cis isomer is recovered in 23 percent yield (about 5.6 g) by a second distillation under the same conditions. The pure trans compound is not recoverable in quantity by this procedure.

Again several repetitions of this procedure may be necessary to give consistently the optimum yield indicated.

C. Properties

The numerical constants of cis and trans sulfanuric fluorides are summarized in Table III. The single ^{19}F resonance (at −72.8 ppm versus $CFCl_3$)[28] for the cis isomer indicates that all three fluorine atoms in this compound are in the same environment and suggests a cis molecular conformation.[26,28] The trans conformation for the other isomer is suggested by two ^{19}F resonances. The infrared data recorded for the cis isomer compare closely with those for the sulfanuric chloride trimers (Table II). It seems probable that non-planar molecular structures also characterize these compounds, but definitive data are lacking.

* The cis-trans designation is that proposed by Seel and Simon on the basis of ^{19}F nuclear magnetic resonance data (p. 72). This designation is used here in the absence of definitive structural data.

TABLE III

Properties of Sulfanuric Fluorides[26]

Property	cis $N_3S_3O_3F_3$	trans $N_3S_3O_3F_3$
Melting point, °C	17.4	−12.5
Vapor pressure at 25°C, mm Hg	9	10
Vapor pressure at 100°C, mm Hg	227	279
Boiling point at 760 mm Hg, °C	138.4	130.3
Enthalpy of vaporization, kcal $mole^{-1}$	9.6	9.8
Entropy of vaporization, cal deg^{-1} $mole^{-1}$	23.3	24.4
Density, g ml^{-1}	1.92	1.92
Refractive index, n_D at 25°C	1.4166	1.4169
^{19}F nuclear magnetic resonance parameter W_F, ppm (in CH_3CN)	−72.25	−71.86, −72.76
Strongest infrared bands, cm^{-1}	518	522
	776	553
	875	781
	1168	798
	1395	899
		1172
		1389

cis-Sulfanuric fluoride is soluble in diethyl ether, benzene, and acetonitrile. The compound is hydrolytically stable. Mono- and diphenyl derivatives are obtainable by reaction with phenyl lithium, triphenyl derivatives by reaction with benzene in the presence of aluminum chloride, and diamino derivatives by reaction with various amines.[23] Comparable data for the trans isomer are lacking.

D. Derivatives of Sulfanuric Fluoride Trimers

Although the derivative chemistry of α-sulfanuric chloride is extremely limited, that of cis-sulfanuric fluoride is varied, and presumably very extensive, probably as a consequence of enhanced ring stability promoted by the more electronegative fluorine atoms. Furthermore, the increased strength of the sulfur-fluorine bond allows both for the handling of the material under more rigorous experimental conditions and for the removal of the halogen atom in a less exothermic fashion. Hydrolysis and ammonolysis reactions have not been investigated, but a limited number of aminolysis and phenylation reactions have.[23]

1. Amino Derivatives

Several derivatives of secondary amines have been obtained by direct aminolysis.[23] In every instance, only disubstituted products result, irrespective of the mole ratio of reactants used. The yields of crude products are good, but separation into isomeric products reduces final yields substantially. The procedure, as follows, for the syntheses of dimorpholino derivatives is typical:

To a solution of cis-sulfanuric fluoride (1.8 g, 0.0075 mole) in 30 ml of acetonitrile cooled to 0°C is added, dropwise, with stirring, and over a period of 1 hr, a solution of 5.3 g (0.06 mole) of morpholine in 30 ml of acetonitrile. The system is allowed to warm to room temperature and to stand at this temperature for 2 hr. The precipitate that forms is removed, and the clear filtrate evaporated at 60°C and 5 mm Hg. The solid residue is extracted fractionally with 10-ml volumes of absolute ethanol. Fractions 1 to 5 contain largely the lower-melting isomer; fractions 6 to 12 the higher-melting isomer. The pure isomers are obtained by recrystallization from absolute ethanol. Yields are 18 percent (mp 146°C) and 32 percent (mp 195°C).

Melting points and more important bands from the infrared spectra are included in Table IV. The vibrational spectra resemble those already discussed for other trimeric sulfanuric compounds, but absolute assignments can only be tentative. Strong absorptions in the 1250 to 1300 cm^{-1} region

TABLE IV

Properties of Sulfanuric Fluoride Derivatives[23]

Compound*	Melting point (°C)	Strongest infrared absorptions (cm^{-1})
$N_3S_3O_3F(C_4H_8ON)_2$	146	520, 750, 940, 1115, 1125, 1160, 1290, 1355
	195	515, 720, 840, 940, 955, 1115, 1140, 1170, 1290, 1355
$N_3S_3O_3F(C_6H_{12}ON)_2$	129	750, 1010, 1085, 1130, 1170, 1290, 1360,
	166	760, 1015, 1085, 1130, 1285, 1295, 1350
$N_3S_3O_3F(C_6H_{10}N)_2$	114	710, 725, 835, 935, 1060, 1135, 1280, 1290, 1355
	134	735, 840, 940, 955, 1065, 1275, 1290, 1355
$N_3S_3O_3F(C_4H_8N)_2$	127	595, 740, 845, 1020, 1130, 1170, 1290, 1360
	148	525, 585, 655, 740, 845, 1025, 1115, 1130, 1145, 1350
$N_3S_3O_3F_2(C_6H_5)$	95	745, 810, 1140, 1292, 1385
$N_3S_3O_3F(C_6H_5)_2$	107[19]	528, 560, 708, 733, 812, 823, 1080, 1125, 1270, 1315, 1345, 1445
	119	730, 1135, 1265, 1280, 1360
$N_3S_3O_3(C_6H_5)_3$	148	565, 685, 715, 755, 830, 1120, 1265
	177	535, 560, 715, 830, 1120, 1135, 1265

* C_4H_8ONH = morpholine
$C_6H_{12}ONH$ = 2,6-dimethylmorpholine
$C_5H_{10}NH$ = piperidine
C_4H_8NH = pyrrolidine

probably reflect the S—O stretching mode. Ring stretching modes are probably indicated by strong absorptions in the 1100 to 1200 and 750 cm^{-1} regions. Major differences between isomers for a particular amine appear in these regions, as may be expected if the ring atoms find themselves in different geometrical environments. Fine structure in the 730 to 770 cm^{-1} regions suggest that exocyclic S—N vibrations may also contribute to these absorptions.

Interpretation of ^{19}F nuclear magnetic resonance data (Table V), on the basis of assignments of a chemical shift of about -73 ppm to an axial fluorine atom and a shift of about -77 ppm to an equatorial fluorine atom, suggests a cis type of molecular conformation for each higher melting isomer and a trans type for each lower melting isomer.[28]

2. *Phenyl Derivatives*

Monophenyl and diphenyl derivatives can be obtained from cis-sulfanuric fluorides by reaction with phenyl lithium in appropriate stoichiometry, using diethyl ether as a reaction medium[23]:

$$N_3S_3O_3F_3 + C_6H_5Li \longrightarrow N_3S_3O_3F_2(C_6H_5) + LiF(s)$$

$$N_3S_3O_3F_3 + 2C_6H_5Li \longrightarrow N_3S_3O_3F(C_6H_5)_2 + 2LiF(s)$$

This kind of reaction has proved particularly useful for the preparation of phenylphosphonitrilic fluorides,[29] but contrary to what is observed with the latter series of compounds, the completely phenylated derivative cannot be made in this way. This compound does result, in two isomeric forms, in a Friedel-Crafts reaction.[23] An isomeric diphenyl derivative results when diphenylsulfanuric chloride is treated with potassium fluoride.[19]

TABLE V

^{19}F Nuclear Magnetic Spectral Data[28]

Compound	Melting point, °C	Chemical shift, ppm versus $CFCl_3$
$N_3S_3O_3F_3$ (cis)	16	-72.8
$N_3S_3O_3F(C_4H_8ON)_2$	146	-73.0
	195	-73.2
$N_3S_3O_3F(C_6H_{12}ON)_2$	129	-72.6
	166	-78.1
$N_3S_3O_3F(C_5H_{10}N)_2$	114	-72.4
	134	-78.8
$N_3S_3O_3F(C_4H_8N)_2$	127	-72.3
	148	-76.9
$N_3S_3O_3F_2(C_6H_5)$	95	-77.2
$N_3S_3O_3F(C_6H_5)_2$	119	-72.0

Useful procedures for synthesis are as follows:

a. Monophenyl sulfanuric fluoride.[23] To a solution of 4.9 g (0.02 mole) of cis-sulfanuric fluoride in 100 ml of diethyl ether and cooled to −70°C, 62 ml of a 3.4*M* solution (0.02 mole) of phenyl lithium in diethyl ether is added during a period of 1 hr with stirring. The suspension is allowed to warm to room temperature and after 1 hr at this temperature is filtered to remove precipitated lithium fluoride. The solvent is removed *in vacuo*, and the product recrystallized twice from *n*-heptane. The yield is about 25 percent.

*b. Diphenyl sulfanuric fluoride (mp 119°*C).[23] The same procedure is followed, using 3.75 g (0.013 mole) of cis-sulfanuric fluoride and 0.04 mole of phenyl lithium. The yield is 62 percent.

c. Diphenyl sulfanuric fluoride (mp 107°C).[19] Diphenyl sulfanuric chloride (1.6 mmoles) is treated with 3.2 mmoles of anhydrous potassium fluoride in 10 ml of dry acetonitrile containing 0.01 ml of water. After being stirred for 21 hr, the precipitated potassium chloride is removed and the filtrate evaporated *in vacuo*. The residue is extracted with several 10-ml portions of dry *n*-heptane, and the product crystallized by cooling. The yield is about 71.4 percent.

d. Triphenyl sulfanuric fluorides.[23] A solution of 1.8 g (0.0075 mole of cis-sulfanuric fluoride in 40 ml of benzene is treated with 6 g (0.045 mole) of anhydrous aluminum chloride and heated at reflux for 2 days. The unreacted benzene is removed at 60°C and 5 mm Hg, and the residue extracted repeatedly with hot absolute ethanol. The resulting solution is allowed to crystallize for 2 days in a refrigerator. The solid is extracted repeatedly with 10-ml volumes of absolute ethanol, with activated charcoal added to remove color. Crystallization of the first two extracts yields largely the lower-melting isomer (Table IV); of the remaining extracts, the higher-melting isomer. Each is purified by recrystallization from absolute ethanol. The yields are about 7 percent (mp 148°C) and 10 percent (mp 177°C).

Melting-point and vibrational-spectral data are given in Table IV. These compounds are soluble in hot ethanol, *n*-heptane, and *n*-hexane and in benzene, toluene, and acetonitrile. Diphenyl sulfanuric fluoride (mp 107°C) sublimes readily *in vacuo*.[19] Thermal decomposition and fragmentation patterns are comparable with those of diphenyl sulfanuric chloride.[19] Reactions of phenyl lithium with cis-sulfanuric fluoride do not yield isomeric products. It is probable that the two diphenyl derivatives[19,23] bear the same kind of structural relationship to each other as noted for other isomeric derivatives, but no definitive evidence is yet available. The same situation probably exists with the triphenyl derivatives.

A detailed single-crystal x-ray analysis of monophenyl sulfanuric fluoride

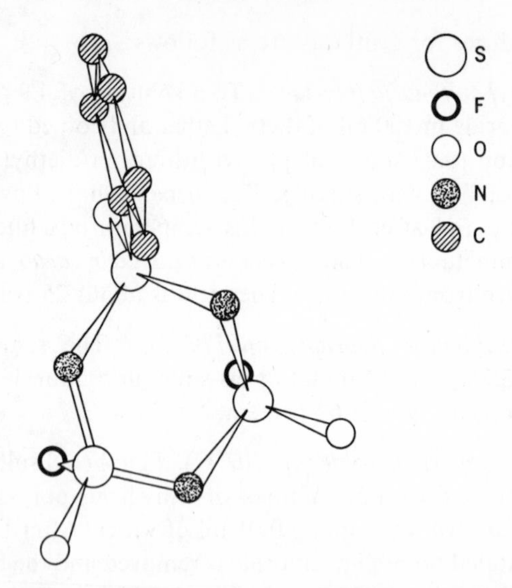

Figure 2. Molecular structure of $N_3S_3O_3F_2$ (C_6H_5).[30]

trimer indicates a triclinic crystal structure, space group $P\bar{1}$, with unit cell constants $a = 7.59(2)$ Å., $b = 8.25(3)$ Å., $c = 9.96(3)$ Å., $\alpha = 109.6(1)°$, $\beta = 108.0(2)°$, $\gamma = 93.3(2)°$.[30] There are two molecules per unit cell. The molecular structure (Figure 2) amounts to six-membered S—N ring in a somewhat irregular chair conformation, with the two fluorine atoms and one oxygen atom in axial positions and the phenyl group and the other two oxygen atoms in equatorial positions. Important bond distances and angles are S—N, 1.52 to 1.64 Å, with a mean of 1.56 Å; S—O, average 1.41 Å; S—F, 1.55 Å; N—S—N, 112.1°, 112.9°, 116.8°; S—N—S, 118.9°, 119.8°, 125.5°; N—S—F, 100.8°, 103.2°, 103.5°; O—S—F, 105.6°, 107.6°; N—S—C, 103.5°, 103.8°; O—S—C, 111.4°. An interpretation of the single ^{19}F resonance (Table V) suggests that in solution the two fluorine atoms and the phenyl group in this compound are in equatorial positions.[28] Whether the difference between the structure in the solid state and in solution results from a temperature-dependent isomerization process cannot be determined until a nuclear-magnetic-resonance study as a function of temperature has been completed.[31]

IV. MIXED SULFANURIC-PHOSPHONITRILIC CHLORIDES

An extension of the isoelectronic concept predicts the existence of two mixed sulfanuric-phosphonitrilic trimers. Only one of these, $(NSOCl)_2$-$(NPCl_2)$ (V), has been described.[32–34] This compound, for which the name cyclo-tri-μ-nitrido-dichlorophosphorusbis(oxochlorosulfur) has been proposed,[33] was first obtained as a product of the decomposition of trichlorophosphazosulfonyl chloride[11,14] at 8 mm Hg and 146°C while under ultraviolet radiation.[32] Attempts to repeat this procedure have been either unsuccessful[25] or productive of only a very small yield.[34] An alternative procedure, based upon a Kirsanov reaction between a linear phosphonitrile, such as $[Cl_3PNPCl_3]^+[PCl_6]^-$ or $[Cl_3PNPCl_2NPCl_3]^+[PCl_6]^-$, and sulfamic acid, gives a larger yield without contamination by α- and β-sulfanuric chlorides.[34]

A. Preparative Procedure for $(NSOCl)_2(NPCl_2)$

In a typical procedure,[34] 120 g (0.23 mole) of the compound $[Cl_3PNCl_3]^+[PCl_6]^-$, prepared by the method of Becke-Goehring and Lehr,[35] is heated with 21.8 g (0.23 mole) of sulfamic acid at 100°C in a flask equipped with a condenser until the mixture liquefies and the evolution of hydrogen chloride gas ceases (about 6 hr). The pale yellow viscous liquid is freed from particles of unreacted sulfamic acid by filtration under dry nitrogen. The filtrate and washings, using dry carbon tetrachloride, are then heated at about 1 mm Hg in a flask equipped with a Vigreux column to remove carbon tetrachloride and phosphorus(V) oxotrichloride from the initial reaction. The temperature is raised to about 155°C (oil-bath), and the flask is held at this temperature for about 32 hr to complete the pyrolysis reaction. The reaction vessel is brought to 1 atm pressure with dry nitrogen, and its contents extracted with dry carbon tetrachloride. The solvent is removed, and the crude product is recrystallized from *n*-hexane, cyclohexane, or *n*-heptane. The yield is 15 percent.

B. Properties of $(NSOCl)_2(NPCl_2)$

The compound is obtained as colorless, needlelike crystals that melt at 96.5°C without decomposition. It is soluble in carbon tetrachloride, cyclohexane, *n*-hexane, *n*-heptane, and acetonitrile. Hydrolytic stability in moist air has been reported. Crystals suitable for x-ray examination can be grown from petroleum ether (80 to 100°C). A single-crystal investigation[33] indicated a monoclinic unit cell of space group $P2_1/n$ with $a = 11.649(3)$ Å, $b = 7.705(3)$ Å, $c = 11.092(3)$ Å, and $\beta = 101.37(13)°$. There are four molecular units per unit cell. The molecular structure (Figure 3) amounts to a deformed chairlike cyclic arrangement of atoms in the sequence —S—N—S—N—P—N—, with one chlorine atom bonded to each sulfur atom, two chlorine atoms to the phosphorus atom, and one oxygen atom to each sulfur atom. The arrangement about each sulfur atom and the phosphorus atom is roughly tetra-

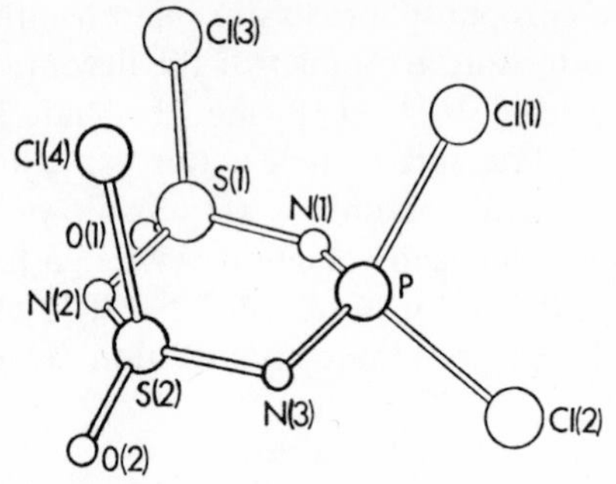

Figure 3. Molecular structure of $(NSOCl)_2$ $(NPCl_2)$.[33]

hedral. Mean bond lengths are N—P = 1.585(13) Å, P—Cl = 1.957(6) Å, S—Cl = 2.018(6) Å, and S—O = 1.421(13) Å. The N—S bond lengths average to 1.578(13) Å for the S—N—S group and 1.540(13) Å for the others. Endocyclic angles at P and S are 115.3(7)° and 115.0(7)°, respectively. Other angles are S—N—S = 120.3(8)°; P—N—S = 123.5(8°) and 120.6(8)°. The molecular structure is closely related to other sulfanuric molecular structures[17,30] and that of trimeric phosphonitrilic chloride.[36]

Strong infrared absorptions at 1348, 1335, 1177, 1142, 720, 624, 558, and 537 cm^{-1} (cyclohexane solution)[34] are in agreement with a mixed-ring molecular structure. The ^{35}Cl nuclear quadrupole resonance spectrum also supports the molecular geometry obtained for the crystalline compound.[34]

References

1. E. C. Franklin, *The Nitrogen System of Compounds*, Reinhold, New York, 1935.
2. L. F. Audrieth, M. Sveda, H. H. Sisler, and M. J. Butler, *Chem. Rev.*, **26**, 49 (1940).
3. W. Traube, *Ber.*, **25**, 2472 (1892).
4. A. Hantzsch and A. Holl, *Ber.*, **34**, 3430 (1901).
5. A. Hantzsch and B. C. Stuer, *Ber.*, **38**, 1022 (1905).
6. G. Heinze and A. Meuwsen, *Z. Anorg. Allgem. Chem.*, **275**, 49 (1954).
7. R. Appel and M. Goehring, *Z. Anorg. Allgem. Chem.*, **271**, 171 (1953).
8. M. Becke-Goehring, in *Advances in Inorganic Chemistry and Radiochemistry*, H. J. Emeléus and A. G. Sharpe (eds.), vol. 2, pp. 185–190, Academic, New York, 1960.
9. M. Becke-Goehring, *Angew. Chem.*, **73**, 589 (1961); *Inorg. Macromol. Rev.*, **1**, 17 (1970).
10. O. Glemser, *Angew. Chem. Intern. Ed. Engl.*, **2**, 530 (1963).
11. A. V. Kirsanov, *J. Gen. Chem. U.S.S.R.*, **22**, 93, 101 (1952).
12. M. Goehring, J. Heinke, H. Malz, and G. Roos, *Z. Anorg. Allgem. Chem.*, **273**, 200 (1953).
13. M. Goehring and H. Malz, *Z. Naturforsch.*, **9b**, 567 (1954).
14. A. Vandi, T. Moeller, and T. L. Brown, *Inorg. Chem.*, **2**, 899 (1963).
15. A. J. Banister and A. C. Hazell, *Proc. Chem. Soc.*, 282 (1962).

16. G. A. Wiegers and A. Vos, *Proc. Chem. Soc.*, 387 (1962).
17. A. C. Hazell, G. A. Wiegers, and A. Vos, *Acta Cryst.*, **20**, 186 (1966).
18. D. P. Craig, *J. Chem. Soc.*, 997 (1959).
19. R. L. McKenney, Jr., and N. R. Fetter, *J. Inorg. Nucl. Chem.*, **30**, 2927 (1968).
20. H. Malz, doctoral dissertation, Heidelberg (1954).
21. M. Becke-Goehring, in *Developments in Inorganic Polymer Chemistry*, M. F. Lappert and G. J. Leigh (eds.), p. 116, Elsevier, New York, 1962.
22. A. Failli, M. A. Kresge, C. W. Allen, and T. Moeller, *Inorg. Nucl. Chem. Letters*, **2**, 165 (1966).
23. T. Moeller and A. Ouchi, *J. Inorg. Nucl. Chem.*, **28**, 2147 (1966).
24. M. Goehring, H. Hohenschutz, and R. Appel, *Z. Naturforsch.*, **9b**, 678 (1954).
25. T. Moeller, *Final Report*, Army Research Office (Durham), Contract DA-31-124-ARO(D)-35, Jan. 31, 1969.
26. F. Seel and G. Simon, *Z. Naturforsch.*, **19b**, 354 (1964).
27. F. Seel and G. Simon, *Angew. Chem.*, **72**, 709 (1960).
28. T. H. Chang, T. Moeller, and C. W. Allen, *J. Inorg. Nucl. Chem.*, **32**, 1043 (1970).
29. T. Moeller and F. Tsang, *Chem. Ind.*, 361 (1962).
30. D. E. Arrington, T. Moeller, and I. C. Paul, *J. Chem. Soc.*, in press.
31. C. W. Allen, private communication.
32. J. C. van de Grampel and A. Vos, *Rec. Trav. Chim.*, **82**, 246 (1962).
33. J. C. van de Grampel and A. Vos, *Acta Cryst.*, **B25**, 651 (1969).
34. R. Clipsham, R. M. Hart, and M. A. Whitehead, *Inorg. Chem.*, **8**, 2431 (1969).
35. M. Becke-Goehring and W. Lehr, *Chem. Ber.*, **94**, 1591 (1961); *Z. Anorg. Allgem. Chem.*, **325**, 287 (1963).
36. A. Wilson and D. F. Carroll, *J. Chem. Soc.*, 2548 (1960).

Heterocyclic Compounds of the Group IV Elements

C. H. YODER

Department of Chemistry
Franklin & Marshall College, Lancaster, Pennsylvania

and

J. J. ZUCKERMAN

Department of Chemistry
State University of New York at Albany
Albany, New York

I. INTRODUCTION

In this chapter we discuss the methods for preparing heterocyclic compounds of the fourth-group elements excluding carbon. During recent years the syntheses of cyclic compounds in which atoms of silicon, germanium, tin, and lead are members of the ring has attracted wide attention. This may in part be due to the possibilities that these compounds present as precursors of polymeric materials or due to the comparisons that they allow with the analogous organic systems. The first of these compounds, silacyclohexane, was prepared independently in 1915 by Bygden[1] and by Grüttner and Wiernik[2] from a Grignard route. Today a very large number of compounds of various kinds are known, although no systematic review of the entire field has been attempted. Andrianov and Khananashvili published a review of cyclic organosilicon compounds in 1967.[3] A comprehensive monograph encompassing the whole field of inorganic ring systems has been written by Haiduc.[3a]

In this survey of the preparative methods used to close rings containing silicon, germanium, tin, and lead, we exclude those reactions which do not form a bond directly to these atoms; for example, we exclude from consideration ring-closing reactions taking place at remote locations in the molecule. We also exclude *homocyclic* compounds[4] constituted from a simple repeat unit

$$-\left[-\overset{|}{\underset{|}{M}}-\right]_n-$$

and *pseudoheterocyclic* compounds constituted from a simple alternating repeat unit

$$-\left[-\overset{|}{\underset{|}{M}}-E-\right]_n-$$

or those rings that are formed as transient intermediates and have not been isolated, or those ring systems formed by a donor-acceptor linkage to the fourth-group atom. The chemical and physical properties of the resulting heterocycles are not dealt with. We deviate from these exclusions, however, when an interesting method or example is at hand.

It will be apparent from the survey of methods that there is little of a systematic nature in the development of this field, and the topic-by-topic mode of presentation reflects the overall lack of coordinating principles. It is also difficult in most cases to discuss the earlier investigations critically, since structural information is largely lacking, and as a rule only analytical data are reported. Critical molecular-weight data that could distinguish between the monocyclic systems and their higher oligomers and polymers are, unfortunately, also largely lacking.

The method chosen for classifying the work is based on the kind of linkage found at the silicon, germanium, tin, or lead atom. Thus the large number of group IV–group IV carbocyclic rings are gathered into one category, followed by group IV–group V and group IV–group VI heterocycles. Recently there has been much activity in the field of group IV–transition-metal compounds, and several heterocyclic systems are now known. Methods of preparing these compounds are treated in a final section.

It will be seen that silicon compounds are by far the most numerous, and this no doubt reflects the major technical interest associated with the chemistry of the silicones. No effort is made to include all the possible examples of the application of each method.

II. SYNTHETIC PROCEDURES

A. Group IV–Group IV Bonds

1. The Direct Synthesis

The direct synthesis of organosilicon compounds was discovered by E. G. Rochow in 1940 and is now the basic process for the manufacture of starting materials for the silicone industries in all countries. Rochow later extended its application to germanium and tin.[5] The use of polyfunctional reactants has been widely explored with silicon, where heterocyclic compounds have been isolated from the products. The reaction of 1,4-dichlorobutane, for example, produces two heterocycles at 300°:

$$Cl(CH_2)_4Cl + Si \xrightarrow{Cu} \text{(1,1-dichlorosilacyclopentane)}^{6,7} + Cl_2Si\langle\text{ring}\rangle SiCl_2^{\,8} \tag{1}$$

The silacyclopentane was reported to constitute 30 percent of the yield.[6,7] The unsaturated 1,2-butene also gives a silacyclopentene in 15.6 percent yield[6,9]:

$$H_2C{=}CHCH(Cl)CH_2Cl + Si \xrightarrow[275\text{–}290°]{Cu} \text{(1,1-dichlorosilacyclopent-3-ene)} \tag{2}$$

since an allylic rearrangement to 1,4-dichlorobutene-2 takes place under the conditions of the reaction.

Elemental germanium has been shown to react with 2,2′-diiodoperfluorobiphenyl to give a spiro derivative[10]:

$$2\ \text{(2,2'-diiodoperfluorobiphenyl)} + M \longrightarrow \text{(spiro-bis(perfluorobiphenylene)M)} \qquad M = Ge, Sn \tag{3}$$

and tin metal reacts similarly at 230°.[11]

2. *Ring Closure by Alkylation*

a. DiGrignard reagents. In order of general importance the diGrignard reagents must be presented first. This method is not only by far the most frequently used, but the first silicon-containing carbocycle was prepared in this way in 1915 from the diGrignard reagent of 1,5-dibromopentane[1,2]:

$$BrMg(CH_2)_5MgBr + MCl_4 \longrightarrow \text{(1,1-dichloro-1-metallacyclohexane, } MCl_2\text{)} \tag{4}$$

and the analogous germanium,[12] tin,[13] and lead[14,15] compounds were prepared similarly. Although the element dichlorides react only slowly with additional diGrignard reagent, the spirocyclic compounds of silicon,[16,17] germanium,[16,18] and tin[16] have been prepared by this route:

$$BrMg(CH_2)_5MgBr + MCl_4 \longrightarrow \text{spiro-}M(C_5H_{10})_2 \quad M = Si, Ge, Sn \tag{5}$$

The lead compound has been made from the diGrignard reagent in two ways:

$$BrMg(CH_2)_5MgBr + (NH_4)_2PbCl_6 \xrightarrow{19} \text{spiro-}Pb(C_5H_{10})_2 + 2NH_4Br + 2MgBrCl \tag{6}$$

$$ClMg(CH_2)_5MgCl + PbCl_2 \xrightarrow{20} \text{spiro-}Pb(C_5H_{10})_2 + Pb + 4MgCl_2 \tag{7}$$

West studied the synthesis of cyclic organosilicon compounds from di-Grignard reagents of varying chain length and found that the yields increased: seven-membered < five-membered < six-membered ring and also with the number of chlorine atoms attached to silicon.[21]

Aralkyl diGrignard reagents have also been used to produce fused ring systems[22,23]:

$$o\text{-}C_6H_4[(CH_2)_3MgBr](MgBr) + R_2SiCl_2 \longrightarrow \text{benzo-fused silacyclohexane } (SiR_2) \tag{8}$$

which have also been prepared from diaryl diGrignard reagents[24]:

$$(o\text{-}BrMgC_6H_4CH_2)_2 + (C_6H_5)_2SiCl_2 \longrightarrow \text{dibenzosilepin with } Si(C_6H_5)_2 \tag{9}$$

The general reaction

$$ClSiSiCl + BrMg(CH_2)_nMgBr \xrightarrow{THF} \text{cyclic } \text{-Si—Si-}(CH_2)_n \tag{10}$$

where $n = 4$, 5, and 6, has been used to produce a series of 1,1,2,2-tetramethyl-1,2-disilacycloalkanes. The yields for the $n = 4, 5$, and 6 reactions were 66, 65, and >3 percent, respectively.[25]

The Barbier reaction has been used with trifunctional chlorosilanes to form bridgehead bicyclic systems[26,27]:

$$HC(CH_2Cl)_2(CH_2SiCl_3) + Mg \xrightarrow{\text{ether}} \text{(bicyclic)}\ SiCl \quad (11)$$

$$HC(CH_2{-}Cl)_2((CH_2)_2SiCl_3) + Mg \xrightarrow{\text{ether}} \text{(bicyclic)}\ SiCl \quad (12)$$

The disilanyl heterocycle below has been prepared by an intermolecular ring closure involving magnesium[28,29]:

$$\text{cyclo-}[(CH_3)_2Si{-}Si(CH_3)_2{-}CH_2{-}(CH_3)_2Si{-}Si(CH_3)_2{-}CH_2]$$

Intramolecular ring closure has been effected through the Barbier reaction on γ-halopropylhalosilanes to give the silacyclobutane[30]:

$$(CH_3)_3Si(CH_2)_3Br \xrightarrow[(2)H_2O]{(1)H_2SO_4} [Br(CH_2)_3Si(CH_3)_2]_2O \xrightarrow[H_2SO_4]{NH_4Cl}$$

$$Br(CH_2)_3Si(CH_3)_2Cl \xrightarrow[\text{ether}]{Mg} \square{-}Si(CH_3)_2 \quad (13)$$

and many derivatives are now known.[31] Cyclization of bromo-(3-bromo-3-phenylpropyl)diphenylsilane failed to take place in ether and required THF.[32] The four-membered ring containing two silicon atoms has been prepared by two routes involving a Barbier alkylation—first by way of an intramolecular closure[33]:

$$FSi(CH_3)_2CH_2Si(CH_3)_2CH_2Cl + Mg \xrightarrow{\text{ether}} (CH_3)_2Si\langle{}^{CH_2}_{CH_2}\rangle Si(CH_3)_2 \quad (14)$$

and later by an intermolecular route[34,35]:

$$2ClCH_2Si(CH_3)_2Cl + 2Mg \xrightarrow{\text{THF}} (CH_3)_2Si\langle{}^{CH_2}_{CH_2}\rangle Si(CH_3)_2 \quad (15)$$

In the preparation of compounds unsymmetrically substituted at silicon an added complication is the presence of the two symmetrical species.[36] The mixed carbon, silicon, germanium cyclobutane has been made similarly intermolecularly[37]:

$$(CH_3)_2Si(CH_2Cl)_2 + (CH_3)_2GeCl_2 + 2Mg \xrightarrow{THF} (CH_3)_2Si \diamond Ge(CH_3)_2 \quad (16)$$

or through an intramolecular route[38]:

$$\underset{X}{(CH_3)_2SiCH_2}\underset{CH_2X}{Ge(CH_3)_2} + Mg \xrightarrow{THF} (CH_3)_2Si \diamond Ge(CH_3)_2 \quad (17)$$

The pseudo silacyclobutene, 2,3-benzo-1,1-diphenyl-1-silacyclobut-2-ene, has been reported by three kinds of organomagnesium synthesis[39]:

$$o\text{-}CH_3C_6H_4Si(C_6H_5)_2H \xrightarrow{NBS} o\text{-}BrCH_2C_6H_4Si(C_6H_5)_2Br + Mg \xrightarrow{THF} \text{2,3-benzo-1-silacyclobut-2-ene-}Si(C_6H_5)_2 \quad (18)$$

$$o\text{-}BrC_6H_4CH_2Si(C_6H_5)H \xrightarrow{PCl_5} o\text{-}BrC_6H_4CH_2Si(C_6H_5)_2Cl + Mg \xrightarrow{THF} \text{2,3-benzo-1-silacyclobut-2-ene-}Si(C_6H_5)_2 \quad (19)$$

$$o\text{-}BrC_6H_4CH_2MgBr + (C_6H_5)_2SiCl_2 \xrightarrow{THF} o\text{-}BrC_6H_4CH_2Si(C_6H_5)_2 \xrightarrow[THF]{Mg} \text{2,3-benzo-1-silacyclobut-2-ene-}Si(C_6H_5)_2 \quad (20)$$

The silacyclopentene and silacyclohexene systems have been prepared by way of diGrignard cyclization of the appropriate open-chain chlorosilane[40–43]:

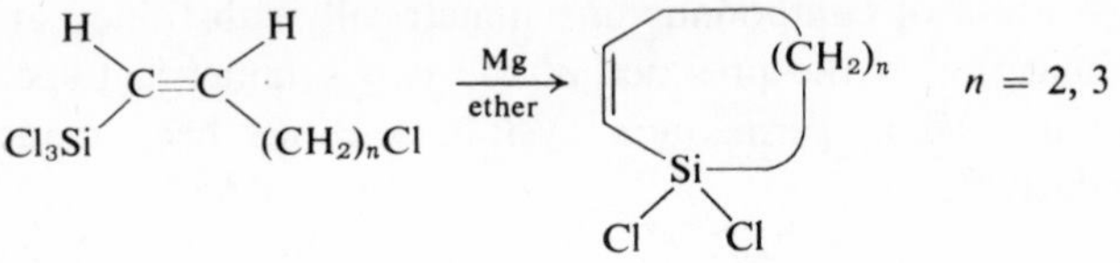

$$\mathrm{Cl_3Si(H)C{=}C(H)(CH_2)_nCl} \xrightarrow[\text{ether}]{\text{Mg}} \text{cyclo-}[\mathrm{CH{=}CH(CH_2)_nSiCl_2}] \qquad n = 2, 3 \tag{21}$$

b. Dilithium reagents. Several examples attest to the more powerful nature of the organolithium derivatives over the analogous diGrignard reagents. West and Rochow noted that the alkyldilithium reagents reacted readily with carbocyclic chlorosilanes to produce the spiro compounds,[44] although the diGrignard synthesis proceeds only slowly.[21] The enhanced reactivity of the organolithium compounds is used in the synthesis of sterically hindered heterocycles; for example, dilithiocarboranes have been used to prepare a number of cyclic silicon[45,46] and tin[47] compounds:

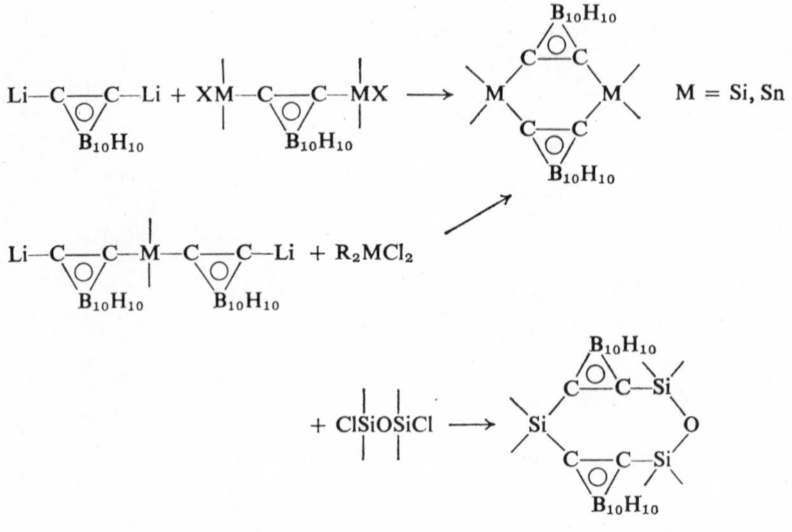

$$\mathrm{Li{-}C(B_{10}H_{10})C{-}Li + XM{-}C(B_{10}H_{10})C{-}MX} \longrightarrow \text{cyclo-}[\mathrm{M{-}C(B_{10}H_{10})C{-}M{-}C(B_{10}H_{10})C}] \qquad \mathrm{M = Si, Sn}$$

$$\mathrm{Li{-}C(B_{10}H_{10})C{-}M{-}C(B_{10}H_{10})C{-}Li + R_2MCl_2} \nearrow \tag{22}$$

$$+ \ \mathrm{ClSiOSiCl} \longrightarrow \text{cyclo-}[\mathrm{Si{-}C(B_{10}H_{10})C{-}Si{-}O{-}Si{-}C(B_{10}H_{10})C}]$$

Silicon-substituted ferrocenes have likewise been prepared from 1,1′-dilithioferrocene.[48] [See equations (23), page 89.]

The organodilithium reagents are formed in a separate step or *in situ* by use of a trans-metalating alkyl- or aryllithium agent or by use of lithium metal. Coupling by alkali metals is treated in the concluding part of this section. The initial work of West on carbocyclic silanes[21,44,49] has been extended by others to the formation of silacyclobutanes,[17] pentanes,[50,51] and hexanes[51]: [See equation (24).]

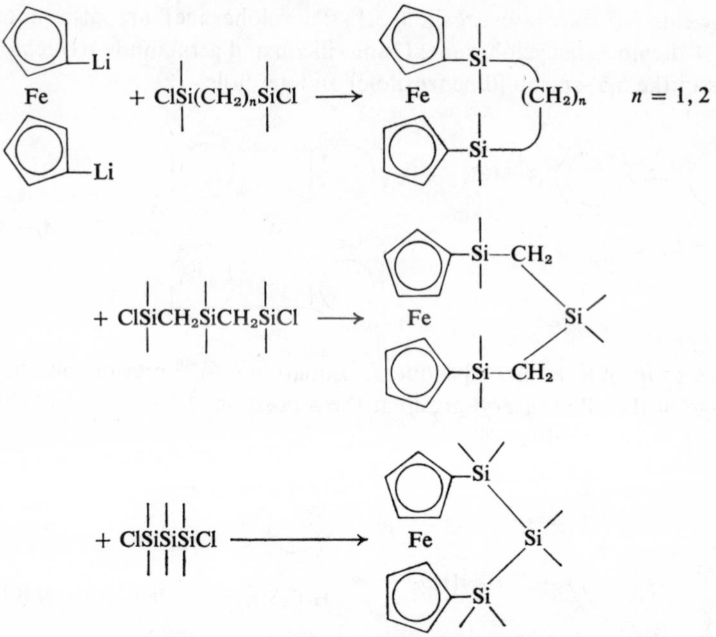

$$\text{(1,1'-dilithioferrocene)} + \text{ClSi(CH}_2)_n\text{SiCl} \longrightarrow \text{Fe}[\text{C}_5\text{H}_4\text{Si}]_2(\text{CH}_2)_n \qquad n = 1, 2$$

$$+ \text{ClSiCH}_2\text{SiCH}_2\text{SiCl} \longrightarrow \text{Fe}[\text{C}_5\text{H}_4\text{Si}-\text{CH}_2]_2\text{Si} \qquad (23)$$

$$+ \text{ClSiSiSiCl} \longrightarrow \text{Fe}[\text{C}_5\text{H}_4\text{Si}]_2\text{Si}$$

$$\text{Li(CH}_2)_n\text{Li} + \text{X}_2\text{SiR}_2 \longrightarrow \text{(CH}_2)_n\text{SiR}_2 \text{ (cyclic)} + 2\text{LiX} \qquad n = 3, 4, 5 \qquad (24)$$

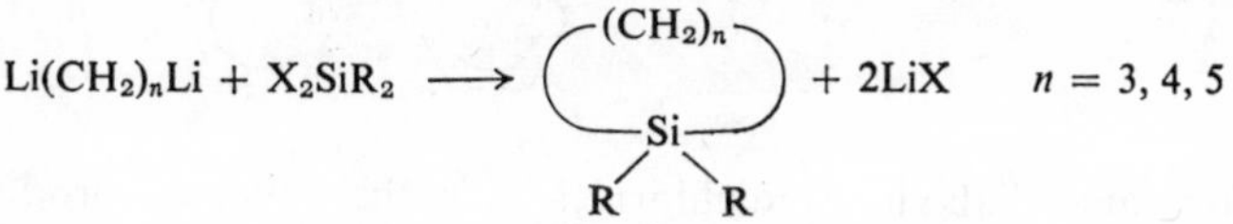

Use of *o*-phenylenedilithium results in the silicon analog of 9,10-dihydroanthracene[52,53]:

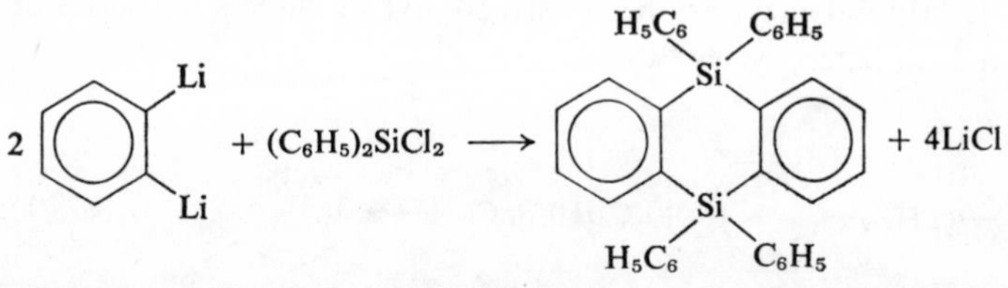

$$2\ o\text{-C}_6\text{H}_4\text{Li}_2 + (\text{C}_6\text{H}_5)_2\text{SiCl}_2 \longrightarrow (\text{C}_6\text{H}_4)_2[\text{Si}(\text{C}_6\text{H}_5)_2]_2 + 4\text{LiCl} \qquad (25)$$

The analogous octafluorotetramethyldisilanthrene has recently been prepared from 1,2,3,4-tetrafluorobenzene after lithiation.[54]

o,o'-Dilithiobiphenyl gives the dibenzosilole in low yield with diphenyldichlorosilane[55,56] or -germane,[56] but more complete conversion is achieved when the number of chlorine atoms on silicon is increased[57]:

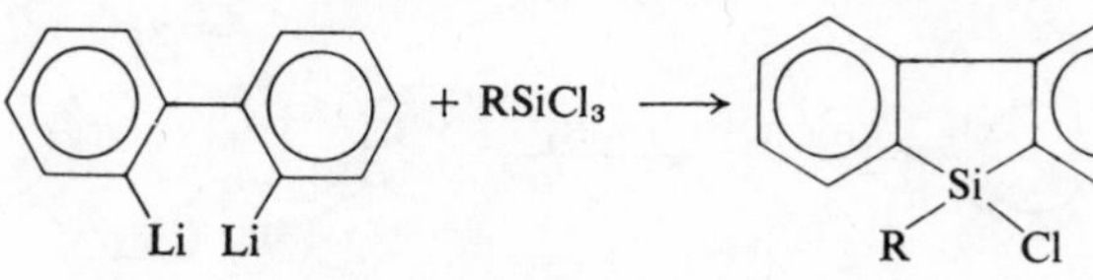

$$o,o'\text{-LiC}_6\text{H}_4\text{C}_6\text{H}_4\text{Li} + \text{RSiCl}_3 \longrightarrow (\text{C}_6\text{H}_4)_2\text{Si(R)Cl} \qquad (26)$$

Low yields of spiro[dibenzosilole-5,1′-silacyclohexane] are also obtained with 1,1-dichlorosilacyclohexane. Using silicon and germanium tetrachlorides results in the 5,5′-spirobi[dibenzosilole] and germole:

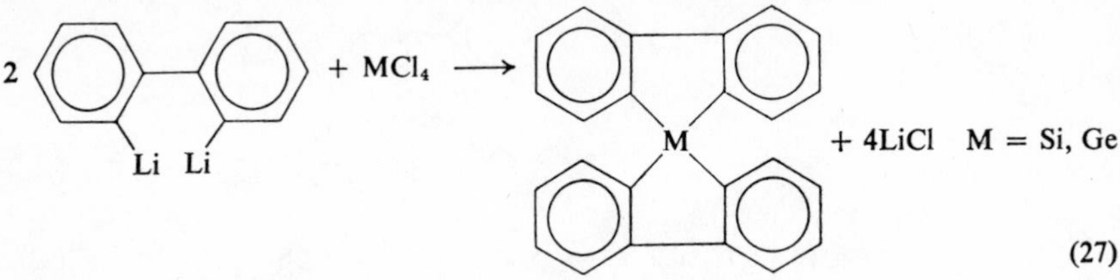

$$2\ (o,o'\text{-}Li_2C_{12}H_8) + MCl_4 \longrightarrow \text{spiro product} + 4LiCl \quad M = Si, Ge \tag{27}$$

which also form as a minor product in Equation (26),[58] presumably through cleavage of the alkyl or aryl group at the 5 position.[58] Use of ethyldichlorosilane produces the dimer[59]:

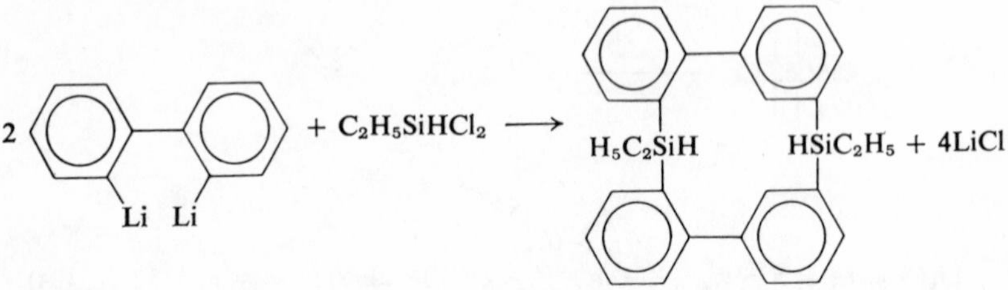

$$2\ (o,o'\text{-}Li_2C_{12}H_8) + C_2H_5SiHCl_2 \longrightarrow H_5C_2SiH \cdots HSiC_2H_5 + 4LiCl \tag{28}$$

Diphenyl-[60] and dialkyltin[61] dichlorides give the expected product, and bis(o,o'-octafluorobiphenylene)germane has been prepared by the route in Equation (27).[62,63]

o,o'-Dilithiobibenzyl has been used to give an interesting series of seven-membered tin heterocycles[64]:

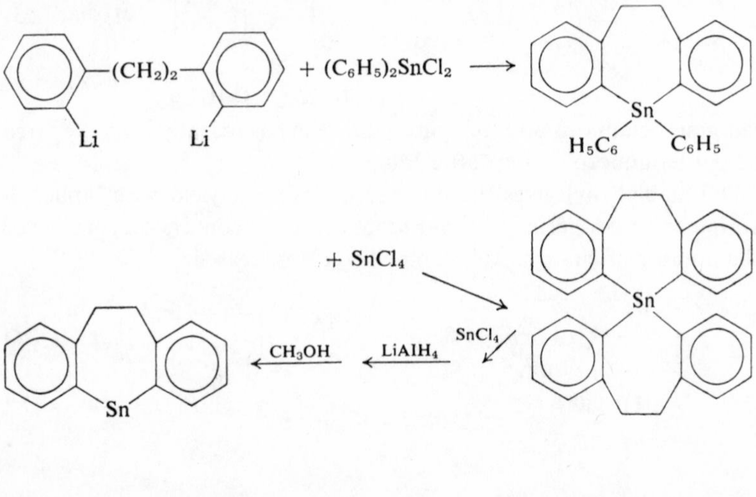

$$o\text{-}LiC_6H_4\text{-}(CH_2)_2\text{-}C_6H_4Li\text{-}o + (C_6H_5)_2SnCl_2 \longrightarrow \text{Sn}(C_6H_5)_2 \text{ heterocycle}$$

$$+ SnCl_4 \longrightarrow \text{spiro-Sn} \xrightarrow{SnCl_4} \xrightarrow{LiAlH_4} \xrightarrow{CH_3OH} \text{Sn heterocycle} \tag{29}$$

1,4-Dilithio-1,2,3,4-tetraphenylbutadiene has been used in an apparently general reaction with fourth-group di- and tetrahalides to yield monocyclic and spiro heterocyclopentadienes (metalloles), respectively. The dark red dilithium derivative is prepared by the dimerization of diphenylacetylene (tolan) by lithium in ether.[65] The general equation below has been used to prepare several silicon,[66–75] germanium,[66–68,76] tin,[66,68,76,77] and lead[76] derivatives of the respective metalloles:

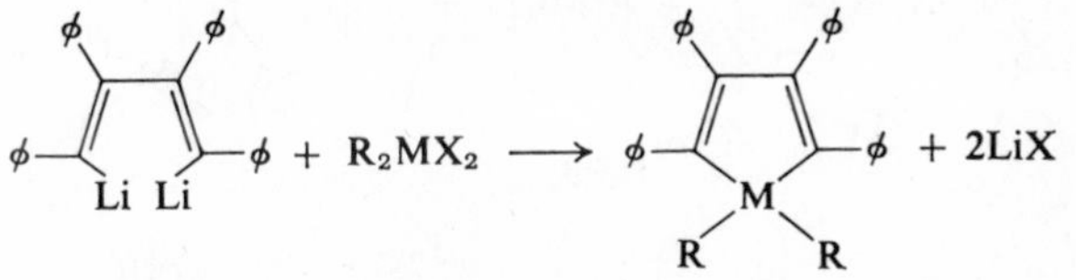

$$\phi_4C_4Li_2 + R_2MX_2 \longrightarrow \phi_4C_4MR_2 + 2LiX \qquad (30)$$

when the organodilithium reagent is added to the halide. In the reverse addition the organodilithium reagent is always present in excess, and the tin synthesis is affected in that cleavage of groups from tin takes place to form the octaphenyl-1,1′-spirobistannole:

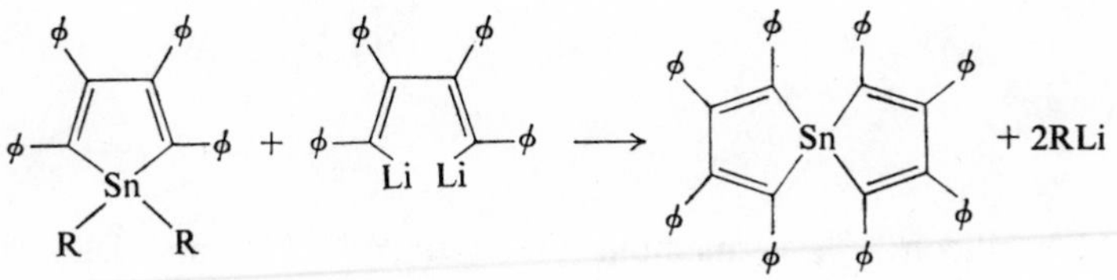

$$\phi_4C_4SnR_2 + \phi_4C_4Li_2 \longrightarrow (\phi_4C_4)_2Sn + 2RLi \qquad (31)$$

where the tendency toward the behavior in Equation (31) is methyl and ethyl > vinyl > phenyl, just the reverse of the usual cleavage series in acidic media at tin. Diorganosilicon, -germanium, and -lead dichlorides give the expected monocyclic products, no matter what the sequence of addition.[7]

The reaction of diphenylacetylene with *n*-butyllithium produces a dilithium intermediate that offers a facile entry to the metalloindenyl systems[77,78]:

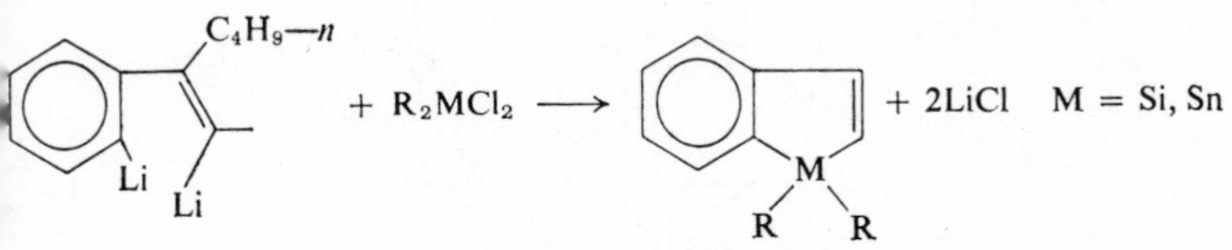

$$\text{(dilithio intermediate, } C_4H_9\text{—}n\text{)} + R_2MCl_2 \longrightarrow \text{(metalloindene, } MR_2\text{)} + 2LiCl \quad M = Si, Sn \qquad (32)$$

$$+ MCl_4 \longrightarrow \quad M = Si, Ge, Sn$$

$$+ (CH_3{=}CH)_2SnCl_2 \longrightarrow \text{(spiro-bis-metalloindene, M)} \qquad (33)$$

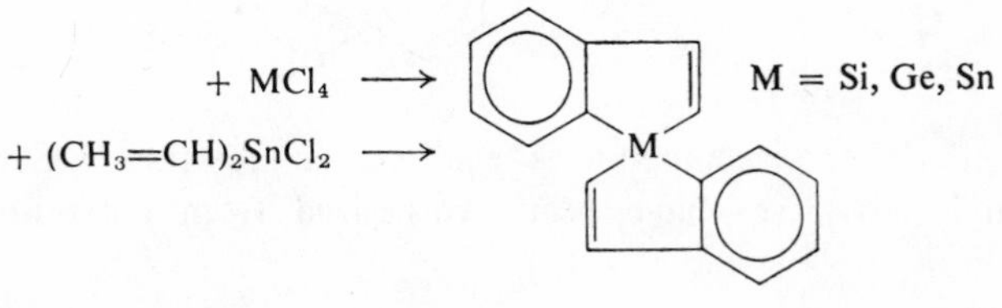

Six-membered rings where a group IV element is para to nitrogen, oxygen, or sulfur have been prepared by way of the intermediacy of an organodilithium reagent, as in the synthesis of the 5,10-dihydroacridine analogs:

R N Li Li $+ R_2MCl_2 \longrightarrow$ R N M R R (34)

$+ MCl_4 \longrightarrow$ R N M N R (35)

where M = silicon[79–87], germanium [80,84,87] tin,[80,84,86–88] and lead.[80,84,87] Likewise 10,10-disubstituted phenoxsilanes[89,90] and phenoxastannepins[91] have been reported:

O Li Li $+ R_2MCl_2 \longrightarrow$ $+ R_2SiH_2 \longrightarrow$ O M R R M = Si, Sn (36)

$+ SiCl_4 \longrightarrow$ O Si O (37)

10,10-Dimethylphenoxastannepin is a dimer in solution.[90] Several perfluorophenyl derivatives have been synthesized from *o,o'*-dilithio intermediates[10]:

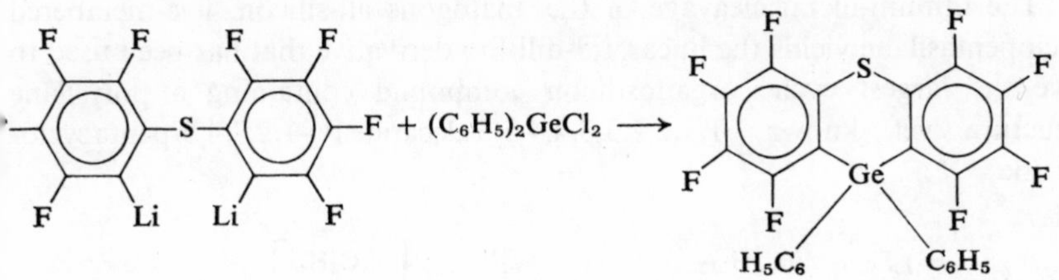

$$\text{(pentafluorophenyl sulfide dilithio)} + (C_6H_5)_2GeCl_2 \longrightarrow \text{(octafluoro-10,10-diphenylphenothiagermin)} \qquad (38)$$

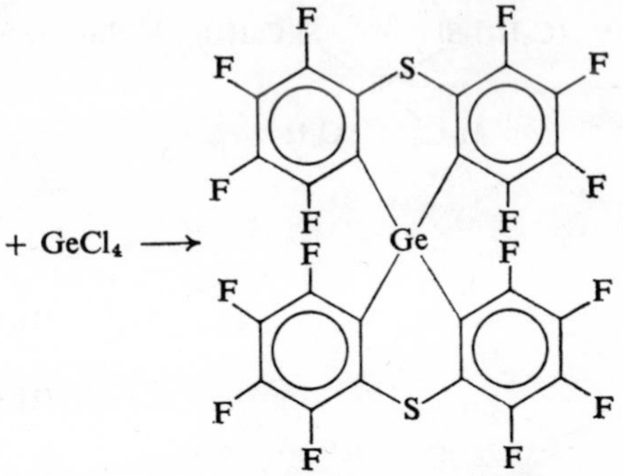

$$+ GeCl_4 \longrightarrow \text{(spiro-bis(octafluorophenothiagermin))} \qquad (39)$$

$$\text{(bis(dilithiotetrafluorophenyl)diphenylgermane)} + (C_6H_5)_2GeCl_2 \longrightarrow \text{(octafluoro-tetraphenyl-digermanthrene)} \qquad (40)$$

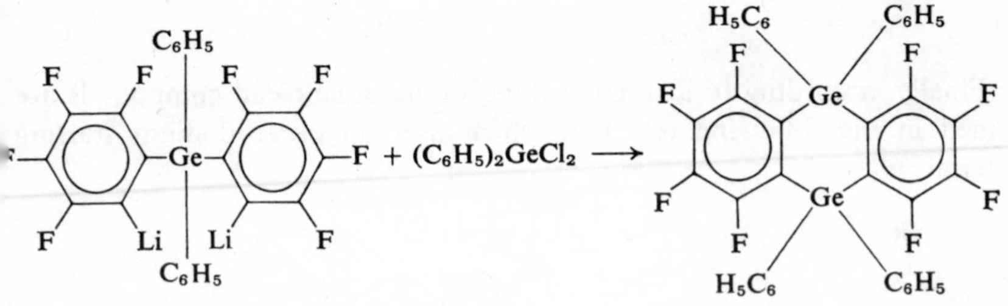

The cleavage of the homocyclic cyclotetrasilane with lithium metal yields a 1,4-dilithio derivative that can be used in subsequent cyclization reactions to give heterocycles[92]:

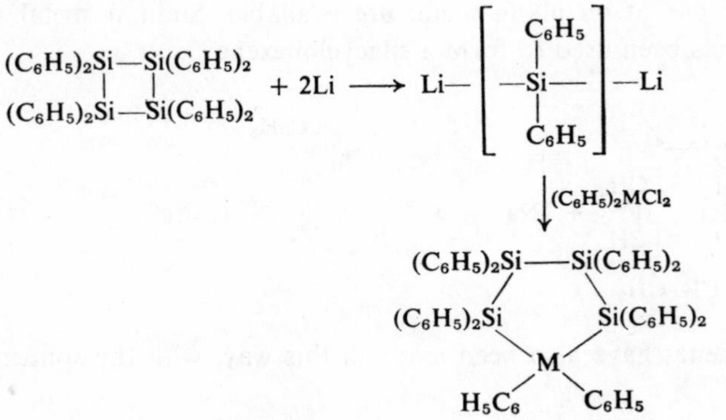

$$[(C_6H_5)_2Si]_4 + 2Li \longrightarrow Li\text{—}[Si(C_6H_5)_2]_4\text{—}Li \xrightarrow{(C_6H_5)_2MCl_2} \text{cyclo-}[(C_6H_5)_2Si]_4M(C_6H_5)_2 \qquad (41)$$

$M = Ge, Sn$

The lithium-metal cleavage of the analogous all-silicon, five-membered cyclopentasilane yields the linear 1,5-dilithio derivative that has been used to give the largest cyclic organosilicon compound containing a polysilane structure yet known, 1,1,2,2,3,3,4,4,5,5-decaphenyl-1,2,3,4,5-pentacyclooctane[93,94]:

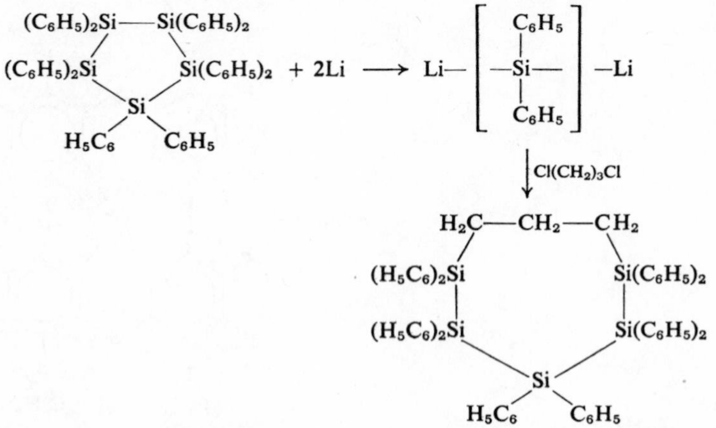

$$\text{cyclo-}[Si(C_6H_5)_2]_5 + 2Li \longrightarrow Li{-}[Si(C_6H_5)_2]_5{-}Li \xrightarrow{Cl(CH_2)_3Cl} \text{cyclo-}[Si(C_6H_5)_2]_5(CH_2)_3 \quad (42)$$

Finally, according to a recent patent, cyclic organolead compounds are formed in the following reaction, which uses an organosodium starting material[95]:

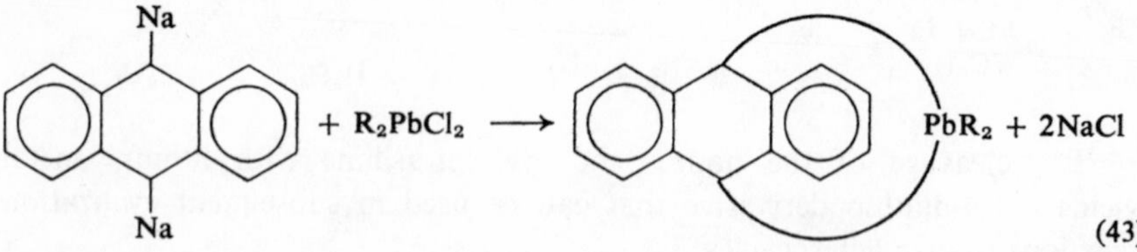

$$\text{9,10-disodio-9,10-dihydroanthracene} + R_2PbCl_2 \longrightarrow \text{(anthracene-9,10-diyl)}PbR_2 + 2NaCl \quad (43)$$

c. *The Wurtz reaction.* Several interesting examples of intramolecular cyclization by the use of an alkali metal are available. Sodium metal in refluxing toluene has been used to form a silacyclohexene[96]:

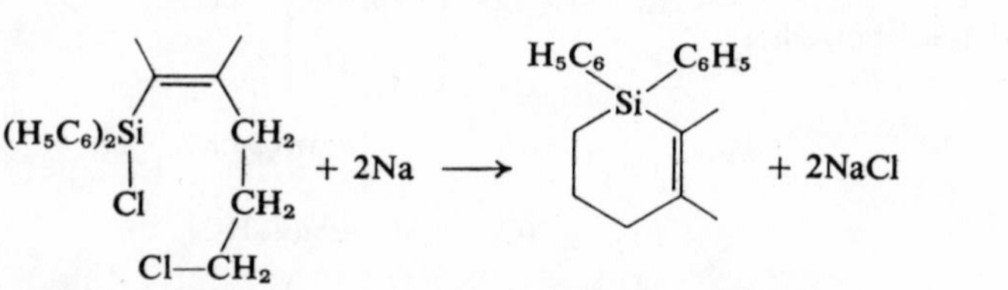

$$(H_5C_6)_2Si(Cl){-}C(CH_3){=}C(CH_3){-}CH_2CH_2CH_2{-}Cl + 2Na \longrightarrow \text{1,1-diphenyl-2,3-dimethyl-1-silacyclohex-2-ene} + 2NaCl \quad (44)$$

The benzocycloalkenes have also been made in this way, with the splitting out of HCl[22,97]:

$$o\text{-}C_6H_4(Cl)(CH_2)_nSiR_2H + Na \longrightarrow \text{[benzo-fused ring: } C_6H_4(CH_2)_nSiR_2\text{]} \qquad n = 2, 3, 4, 5 \quad (45)$$

Lithium in THF gives the $n = 3$ product in small yield, presumably because of silicon-carbon bond cleavage which interferes.[98] Although *o*-chlorophenylsilane is inert to magnesium metal in THF, intermolecular coupling takes place with sodium metal in refluxing toluene[99]:

$$2\ o\text{-}ClC_6H_4SiR_2H \xrightarrow{Na} \text{[dibenzo ring: } C_6H_4(SiR_2)_2C_6H_4\text{]} \quad (46)$$

In Equation (45), where $n = 1$, the fused silabutene was not isolated; the dimeric 2:3,6:7-dibenz-1,1,5,5-tetraphenyl-1,5-disilaoctadiene-2,6 was formed instead[39,97]:

$$o\text{-}ClC_6H_4CH_2SiH(C_6H_5)_2 \xrightarrow{Na} \text{[ring: } C_6H_4\text{–}Si(C_6H_5)_2\text{–}CH_2\text{–}C_6H_4\text{–}Si(C_6H_5)_2\text{–}CH_2\text{]} \quad (47)$$

The intramolecular closure of γ-chloropropyl derivatives of silicon does not proceed with either lithium or sodium metal and gives only low yields of the silacyclobutanes with unactivated magnesium. Satisfactory yields have been obtained through the use of magnesium metal activated with iodine vapor[31]:

$$Cl_2RSi(CH_2)_3Cl + Mg \longrightarrow \text{[}(CH_2)_3Si(Cl)R\text{ ring]} + MgCl_2 \quad (48)$$

o-Dichlorobenzene does react in an intermolecular reaction with dimethylmethoxychlorosilane in the presence of sodium metal in refluxing toluene to give several products, among which are[100]:

$$o\text{-}C_6H_4Cl_2 + CH_3O(Cl)Si(CH_3)_2 \xrightarrow{Na} (\text{dibenzo-silacycle with } Si(CH_3)_2) + (\text{benzo-disiloxane ring } [Si(CH_3)_2\text{-}O\text{-}Si(CH_3)_2]) \quad (49)$$

The cyclic disilanyl dimer was obtained when bis(chlorodimethylsilyl)-methane was treated with metallic sodium[27]:

$$2ClSi(CH_3)_2CH_2Si(CH_3)_2Cl + 4Na \longrightarrow \text{cyclo-}[(H_3C)_2Si\text{-}Si(CH_3)_2\text{-}CH_2\text{-}Si(CH_3)_2\text{-}Si(CH_3)_2\text{-}CH_2] \quad (50)$$

By using 1-(dimethylchlorosilyl)-3-(methylphenylchlorosilyl)propane with metallic sodium, intramolecular cyclization is achieved to give a disilanyl-cyclopentane[101]:

$$ClSi(CH_3)_2(CH_2)_3Si(CH_3)(C_6H_5)Cl + 2Na \longrightarrow \text{cyclo-}[(CH_2)_3Si(CH_3)(C_6H_5)Si(CH_3)_2] \quad (51)$$

Likewise were prepared other 1,1,2,2-tetramethyl-1,2-disilacycloalkanes[29]:

$$ClSi(CH_3)_2(CH_2)_nSi(CH_3)_2Cl + Na/K \xrightarrow[n\text{-heptane}]{\text{benzene}} (CH_3)_2Si\text{—}Si(CH_3)_2 \text{ (ring closed by } (CH_2)_n) \qquad n = 3, 4, 5, 6 \quad (52)$$

Organocyclooxasilaalkanes with a single oxygen atom in the ring have been prepared by the condensation of organo(chloroalkoxy)chlorosilanes in the presence of sodium or lithium[102–104]:

$$R_2ClSiO(CH_2)_nCl + 2M \longrightarrow R_2Si\text{(ring with } (CH_2)_n \text{ and } O) + 2MCl \qquad n = 3, 4, 5;\ M = Li, Na \quad (53)$$

Oxadicarbacyclohexasilanes have also been synthesized by way of an inter-molecular route involving bis(chloromethyl)tetramethyldisiloxane and a dichlorosilane with sodium metal in refluxing toluene[105]:

$$ClCH_2Si(CH_3)_2OSi(CH_3)_2CH_2Cl + (CH_3)_2SiCl_2 + 4Na \longrightarrow 4NaCl + \text{cyclo-}[(H_3C)_2Si{-}O{-}Si(CH_3)_2{-}CH_2{-}Si(CH_3)_2{-}CH_2] \quad (54)$$

Phenoxasilin compounds have also been prepared by a Wurtz coupling of an *o,o′*-dibromodiphenyl ether with a diorganodihalosilane[106]:

$$(o\text{-}BrC_6H_4)_2O + X_2SiR_2 \xrightarrow{Na} \text{10,10-}R_2\text{-phenoxasilin} \quad (55)$$

In the germanium series the Wurtz reaction has been used to prepare the germacyclobutane by an intramolecular closure[107,108]:

$$R_2Ge(Cl)(CH_2)_3Cl + 2Na \longrightarrow \text{cyclo-}[(CH_2)_3GeR_2] + 2NaCl \quad (56)$$

Intermolecular closure in xylene is less satisfactory, since the product, which is formed in only small yield (10 percent) is difficult to separate[107,108]:

$$R_2GeCl_2 + Cl(CH_2)_3Cl + 4Na \longrightarrow 4NaCl + \text{cyclo-}[(CH_2)_3GeR_2] \quad (57)$$

A spiro compound has also been synthesized[107,108]:

$$\text{cyclo-}[(CH_2)_4Ge](Cl)(CH_2)_3Cl + 2Na \xrightarrow{xylene} \text{spiro-}[(CH_2)_4Ge(CH_2)_3] + 2NaCl \quad (58)$$

3. *Ring Closures by Addition of Fourth-Group Hydrides to Unsaturated Systems*

a. Hydrosilylation reactions. The silicon-hydrogen bond can add across olefinic and acetylenic bonds in the presence of a catalyst. The barrier to this kind of reaction is reduced as the fourth group is descended; for example, in the presence of platinum metal suspended on finely divided carbon as the catalyst, organosilanes containing vinyl groups and hydrogen joined to the same silicon atom can be polymerized and cyclized to the silethylenes at atmospheric pressure[109–111]:

$$(CH_3)_2Si(H)(CH{=}CH_2) \xrightarrow{Pt/C} (CH_3)_2Si\langle C_4H_8\rangle Si(CH_3)_2 + CH_3\text{-}\langle Si(CH_3)_2\text{-}CH_2CH_2\text{-}Si(CH_3)_2\rangle \quad (59)$$

Likewise the use of allyldimethylsilane with the same catalyst system in refluxing toluene yields the 1,5-disilacyclooctane ring system[112]:

$$(CH_3)_2Si(H)(CH_2CH{=}CH_2) \xrightarrow{Pt/C} (CH_3)_2Si\langle\ \rangle Si(CH_3)_2 \quad (60)$$

The allyl ether of tetramethyldisiloxane also undergoes intramolecular hydrosilylation to give the 1-trimethylsiloxy-1-methyl-1,2-siloxacyclopentane[113]:

$$(CH_3)_3SiOSiCH_3(H)(OCH_2CH{=}CH_2) \xrightarrow{catalyst} \text{cyclo-}[Si(H_3C)(OSi(CH_3)_3)\text{-}O\text{-}CH_2CH_2CH_2] \quad (61)$$

Another variation to this reaction involves the intermolecular addition of dihydridodisiloxanes to variously substituted acetylenes in the presence of $H_2PtCl_6 \cdot 6H_2O$ or Pt/C as the catalyst at 110 to 120° in isopropyl alcohol[114]:

$$HSiR_2OSiR_2H + R'CH{=}CHR'' \longrightarrow \text{cyclo-}[R_2Si\text{-}CHR'\text{-}CHR''\text{-}SiR_2\text{-}O] \quad (62)$$

A heterocycle containing both silicon and boron atoms in a six-membered ring has been prepared by the hydroboration of a diallylsilane[115]:

$$(CH_3)_3N{:}BH_2R + \underset{\textbf{I}}{(CH_2{=}CH)(CH_2{=}C(CH_3))SiR_2} \longrightarrow \text{cyclo-}[SiR_2\text{-}CH_2CH_2\text{-}BR\text{-}CH_2CH_2] \quad (63)$$

b. Hydrogermylation. Polymerization of dimethyldiallygermane using a Ziegler catalyst gives a polymer in which the germacyclohexane structure is present as a repeating unit:

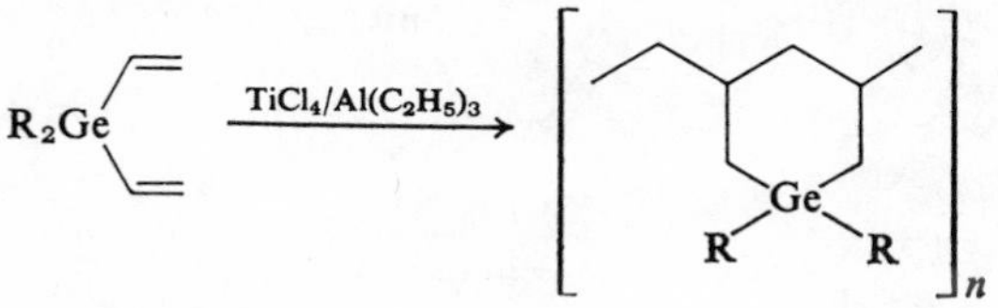

$$R_2Ge(CH_2CH{=}CH_2)_2 \xrightarrow{TiCl_4/Al(C_2H_5)_3} [\ldots]_n \qquad (64)$$

The following compound was also isolated[116]:

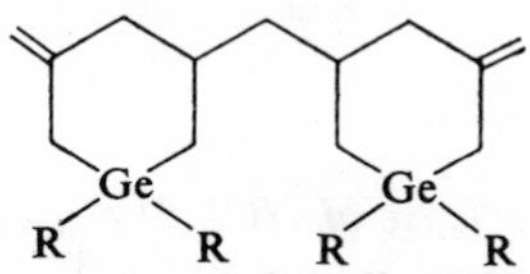

c. *Hydrostannylation.* Addition of the tin-hydrogen bond to olefinic and acetylenic compounds proceeds more readily than in the case of the analogous silanes and germanes. Tin-containing heterocycles have been obtained from the reaction of organotin hydrides with bifunctional unsaturated compounds; for example, heterocycles containing two fourth-group atoms have been prepared from divinyl compounds and diphenyltin dihydride[117]:

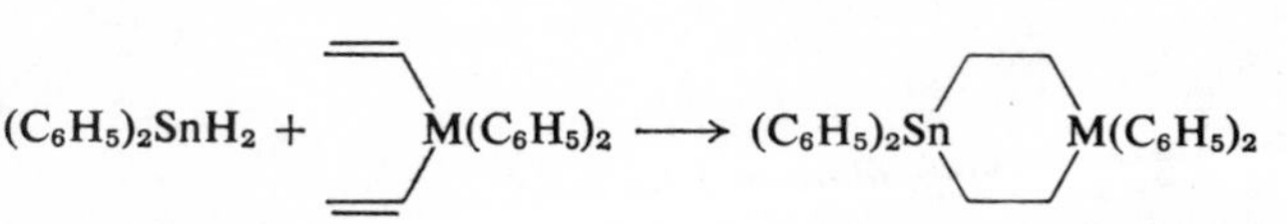

$$(C_6H_5)_2SnH_2 + (CH_2{=}CH)_2M(C_6H_5)_2 \longrightarrow (C_6H_5)_2Sn(CH_2CH_2)_2M(C_6H_5)_2 \quad M = Si, Ge \qquad (65)$$

The distannacyclohexane heterocycle, where M = Sn, could not be made by this procedure,[117–119] nor does it result from the reaction of the dihydride with phenylethyne, since this would involve β,α addition to the acetylene. In fact only β,β addition is observed[120]:

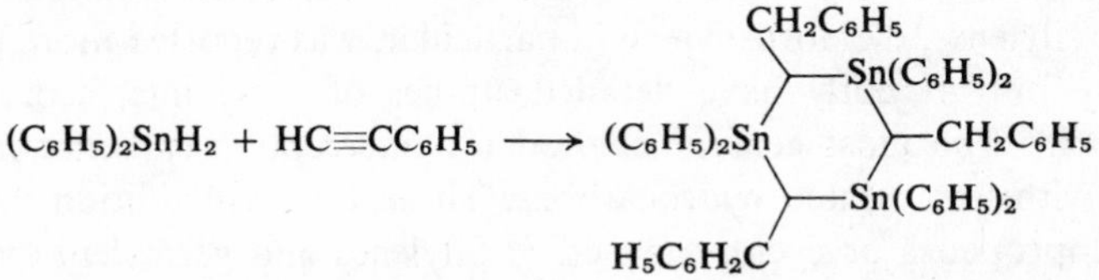

$$(C_6H_5)_2SnH_2 + HC{\equiv}CC_6H_5 \longrightarrow \text{(trimeric ring: } (C_6H_5)_2Sn,\ Sn(C_6H_5)_2,\ Sn(C_6H_5)_2 \text{ with } CH_2C_6H_5 \text{ groups)} \qquad (66)$$

Other cases in which the two unsaturated groups are in a sterically favorable position for ring closure include the *o*-divinyl- and *o*-diethynylbenzenes. Reactions of the latter lead to the cyclic trimer as well[118,121]:

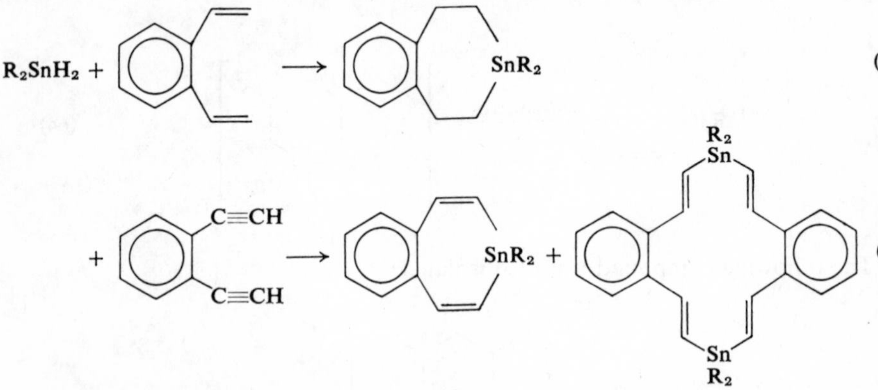

R_2SnH_2 + [o-divinylbenzene] → [benzo-fused ring]SnR_2

\+ [o-diethynylbenzene, C≡CH, C≡CH] → [benzo-fused ring]SnR_2 + [cyclic dimer with two R_2Sn bridges]

d. Hydrosilylation with release of HCl. The tricyclic system, 1,1-dichloro-1-silaacenaphthalene, has been prepared by the gas-phase reaction of α-methylnaphthylene with trichlorosilane or methyldichlorosilane at 600 to 700°[122]:

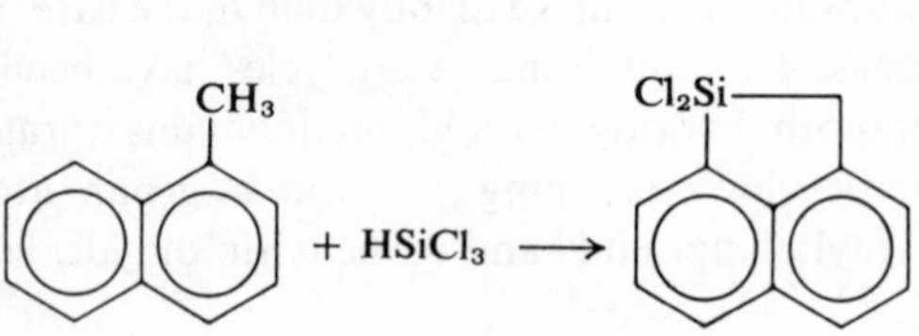

$$\text{CH}_3\text{-naphthalene} + HSiCl_3 \longrightarrow \text{Cl}_2\text{Si-acenaphthene} \tag{69}$$

4. *1,2 Cycloaddition Reactions*

a. Reactions with olefins. In 1960 and 1961 two groups of scientists in the Soviet Union discovered ring-forming reactions that proceeded through the intermediate formation of divalent silicon and germanium species, presumably analogs of the organic carbenes.[123–130] It is known that the carbene intermediates react with unsaturated compounds by way of a cycloaddition route. Although the preparation and spectroscopic observation of the inorganic silylenes, dichlorosilylene in particular, was reported more than 30 years ago, only recently have detailed studies of these intermediates appeared.[131,132] The most general method for intercepting carbenes involves trapping with unsaturated hydrocarbons. These 1,2 cycloaddition reactions yield cyclopropanes or cyclopropenes.[123] Silylenes and germylenes seem to react with a variety of unsaturated organic compounds as well, but to date no stable, isolatable sila- or germacyclopropanes or cyclopropenes are known.[3]

The vapor-phase reaction of dimethylsilylene with ethylene yields vinyldimethylsilane[133]:

$$[:Si(CH_3)_2] + H_2C{=}CH_2 \longrightarrow \left[\text{cyclo-}(CH_2)_2Si(CH_3)_2 \right] \longrightarrow H_2C{=}CH{-}SiH(CH_3)_2 \tag{70}$$

The analogous organic carbene intermediates R_2C: often react similarly.[123] The condensed-phase reaction leads to several cyclic products in addition to polymers that are thought to arise from the silacyclopropane intermediate[131]:

$$\left[\text{cyclo-}(CH_2)_2Si(CH_3)_2 \right] + H_2C{=}CH_2 \longrightarrow \text{1,1-dimethylsilacyclopentane} \tag{71}$$

$$\xrightarrow{\text{dimerization}} \text{1,1,4,4-tetramethyl-1,4-disilacyclohexane} \tag{72}$$

Styrenes, substituted styrenes, stilbenes, vinylsilanes, and the like, have also been used to give the various isomeric 1,1-dimethylsilacyclopentanes.[134–140] When the ethylene is replaced by isobutylene, a tetrasilacyclohexane is the product[131]:

$$[:SiR_2] + (H_3C)_2C{=}CH_2 \longrightarrow \text{cyclo-}[R_2Si{-}SiR_2{-}SiR_2{-}SiR_2{-}C(CH_3)_2{-}CH_2] \tag{73}$$

The reaction of difluorosilylene with ethylene gives two monomeric cyclic products in addition to solid polymers[141]:

$$[:SiF_2] + H_2C{=}CH_2 \longrightarrow \text{cyclo-}[SiF_2{-}SiF_2{-}(CH_2)_2] + \text{cyclo-}[SiF_2{-}SiF_2{-}(CH_2)_4] \tag{74}$$

The reaction of difluorosilylene with cyclohexene may proceed through a three-membered ring intermediate as well[142]:

$$[:SiF_2] + \text{cyclohexene} \longrightarrow \left[\text{bicyclo[4.1.0]heptane-7,7-}SiF_2\right] \longrightarrow \text{dodecahydro-5,10-disila-5,5,10,10-tetrafluoroanthracene} \quad (75)$$

In the germanium series the intermediacy of dimethylgermylene in the formation of 1,1-dimethylgermacyclopentanes can be established in the same way[130]:

$$[:Ge(CH_3)_2] + RCH{=}CH_2 \longrightarrow \text{3,4-R}_2\text{-1,1-(}CH_3)_2\text{-germacyclopentane} \quad (76)$$

However, the trihalogermanes $HGeX_3$, where X = Cl and Br, probably exist in equilibrium with dihalogermylene $:GeX_2$ in ether solution:

$$HGeX_3 \rightleftharpoons H^+ + GeX_3^- \rightleftharpoons :GeX_2 + HX \quad (77)$$

and this inorganic carbene analog,[143–147] like the compound GeI_2[148,149] and the organometallic R_2Ge: species,[150–152] can react with conjugated dienes:

$$[:MX_2] + \rangle C{=}C{-}C{=}C\langle \longrightarrow \text{1,1-}X_2\text{-metallacyclopent-3-ene} \qquad M = Si, Ge \quad (78)$$

Although the metallocyclopentene is formally the product of a 1,4 cycloaddition of the carbene analog, a reaction path involving the formation and thermal isomerization of a vinyl-substituted metallocyclopropane is favored[132]:

$$[:MR_2] + \rangle C{=}C{-}C{=}C\langle \longrightarrow \left[\text{2-vinyl-1,1-}R_2\text{-metallacyclopropane}\right] \longrightarrow \text{1,1-}R_2\text{-metallacyclopent-3-ene} \quad (79)$$

Difluorosilylene reacts with benzene or toluene to give 1,4-hexadiene derivatives with SiF_2 bridges across the 3,6 positions[153]:

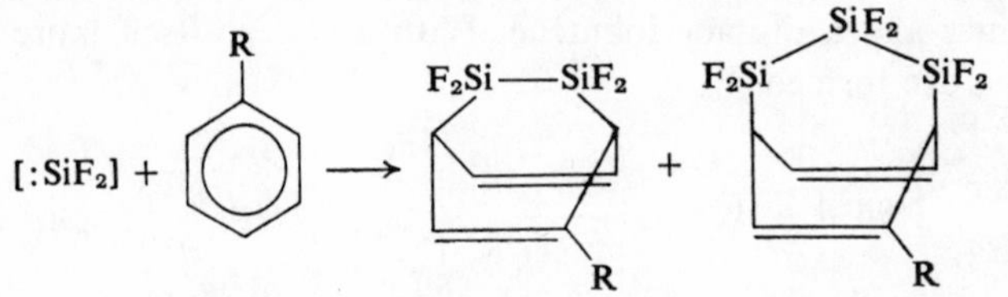

$[:SiF_2]$ + (R-benzene) → (F_2Si—SiF_2 bridged, R) + (F_2Si—SiF_2—SiF_2 bridged, R) (80)

b. Reaction with acetylenes. Since the initial reports that diphenylacetylene could be used as a trapping agent for the silylenes[125,154] and germylenes,[124,126–128,154–156] and the original incorrect assignment of the product as a monomeric three-membered heterocycle was corrected,[157–163] a variety of acetylenes has been used in reactions of this kind. In all cases the product is the corresponding 1,4-disila- or digermacyclohexadiene. It is generally assumed that the product arises by way of a cyclopropene intermediate[131,132]:

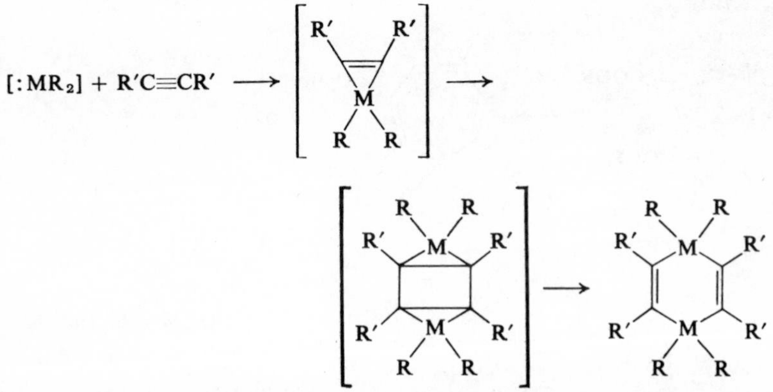

$[:MR_2] + R'C{\equiv}CR' \longrightarrow$ [cyclopropene intermediate] $\longrightarrow$ [bicyclic dimer] $\longrightarrow$ 1,4-dimetallacyclohexadiene (81)

by analogy with the organic carbenes.[123] Reaction of dimethylsilylene with a mixture of dimethyl- and diphenylacetylene, however, gave only three disilacyclohexadienes. The structure of the "mixed" disilacyclohexadiene eliminated from further consideration the π dimerization mechanism for disilacyclohexadiene formation:

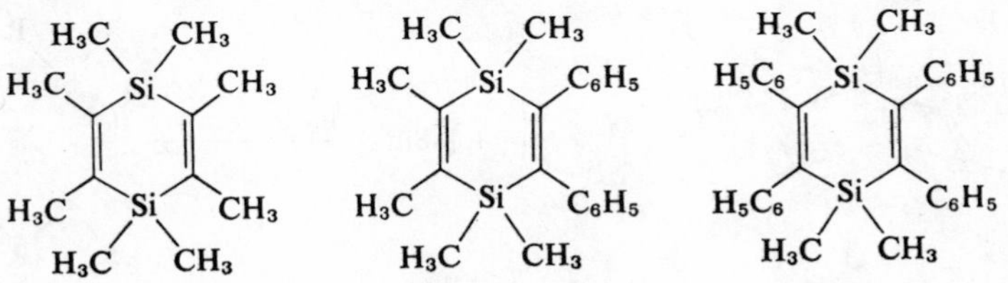

An alternate route involving a rather specific dimerization across the carbon-silicon ring bonds of the silacyclopropene intermediates has been proposed.[152]

Silicon difluoride reacts with alkynes to give a spontaneously inflammable solid polymer and a disilacyclobutene. With acetylene itself more complex ring systems are formed[164]:

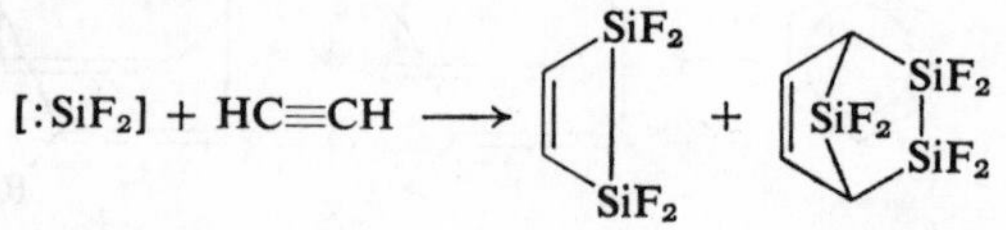

$$[:SiF_2] + HC{\equiv}CH \longrightarrow \text{(disilacyclobutene, } SiF_2/SiF_2\text{)} + \text{(bicyclic, } SiF_2, SiF_2, SiF_2\text{)} \qquad (82)$$

Such apparent 1,4 cycloadditions that result from the ability of $[:SiF_2]$ to give the diradical dimer, trimer, and the like, do not occur with carbenes.[165]

The divalent silicon and germanium species can be prepared by a variety of routes,[131,132] including reduction of the tetrahalide with elemental silicon and germanium or of a diorganodihalo derivative with an alkali metal, thermolysis of hexahalo- or alkoxydisilanes and -germanes, or thermolysis of the 7-sila- and germanorbornadienes. The latter bicyclic compounds undergo rearrangement on standing, but the structure of the product is not known[70,71,166]:

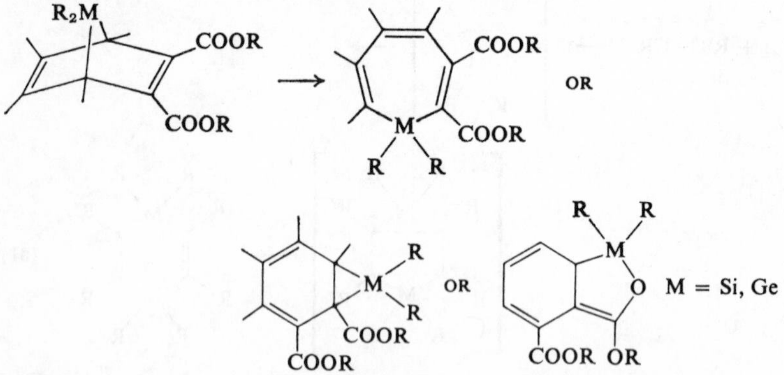

R_2M (7-metallanorbornadiene, COOR, COOR) $\longrightarrow$ (metallacycloheptatriene, M, R, R, COOR, COOR) OR (bicyclic, M, R, R, COOR, COOR) OR (R, R, M, O, COOR, OR) M = Si, Ge (83)

The organic analog of this compound, bicyclo[2.2.1]heptadiene, undergoes thermal isomerization to cycloheptatriene.[167] Thermolysis could lead through the divalent intermediate, and such pyrolyses have been carried out for M = Si in the presence of diphenylacetylene to form the 1,4-silacyclohexadiene[164]:

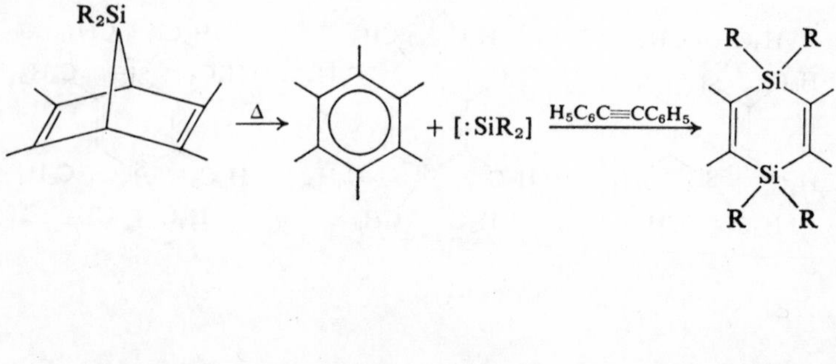

R_2Si (7-silanorbornadiene) $\xrightarrow{\Delta}$ (hexasubstituted benzene) $+ [:SiR_2] \xrightarrow{H_5C_6C{\equiv}CC_6H_5}$ (1,4-disilacyclohexadiene, Si, Si, R, R, R, R) (84)

The 7-germanorbornadiene, on the other hand, in the presence of excess dimethylacetylenedicarboxylate takes a different route[166]:

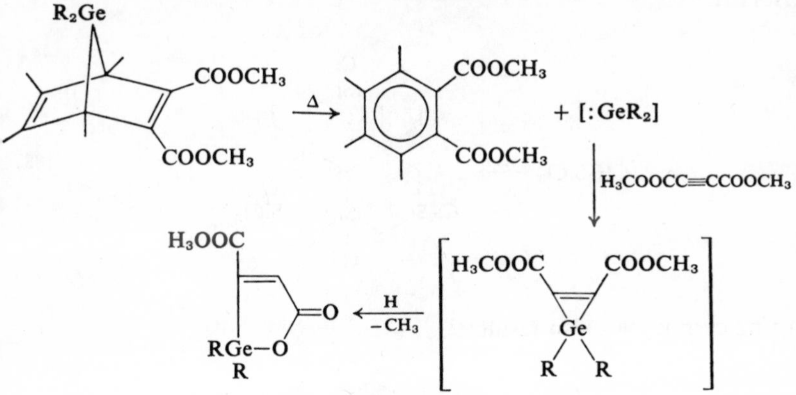

R_2Ge COOCH$_3$ COOCH$_3$ $\xrightarrow{\Delta}$ COOCH$_3$ COOCH$_3$ + [:GeR$_2$]

$H_3COOCC{\equiv}CCOOCH_3$

H$_3$OOC =O RGe—O R $\xleftarrow[-CH_3]{H}$ [H$_3$COOC COOCH$_3$ Ge R R]

(85)

In addition to the germa-γ-butyrolactone, a second heterocycle, which probably results from the reaction of a digermacyclobutene with oxygen, was isolated[166]:

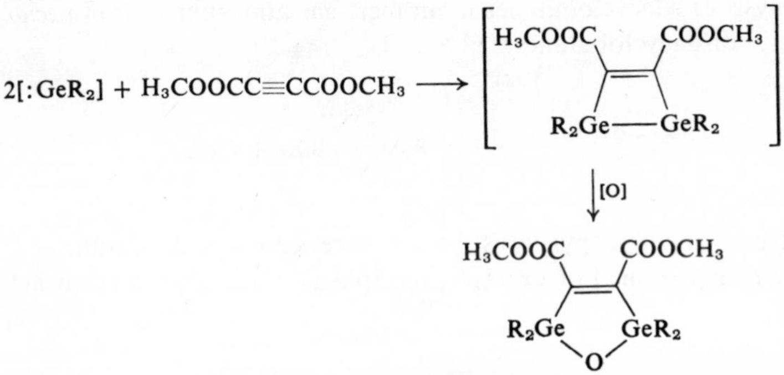

2[:GeR$_2$] + $H_3COOCC{\equiv}CCOOCH_3$ $\longrightarrow$ [H$_3$COOC COOCH$_3$ R$_2$Ge—GeR$_2$]

$\downarrow$[O] (86)

H$_3$COOC COOCH$_3$ R$_2$Ge GeR$_2$ O

5. *Pyrolysis*

The study of the thermal decomposition of simple tetraorganosilanes and organohalosilanes has added a great number of new silicon heterocycles. Although the great majority of these ring systems are pseudoheterocyclic according to our classification, and as such are not discussed, many others are not; for example, the pyrolysis of tetramethylsilane at 720° gives in a complex mixture of products a disilacyclopentene[168] and trisilabicycloheptane[169]:

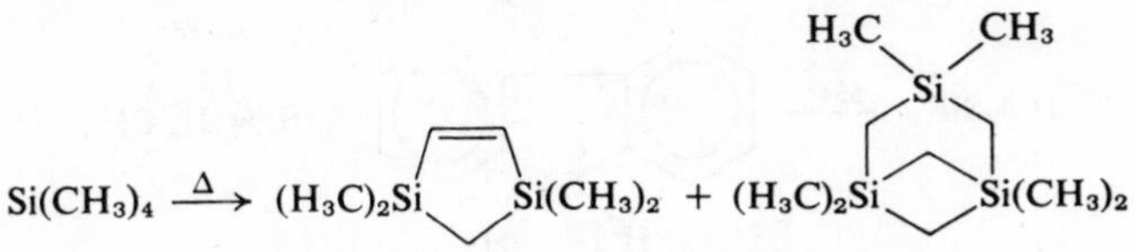

Si(CH$_3$)$_4$ $\xrightarrow{\Delta}$ (H$_3$C)$_2$Si Si(CH$_3$)$_2$ + (H$_3$C)$_2$Si Si(CH$_3$)$_2$ (with H$_3$C CH$_3$ Si bridge)

(87)

The pyrolysis of methyltrichlorosilane at 800° results in a complex mixture of cyclic carbosilanes, from among which the carbocyclic analog of pyrene has been isolated[170]:

$$CH_3SiCl_3 \xrightarrow{\Delta} \qquad (88)$$

Fused ring carbocycles also result[171]:

$$(CH_3)_nSiCl_{4-n} \xrightarrow{\Delta} \qquad (89)$$

The pyrolysis of silacyclobutanes in an inert-gas atmosphere or *in vacuo* is a route to 1,3-disilacyclobutanes[172,173]:

$$R_2Si\square \xrightarrow{500\text{–}700^\circ} R_2Si\square SiR_2 + C_2H_4 \qquad (90)$$

Phenylsilanes undergo pyrolysis to produce compounds containing the benzsilole ring system. For example, tetraphenylsilane gives a spiro heterocycle[174]:

$$(C_6H_5)_4Si \longrightarrow \qquad (91)$$

and diphenylsilane and diphenyldimethylsilane give the dibenzosilole:

$$(C_6H_5)_2SiR_2 \xrightarrow{534^\circ} \qquad R = H, CH_3 \qquad (92)$$

as does $p\text{-}CH_3C_6H_5Si(CH_3)_2H$[175]:

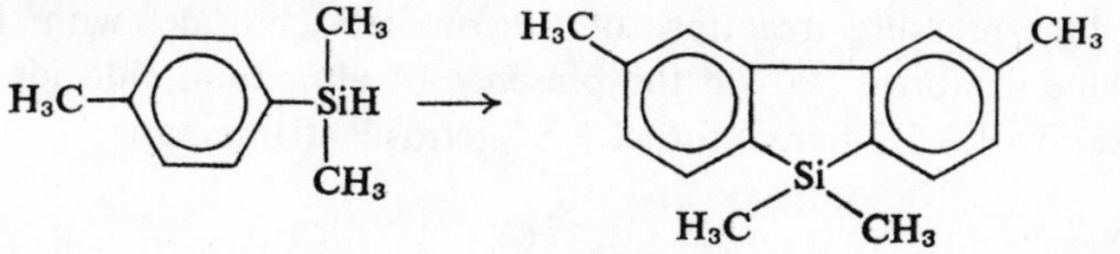

$$p\text{-}H_3C\text{-}C_6H_4\text{-}SiH(CH_3)_2 \longrightarrow \text{(3,7-dimethyl-5,5-dimethyldibenzosilole)} \quad (93)$$

Pyrolysis of the eight-membered carbodisiloxane heterocycle at 700° leads to the isolation of three cyclic products in addition to a high boiling fraction[176]:

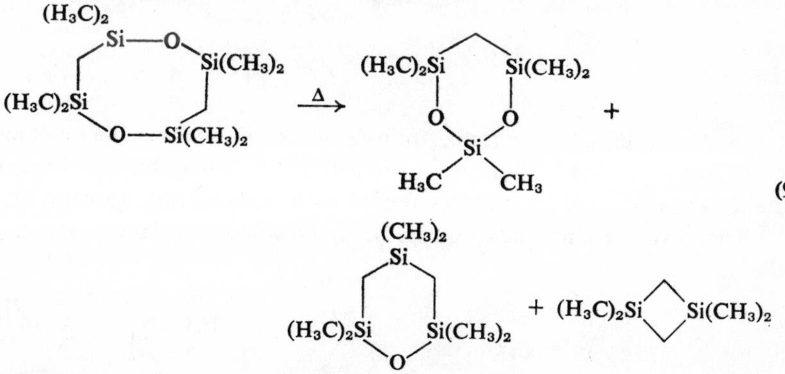

$$\text{(94)}$$

Pyrolysis of 1,1-dichloro-4-bromo-1-silacyclohexane at 560° gives a mixture of silacyclohexenes and silacyclohexadienes[42]:

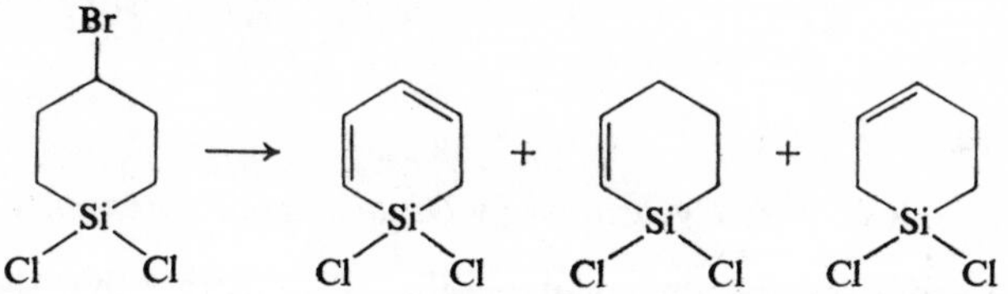

$$\text{(95)}$$

6. *Silent Electrical Discharge*

High-voltage electrical discharge through a volatile organosilane has been used to produce cyclic carbosilanes; for example, among the products found after the condensation of dimethyldichlorosilane in a silent electrical discharge of 24.5 kv is a disilapentane and a trisilaoctane[177]:

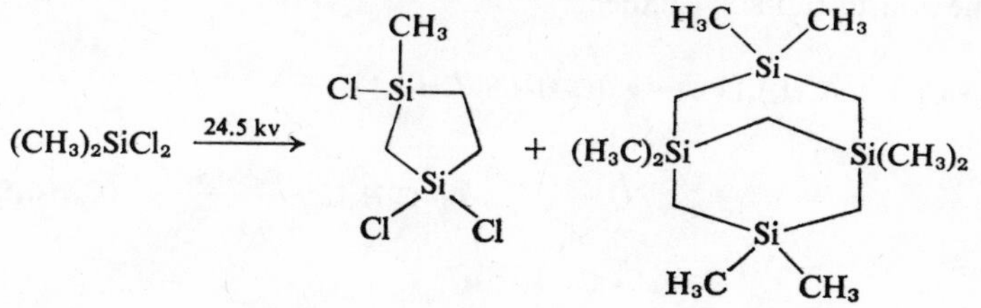

$$(CH_3)_2SiCl_2 \xrightarrow{24.5\text{ kv}} \text{(disilapentane)} + \text{(trisilaoctane)} \quad (96)$$

7. *Aluminum Chloride–Catalyzed Cyclization*

The high-pressure reaction of silicon tetrachloride with trimethylchlorosilane at above 500° in the presence of aluminum chloride has been shown to give 1,3,5,7-tetrachloro-1,3,5,7-tetrasilaadamantane:

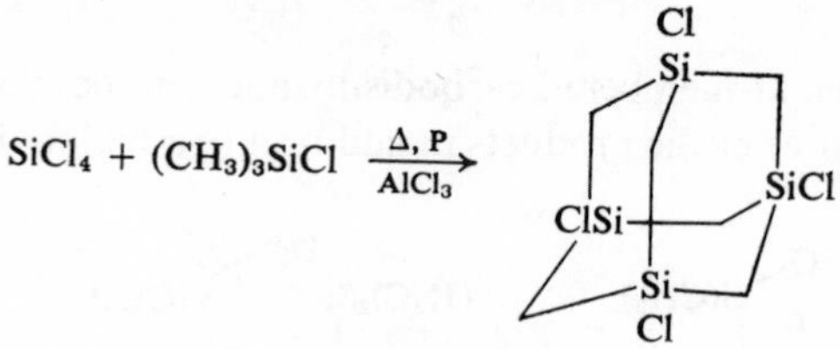

$$SiCl_4 + (CH_3)_3SiCl \xrightarrow[AlCl_3]{\Delta,\ P} \text{1,3,5,7-tetrachloro-1,3,5,7-tetrasilaadamantane (ClSi, SiCl, Si–Cl, Si–Cl)} \quad (97)$$

The product heterocycle is a rigid, strain-free compound whose silicon-chlorine bonds are remarkably resistant to reaction.[178] When hexamethyldisilylethane is heated with aluminum chloride with removal of tetramethylsilane, two heterocyclic compounds were isolated, including a trisilacyclononane[179]:

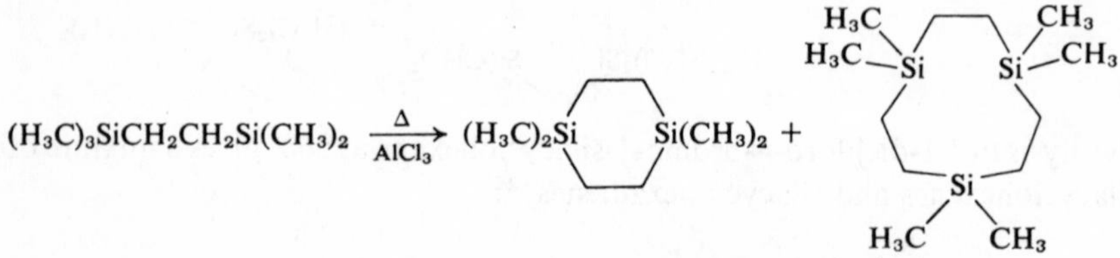

$$(H_3C)_3SiCH_2CH_2Si(CH_3)_2 \xrightarrow[AlCl_3]{\Delta} (H_3C)_2Si\langle\text{ring}\rangle Si(CH_3)_2 + \text{hexamethyltrisilacyclononane} \quad (98)$$

8. *Cleavage-Cyclization by Organometallic Reagents*

Several examples are now available that demonstrate that the formation of small ring compounds by the internal cleavage of a fourth-group carbon bond by an organolithium or Grignard reagent, followed by cyclization, occurs when the organometallic reagent is in excess. For example, attempts to prepare tetramethylene-bis(triphenylsilane) from the reaction of tetramethylenedilithium and triphenylchlorosilane in diethyl ether result in the desired product in only 4.4 percent yield. The main products are a silacyclopentane and tetraphenylsilane[50]:

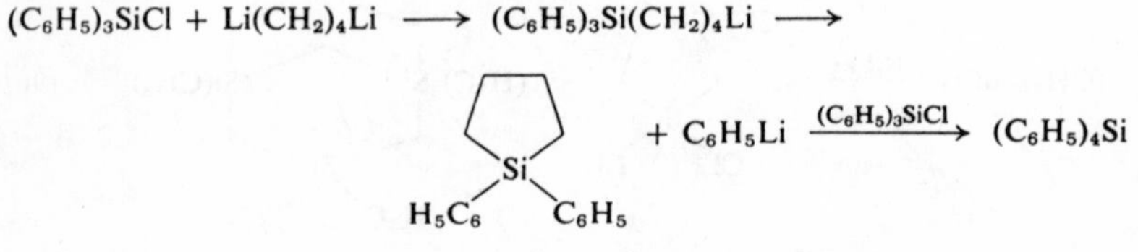

$$(C_6H_5)_3SiCl + Li(CH_2)_4Li \longrightarrow (C_6H_5)_3Si(CH_2)_4Li \longrightarrow$$

$$\text{1,1-}(C_6H_5)_2\text{-silacyclopentane} + C_6H_5Li \xrightarrow{(C_6H_5)_3SiCl} (C_6H_5)_4Si \quad (99)$$

In the dibenzosilole series the use of 5-chloro-5-methyldibenzosilole with 2,2′-dilithiobiphenyl in ether yields equal amounts of the 5,5-dimethyl derivative and the spiroheterocycle[57]:

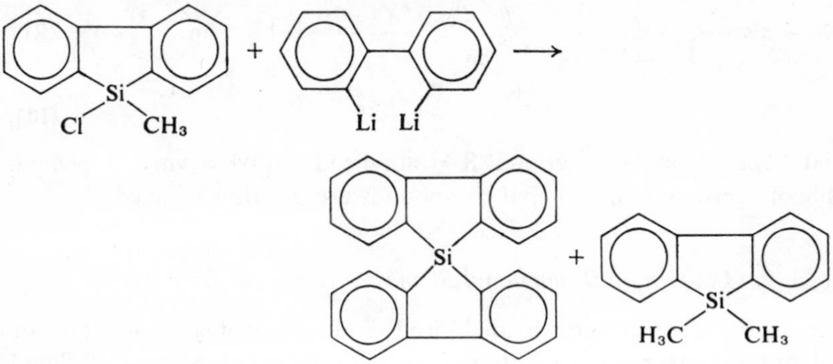

(100)

In another example, this time involving a Grignard reagent, the attempted methylation of 1,1,2,2,3,3,4,5,5,6,7,7,8-tridecachloro 1,3,5,7-tetrasilacyclooctane with CH_3MgCl produces 1,3,3,5,7,7-hexamethyl-1,3,5,7-tetrasilabicyclo[3.3.1]nonane[180]:

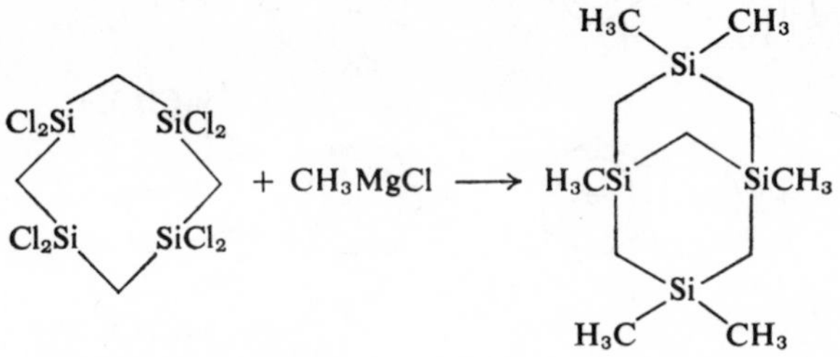

$$Cl_2Si\ldots SiCl_2 + CH_3MgCl \longrightarrow H_3CSi\ldots SiCH_3 \quad (101)$$

Chlorine activates the bridghead atoms in cyclic carbosilanes to rearrangement in the presence of a Grignard reagent. The six-membered, perchloro, pseudoheterocycle $[SiCl_2CCl_2]_3$, for example, undergoes complete rearrangement in the presence of an excess of Grignard reagent to give a complex mixture of products, none of which contain the original ring skeleton[181,182]:

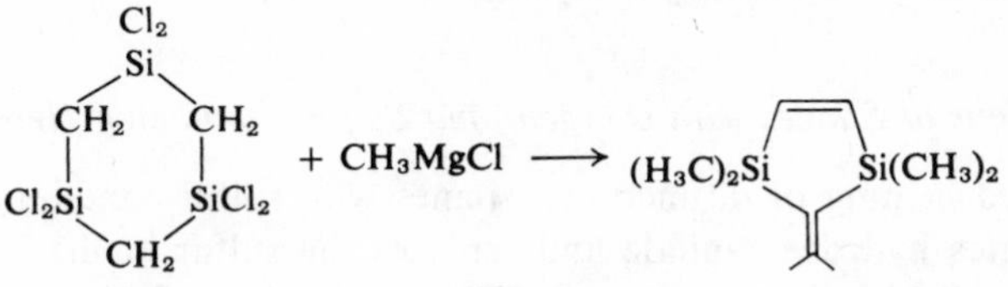

$$+ CH_3MgCl \longrightarrow (H_3C)_2Si \quad Si(CH_3)_2 \quad (102)$$

In the tin series it has recently been reported that in reactions in which 1,4-dilithiotetraphenylbutadiene is in excess over a diorganotin dichloride, the well-known cyclization to the stannole proceeds beyond the formation of the monocyclic ring to give the spiro compound:

R_2SnCl_2 + x's — Li Li → Sn R R —LiLi→ Sn + 2RLi (103)

The liability series in this reaction is R = ethyl and methyl > vinyl > phenyl. The silicon, germanium, and lead compounds are not thus affected.[76]

9. Addition of Difunctional Silanes to Aromatic Dianions

The addition of tetramethyldichlorodisilane to aromatic dianions or radical anions has been used to prepare a number of bridged disilanyl heterocycles:

H_3C CH_3 ClSiSiCl H_3C CH_3 + (=) → $Si(CH_3)_2$ $(H_3C)_2Si$

\+ (=) → $(H_3C)_2$ Si $Si(CH_3)_2$ (104)

\+ (=) → $(H_3C)_2$ Si $Si(CH_3)_2$

The heterocycles were prepared for use in the retrodiene reaction in an attempt to produce tetramethyldisilylene.[183]

10. Reactions of Silanes with Oxygen- and Sulfur-Containing Heterocycles

Prolonged heating of difunctional silanes with sulfur-containing heterocycles liberates hydrogen sulfide and replaces the sulfur in the heterocycle with silicon. Yields are generally low. The mechanism of the reaction is in doubt. Diphenylsilane is usually chosen for its higher boiling point. Examples of this synthetic procedure include cases in which sulfur-containing heterocycles, such as phenoxathin, phenothiazine, 10-ethylphenothiazine, and thianthrene, are used[184,185]:

$$\text{(E, S-heterocycle)} + (C_6H_5)_2SiH_2 \longrightarrow \text{(E, Si}(C_6H_5)_2\text{-heterocycle)} + H_2S$$

E = O, NH, NC_2H_5, S, SO_2, $Si(C_6H_5)_2$ (105)

Silane itself is also effective. The spirocyclopentane has been prepared by passing a mixture of tetrahydrofuran and monosilane through a tube packed with alumina at 375°[186]:

$$\text{C}_4\text{H}_8\text{O} + SiH_4 \xrightarrow{Al_2O_3} \text{spiro-Si(C}_4\text{H}_8)_2 + \text{C}_4\text{H}_8\text{SiH}_2 \qquad (106)$$

11. Ring Contraction by Desulfuration

The desulfuration of 2,2,6,6-tetraethyl-1-thia-2,6-digermacyclohexane by sodium metal in refluxing nonane effects ring contraction by removal of sulfur as sodium sulfide[187]:

$$(H_5C_2)_2Ge(S)Ge(C_2H_5)_2\text{(ring)} + 2Na \longrightarrow Na_2S + (H_5C_2)_2Ge{-}Ge(C_2H_5)_2\text{(ring)} \qquad (107)$$

12. Ring Closure by Pyrolysis of Unstable Mercury Derivatives

The action of diethylmercury on 1,3-bis(diethylgermyl)propane and 1,4-bis(diethylgermyl)butane leads to a mercury derivative that, by pyrolysis, gives a cyclic compound with two adjacent intracyclic germanium atoms. The intermediate mercury-containing material may be a linear polymer or a mercury heterocycle[187]:

$$(H_5C_2)_2Ge(H)(CH_2)_nGe(H)(C_2H_5)_2 + (C_2H_5)_2Hg \xrightarrow{-2C_2H_6} \qquad (108)$$

$$\left[(C_2H_5)_2Ge(CH_2)_nGe(C_2H_5)_2\text{—Hg—}\right]_x \longrightarrow Hg + (C_2H_5)_2Ge{-}Ge(C_2H_5)_2{-}(CH_2)_n\text{(ring)} \qquad n = 3, 4$$

13. *Transition-Metal–Acetylene Complexes with Chlorosilanes*

Diphenyldichlorosilane has been reported to react with $Fe_2(CO)_6$-$(C_6H_5C{\equiv}CC_6H_5)_2$ to give hexaphenylsilole[188]:

$$Fe_2(CO)_6(C_6H_5C{\equiv}CC_6H_5)_2 + (C_6H_5)_2SiCl_2 \longrightarrow (C_6H_5)_4C_4Si(C_6H_5)_2 \quad (109)$$

14. *Ring Expansion Reactions*

a. Carbon-functional silanes with metal halides. In studies of the polymerization of various silacyclopentane derivatives in the presence of metal halides, especially aluminum chloride, it was discovered that polymerization does not occur if the silicon atom is carbon-functional.[189,190] When 1-chloromethyl-1-methyl-1-silacyclopentane was treated with aluminum chloride at room temperature, a vigorous reaction set in to give 1-chloro-1-methyl-1-silacyclohexane[191,192]:

$$(ClCH_2)(H_3C)Si(CH_2)_4 \xrightarrow{AlCl_3} Cl(H_3C)Si(CH_2)_5 \quad (110)$$

An analogous ring expansion reaction occurs with 1-chloromethyl-1-methyl-1-silacyclohexane[190]:

$$(ClCH_2)(H_3C)Si(CH_2)_5 \xrightarrow{AlCl_3} Cl(H_3C)Si(CH_2)_6 \quad (111)$$

The cyclization of 1-chloromethyldimethylsilyl-3-trimethylsilylpropane in the presence of aluminum chloride results in the formation of 1,1-dimethylsilacyclopentane and trimethylchlorosilane[193]:

$$ClCH_2Si(CH_3)_2(CH_2)_3Si(CH_3)_3 \xrightarrow[AlCl_3]{140^\circ} (H_3C)_2Si(CH_2)_4 + (CH_3)_3SiCl \quad (112)$$

presumably through the expansion of the trimethylene group in —$Si(CH_2)_3Si$— to a tetramethylene group found in the product before, during, or after cyclization.

No ring expansion reaction is observed when 1-(γ-chloropropyl)-1-methyl-1-silacyclopentane is treated with aluminum chloride; γ elimination takes place instead to liberate cyclopropane[189]:

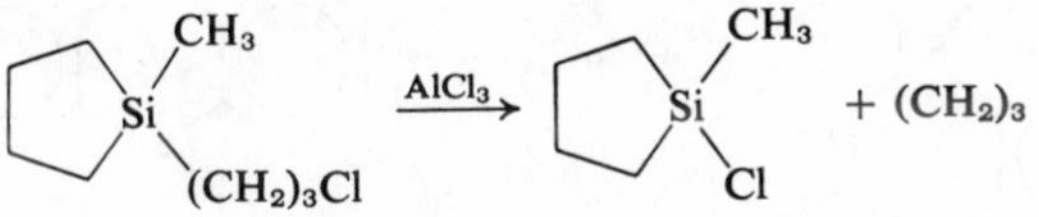

$$\text{1-methyl-1-(CH}_2)_3\text{Cl-silacyclopentane} \xrightarrow{AlCl_3} \text{1-methyl-1-chloro-silacyclopentane} + (CH_2)_3 \qquad (113)$$

b. Insertion of carbenes into the intracyclic silicon-carbon bond. Only one example of the insertion of dichlorocarbene into a silicon-carbon bond is known. This reaction has been observed to take place with the strained silacyclobutane ring system to result in ring expansion to a silacyclopentane[194]:

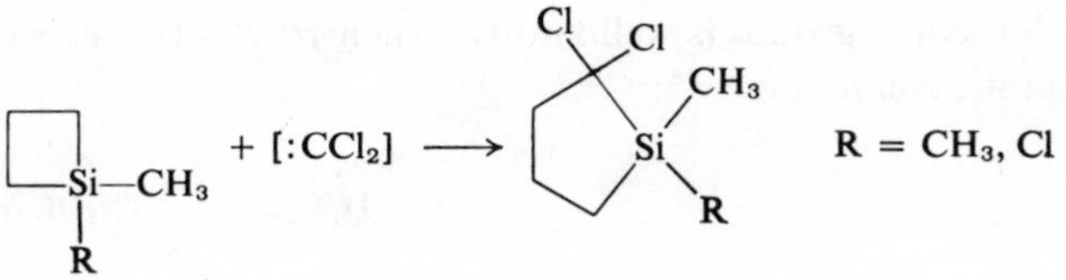

$$\text{1-R-1-CH}_3\text{-silacyclobutane} + [:CCl_2] \longrightarrow \text{2,2-dichloro-1-CH}_3\text{-1-R-silacyclopentane} \qquad R = CH_3, Cl \qquad (114)$$

1,3-Disilacyclobutane is inert under the same conditions.

15. *Transformations of Heterocycles*

Aluminum heterocycles undergo reaction with difunctional or tetrafunctional halostannanes to give tin heterocycles. For example, tris[3,3-dimethylpentamethylene]dialuminum is converted to the corresponding tin heterocycles with tin tetrachloride or dibutyltin dichloride[195–197]:

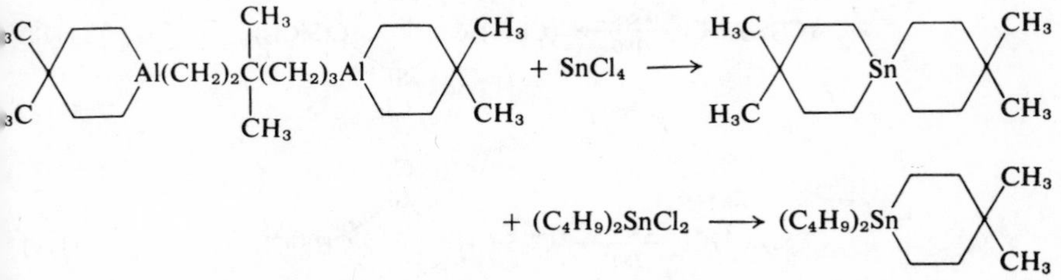

$$\text{(H}_3\text{C)}_2\text{C(CH}_2\text{CH}_2)_2\text{Al(CH}_2)_2\text{C(CH}_3)_2\text{(CH}_2)_2\text{Al(CH}_2\text{CH}_2)_2\text{C(CH}_3)_2 + SnCl_4 \longrightarrow \text{3,3,9,9-tetramethyl-6-stannaspiro[5.5]undecane}$$

$$+ (C_4H_9)_2SnCl_2 \longrightarrow (C_4H_9)_2Sn(CH_2CH_2)_2C(CH_3)_2 \qquad (115)$$

The tin spiro heterocycle containing the phenazastannine ring system can be transformed by refluxing with tin(IV) chloride[88]:

$$\text{spiro-bis(10-R-phenazastannine)} + SnCl_4 \longrightarrow \text{10-R-5,5-dichlorophenazastannine} \quad (116)$$

16. Reaction with Activated Methylene Protons

The enhanced reactivity of the hydrogen atoms on a carbon attached to electron-withdrawing groups is well known. Their reactivity has been used to prepare a stannocyclobutane[198]:

$$(C_4H_9)_2SnCl_2 + H_2C[CH(COOC_2H_5)_2]_2 \longrightarrow (C_4H_9)_2Sn\!\left[C(COOC_2H_5)_2\right]_2CH_2 \quad (117)$$

17. Dechlorination by Fe/Cu

Novel perchlorosilacyclobutanes have been prepared by heating poly-(trichlorosilyl)chloromethanes in the presence of iron-copper alloy[199,200]:

$$(Cl_3Si)_3CCl \xrightarrow[190^\circ]{Fe/Cu} (Cl_3Si)_2C\langle(SiCl_2)_2\rangle C(SiCl_3)_2 \quad (118)$$

$$(Cl_3Si)_2CHCl \xrightarrow[350^\circ]{Fe/Cu} Cl_3SiCH\langle(SiCl_2)_2\rangle CHSiCl_3 \quad (119)$$

B. Group IV–Group V Bonds

1. Amination

The most frequently used method for forming cyclic silicon-nitrogen or germanium-nitrogen linkages is the amine-chloride reaction, exemplified below for the formation of 2-sila-1,3-dimethylimidazolidine[201]:

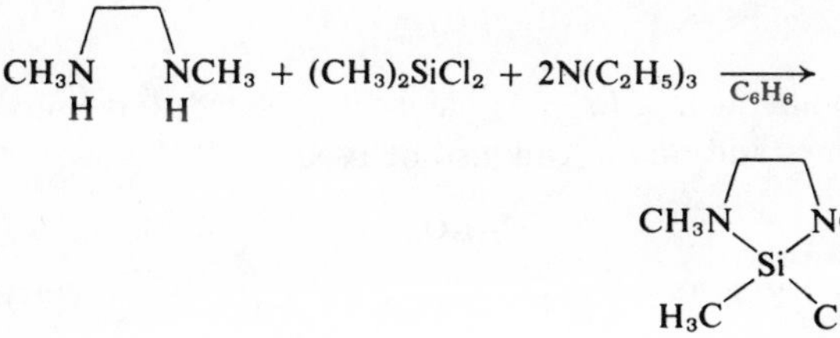

$$CH_3NH-CH_2CH_2-NHCH_3 + (CH_3)_2SiCl_2 + 2N(C_2H_5)_3 \xrightarrow{C_6H_6} \text{[ring: } CH_3N-CH_2CH_2-NCH_3 \text{ bridged by } Si(CH_3)_2] + 2(C_2H_5)_3NHCl \quad (120)$$

The reaction seems to be quite general for silicon with regard to the size and nature of the ring, as is apparent from the following examples:

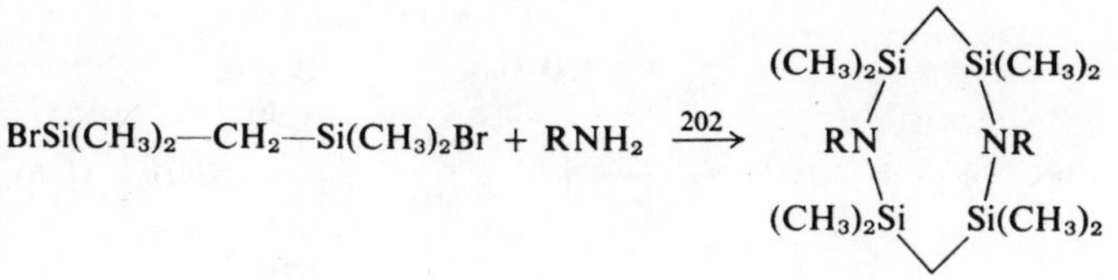

$$BrSi(CH_3)_2-CH_2-Si(CH_3)_2Br + RNH_2 \xrightarrow{202} \text{[8-membered ring: } (CH_3)_2Si(CH_2)Si(CH_3)_2-NR-Si(CH_3)_2(CH_2)Si(CH_3)_2-NR] \quad (121)$$

$$RNH-CH_2CH_2-NHR + MCl_4 \xrightarrow[C_6H_6]{203} \text{[spiro: } (RNCH_2CH_2NR)M(RNCH_2CH_2NR)] + RNH-CH_2CH_2-NHR\cdot HCl$$

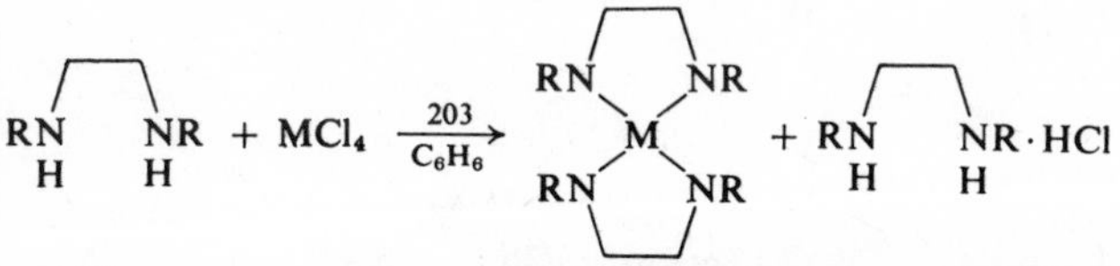

M = Si, Ge

R = CH_3, C_2H_5, p-$CH_3C_6H_4$ (triethylamine used in place of excess diamine) (122)

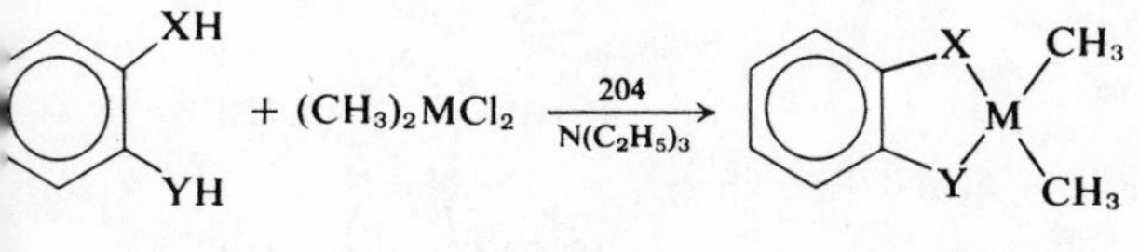

$$o\text{-}C_6H_4(XH)(YH) + (CH_3)_2MCl_2 \xrightarrow[N(C_2H_5)_3]{204} \text{[ring: } C_6H_4(X)(Y)M(CH_3)_2] \quad (123)$$

X = O, S, NH
Y = NH
M = Si, Ge

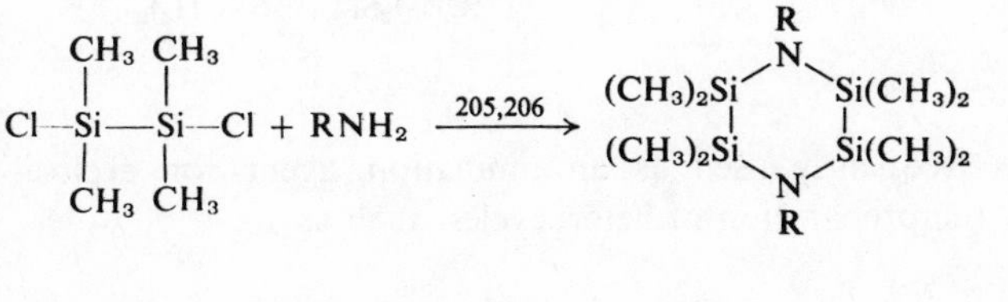

$$Cl-Si(CH_3)_2-Si(CH_3)_2-Cl + RNH_2 \xrightarrow{205,206} \text{[6-membered ring: } (CH_3)_2Si-Si(CH_3)_2-NR-Si(CH_3)_2-Si(CH_3)_2-NR] \quad (124)$$

A variety of substituents on the nucleophilic nitrogen have been used; aromatic groups sometimes lower the yield,[203] and the trimethylsilyl group enhances ring formation.[207,208] The *N*-unsubstituted 2-sila-1,3-diazacyclopentanes ($n = 2$) and hexanes ($n = 3$) apparently cannot be isolated because

$$HN\!-\!(CH_2)_n\!-\!NH\!-\!SiR_2 \ (\text{ring})$$

of polymerization; the heptane analog ($n = 4$) has been isolated, but polymerizes readily.[207] Hydrazines and imines can also be used:

$$C_2H_5O(HN{=})C\!-\!C({=}NH)OC_2H_5 + RR'SiCl_2 \xrightarrow[N(C_2H_5)_3]{209} \text{cyclo-}[C_2H_5O\!-\!C({=}N)\!-\!C(OC_2H_5)({=}N)\!-\!SiRR'] \qquad (125)$$

$$\text{cyclo-}[(CH_3)_2Si(Cl)\!-\!CH_2CH_2\!-\!Si(CH_3)_2Cl] + H_2NNH_2 \xrightarrow[C_6H_6]{210,211} \text{cyclo-}[(CH_3)_2Si\!-\!NH\!-\!NH\!-\!Si(CH_3)_2\!-\!CH_2CH_2] + \text{cyclo-}[(CH_3)_2Si\!-\!N(NH_2)\!-\!Si(CH_3)_2\!-\!CH_2CH_2] \qquad (126)$$

(see also Ref. 212)

$$RR'SiCl_2 + R''NH\!-\!NHR''' \xrightarrow{213,214} \text{cyclo-}[RR'Si\!-\!NR''\!-\!NR'''\!-\!SiRR'\!-\!NR'''\!-\!NR''] \qquad (127)$$

$$2CH_3N(Si(CH_3)_2Cl)_2 + 5H_2NNH_2 \xrightarrow{215} \text{bicyclo-}[HN(Si(CH_3)_2)_2N\!-\!N(Si(CH_3)_2)_2NH] \qquad (128)$$

Ammonia itself is frequently used as an amination agent for cyclosiloxazanes[216] or for the preparation of heterocycles, such as

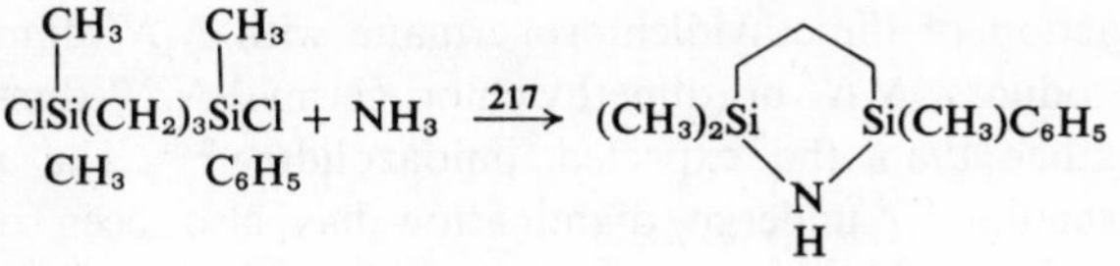

$$ClSi(CH_3)_2(CH_2)_3Si(CH_3)(C_6H_5)Cl + NH_3 \xrightarrow{217} (CH_3)_2Si\overline{(CH_2)_3}Si(CH_3)C_6H_5\,(NH) \quad (129)$$

$$Cl{-}Si(CH_3)_2(CH_2)_8Si(CH_3)_2{-}Cl + NH_3 \xrightarrow{218} \overline{(CH_3)_2Si(CH_2)_8Si(CH_3)_2NH} \quad (130)$$

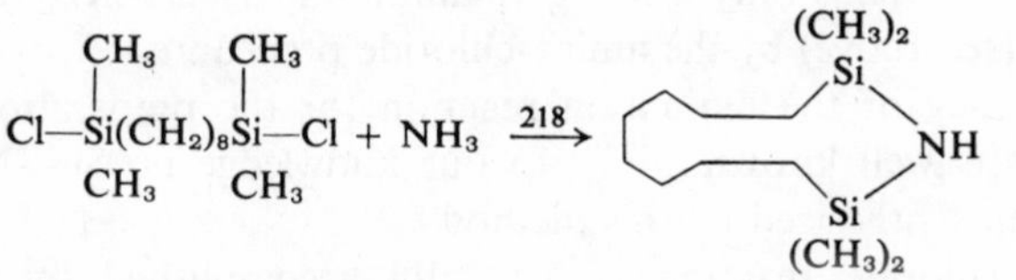

The organic substituent on silicon has generally been methyl; phenyltrichlorosilane and diphenyldichlorosilane do not react with *N*,*N'*-bis(trimethylsilyl)-*o*-phenylenediamine in the presence of triethylamine, while $(CH_3)_2SiCl_2$, CH_3SiCl_3, and $C_6H_{11}SiCl_3$ proceed to[219]

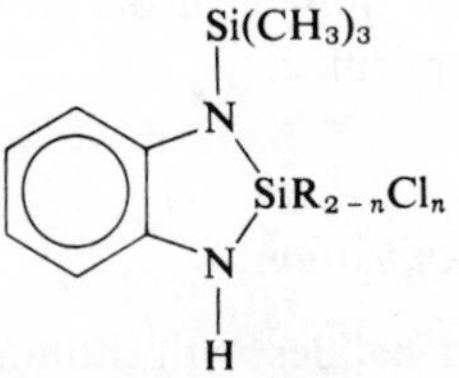

$$C_6H_4[N(Si(CH_3)_3)][NH]SiR_{2-n}Cl_n \qquad R = CH_3 \text{ or } C_6H_{11}$$

The use of chloromethyldimethylchlorosilane permits the introduction of the N—C—Si—N linkage:

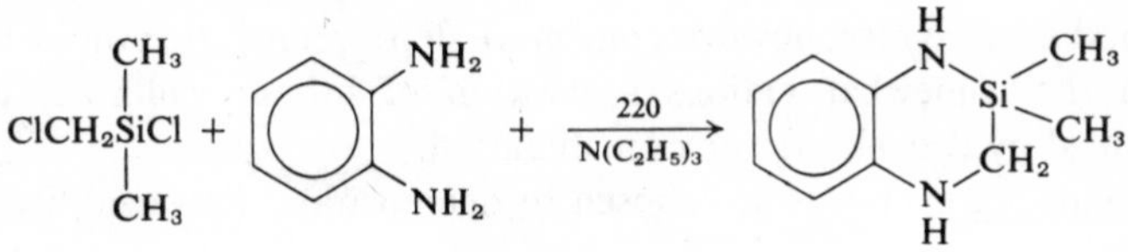

$$ClCH_2Si(CH_3)_2Cl + o\text{-}C_6H_4(NH_2)_2 + \xrightarrow[N(C_2H_5)_3]{220} C_6H_4[NH{-}Si(CH_3)_2{-}CH_2{-}NH] \quad (131)$$

$$BrCH_2Si(CH_3)_2Cl + CH_3NH(CH_2)_2NHCH_3 \xrightarrow{221} \overline{CH_3N{-}Si(CH_3)_2{-}CH_2{-}N(CH_3){-}(CH_2)_2} \quad (132)$$

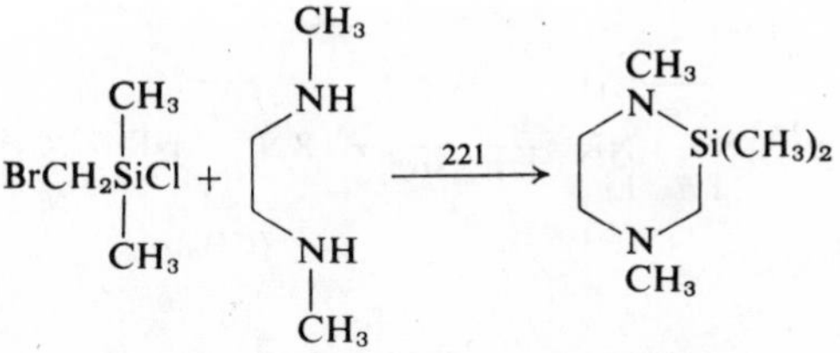

Amination of 1,2-bis(chlorosilyl)-carboranes produces novel boron-containing bicyclic heterocycles[45,222]:

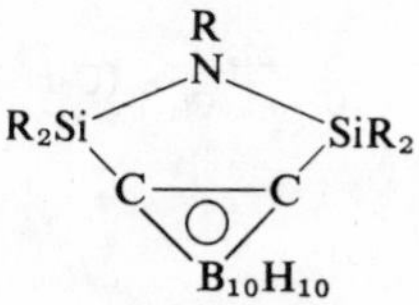

$$R_2Si{-}N(R){-}SiR_2 \text{ bridged by } C{-}C(B_{10}H_{10})$$

The reaction of dimethyldichlorogermane with *N,N'*-dimethylethylenediamine produces *N,N'*-bis(dimethylchlorogermyl)-*N,N'*-dimethylethylenediamine rather than the expected imidazolidine.[223] The reluctance of dichlorogermanes to undergo diamination has also been observed with monoamines.[223,224] Thus only a few germanium-nitrogen heterocycles have been obtained (see above) by the amine-chloride procedure.

The inefficiency of the amination reaction for the preparation of tin-nitrogen bonds is well known[225,226]; to our knowledge no Sn—N heterocycles have been synthesized by this method.

The amine-chloride reaction is generally accomplished by dropwise addition of the chlorosilane or -germane to the amine dissolved in benzene (usually), ether, or petroleum ether. Excess amine or triethylamine, if the starting amine is only weakly basic or in short supply, is used to take up the HCl produced. The addition is performed cold if the reaction is exothermic, followed by a short period of reflux. The amine hydrochloride is then filtered, the solvent removed, and the product purified.

2. *Lithioamination*

The reaction of metal or metalloid halides with lithium, or other alkali metal, salts of amines has two primary advantages over the simple amination procedure: (a) greater reactivity—for example, tin amines can be obtained by this method but not by amination; and (b) conservation of starting materials. The method does involve, however, one more step—preparation of the lithium salt—and the somewhat tedious separation of lithium chloride, usually accomplished by decantation or centrifugation.

The following examples are chosen to demonstrate the versatility of the method:

$$\text{RN(H)CH}_2\text{CH}_2\text{N(H)R} \xrightarrow{n\text{-C}_4\text{H}_9\text{Li}} \text{RN(Li)CH}_2\text{CH}_2\text{N(Li)R} \xrightarrow{(\text{CH}_3)_2\text{MCl}_2} \text{RN}\overline{\text{CH}_2\text{CH}_2\text{NR}}\text{M(CH}_3)_2 \quad \text{M} = \text{Ge, R} = \text{CH}_3{}^{227}; \ \text{M} = \text{Sn, R} = \text{Si(CH}_3)_3{}^{2} \tag{13}$$

$$[(\text{C}_2\text{H}_5)_3\text{Si}](\text{Li})\text{N}-\text{Ge}(\text{CH}_3)_2-\text{N}(\text{Li})[\text{Si}(\text{C}_2\text{H}_5)_3] \xrightarrow[(\text{CH}_3)_2\text{SnCl}_2]{229} (\text{C}_2\text{H}_5)_3\text{SiN}\overline{\text{Ge}(\text{CH}_3)_2\text{NSi}(\text{C}_2\text{H}_5)_3\text{Sn}(\text{CH}_3)_2} \tag{134}$$

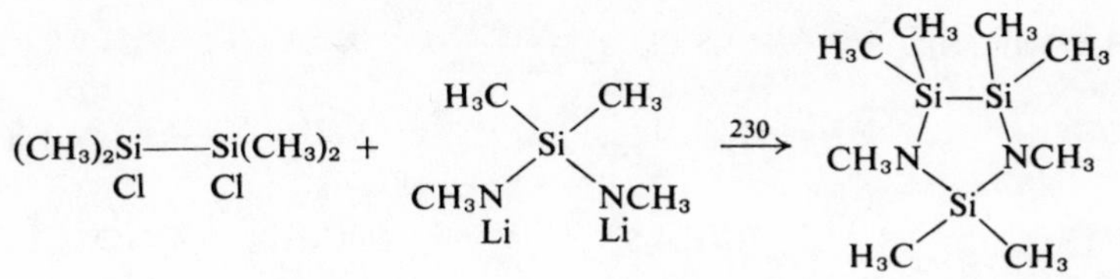

$$(CH_3)_2Si(Cl)\text{—}Si(CH_3)_2Cl + (H_3C)(CH_3)Si(N(Li)CH_3)_2 \xrightarrow{230} \text{[ring: } (CH_3)_2Si\text{—}Si(CH_3)_2,\ CH_3N,\ NCH_3,\ Si(CH_3)_2 \text{]} \qquad (135)$$

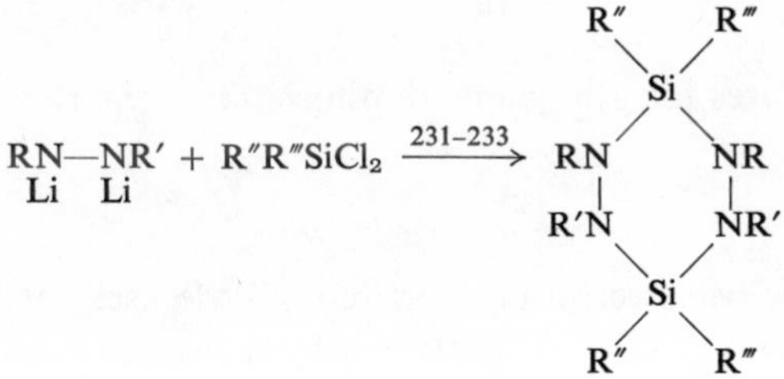

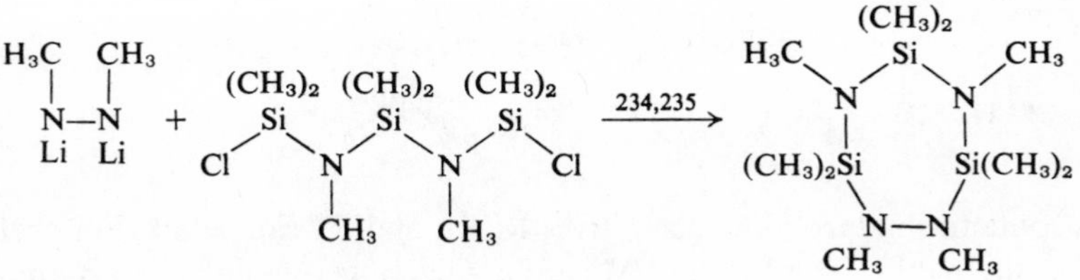

$$RN(Li)\text{—}N(Li)R' + R''R'''SiCl_2 \xrightarrow{231\text{–}233} \text{[ring: } R''R'''Si,\ RN,\ NR,\ R'N\text{—}NR',\ SiR''R''' \text{]} \qquad (136)$$

$$H_3C\text{—}N(Li)\text{—}N(Li)\text{—}CH_3 + Cl\text{—}Si(CH_3)_2\text{—}N(CH_3)\text{—}Si(CH_3)_2\text{—}N(CH_3)\text{—}Si(CH_3)_2\text{—}Cl \xrightarrow{234,235} \text{[ring: } Si(CH_3)_2,\ H_3C\text{—}N,\ N\text{—}CH_3,\ (CH_3)_2Si,\ Si(CH_3)_2,\ N(CH_3)\text{—}N(CH_3) \text{]} \qquad (137)$$

$$C_6H_5CN + CH_3N(H)Li \xrightarrow{236} C_6H_5C(=NLi)\text{—}N(H)CH_3 \xrightarrow[(C_2H_5)_3N]{(CH_3)_2SiCl_2}$$

$$\text{[ring: } C_6H_5\text{—}C{=}N,\ Si(CH_3)_2,\ NCH_3,\ N{=}C\text{—}C_6H_5,\ (CH_3)_2Si,\ N \text{]} \qquad (138)$$

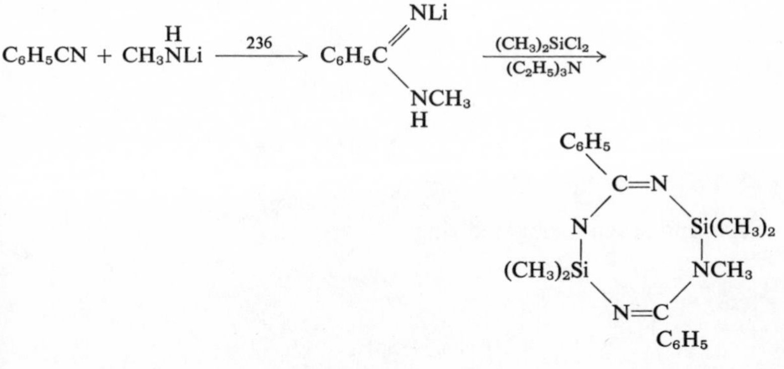

Other examples of ring closure between mono and dilithio amines and chlorosilanes can be found in the work of the Rochow[237–239] and Wannagat groups.[240] Lithioamination has also been applied to the preparation of cyclosiloxazanes.[241]

A silicon-phosphorus heterocycle has been obtained from tetrakis-(sodiumphenylphosphinomethyl)methane according to[242]

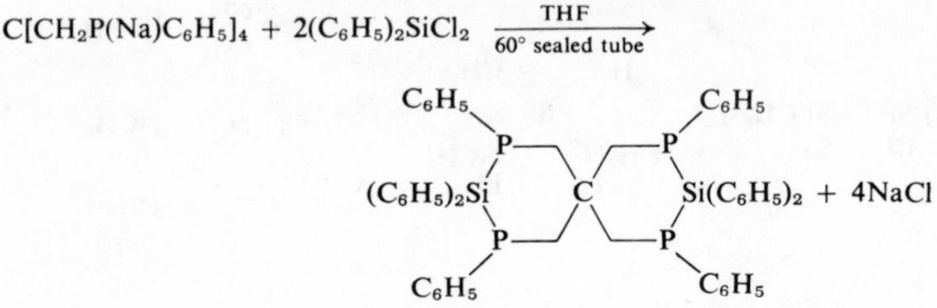

$$C[CH_2P(Na)C_6H_5]_4 + 2(C_6H_5)_2SiCl_2 \xrightarrow[60^\circ \text{ sealed tube}]{\text{THF}}$$

$$(C_6H_5)_2Si[P(C_6H_5)CH_2]_2C[CH_2P(C_6H_5)]_2Si(C_6H_5)_2 + 4NaCl \quad (139)$$

General procedures for lithioamination have been reported.[225–227,239]

3. *Transamination*

This potentially powerful but apparently little used method involves the exchange of amino groups, as illustrated by the reaction

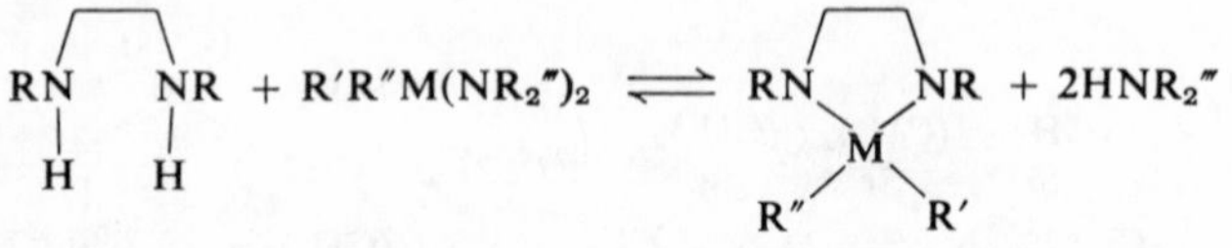

$$RNH\text{—}CH_2CH_2\text{—}NHR + R'R''M(NR_2''')_2 \rightleftharpoons \overline{RN\text{—}CH_2CH_2\text{—}NR}\text{—}M(R'')(R') + 2HNR_2''' \quad (140)$$

In this route to imidazolidine derivatives R, R′, and R″ can be alkyl or aryl, R‴ is usually methyl or ethyl, and M can be silicon, germanium, or tin [only when $R = Si(CH_3)_3$].[201,207,223,227,228,238] The spiro derivative

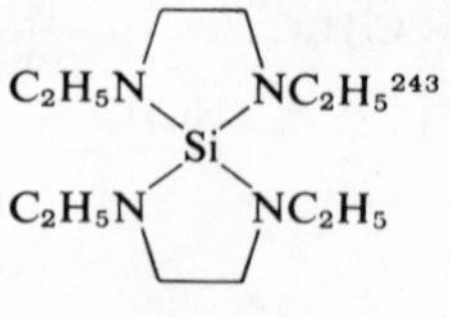

$$[C_2H_5NCH_2CH_2NC_2H_5]_2Si \text{ (spiro)}$$ [243]

and the six- and seven-membered rings

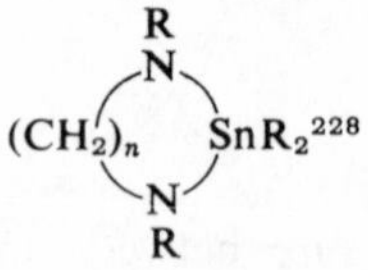

$$(CH_2)_n(NR)_2SnR_2$$ [228]

have also been obtained in this way.

Transamination of silylamines generally requires an acidic catalyst—$(NH_4)_2SO_4$ or amine hydrochloride—and temperatures of 80 to 150°C; germyl- and stannylamines require no catalyst and proceed at lower temperatures.

4. *Reactions of Silanes with Amines*

The reaction of silanes with alkali-metal amine salts, as modified by Fink, provides a simple route to five- and six-membered heterocycles, as illustrated by the reactions[244,245]

$$H_2N—(CH_2)_n—NH_2 + 3C_6H_5RR'SiH \xrightarrow[NaH]{40–80°}$$

$$C_6H_5RR'Si\overset{(CH_2)_n}{N \quad\quad N}SiRR'C_6H_5 \text{ (ring closed by } SiR_2) + 3H_2 + C_6H_6 \qquad n = 2, 3 \qquad (141)$$

$$RNH–CH_2CH_2–NHR + R_2'R''SiH \xrightarrow{NaH} \text{RN–CH}_2\text{CH}_2\text{–NR (ring closed by SiRR'')} \qquad R = \text{alkyl, aryl, } Si(CH_3)_3 \qquad (142)$$

$$o\text{-}C_6H_4(NH_2)_2 + 3(C_6H_5)_2CH_3SiH \xrightarrow{NaH} C_6H_4[N(Si(C_6H_5)_2CH_3)]_2Si(C_6H_5)(CH_3) + 3H_2 + C_6H_6 \qquad (143)$$

The reaction can be carried out in a variety of solvents, and produces excellent yields.

5. *Others*

A silazarophenanthrene has been obtained by photolysis of the parent azide[246]:

$$C_6H_5–C_6H_4–Si(CH_3)_2N_3 \xrightarrow{h\nu} \text{silazarophenanthrene with } N(H)–Si(CH_3)_2 \qquad (144)$$

Pyrolysis of

$$(CH_3)_3Si(\overset{Si(CH_3)_2}{N}Si(CH_3)_2)_2\overset{Si(CH_3)_3}{N}Li$$

produced, among other products,[247]

$$\begin{array}{c} Si(CH_3)_3 \\ | \\ N \\ / \quad \backslash \\ (CH_3)_2Si \text{———} Si(CH_3)_2 \end{array}$$

Cleavage of octaphenylcyclotetrasilane with a halogen or lithium in THF, followed by reaction with $C_2H_5NH_2$, $C_6H_5PCl_2$, or $(C_6H_5)_2MCl_2$, where M = Si, Ge, Sn, produces heterocycles of the type[248]

$$\begin{array}{c} (C_6H_5)_2Si\text{—}Si(C_6H_5)_2 \\ | \qquad\quad | \\ (C_6H_5)_2Si\text{—}Si(C_6H_5)_2 \end{array} \xrightarrow[(2)\ C_6H_5PCl_2]{(1)\ Li} \begin{array}{c} (C_6H_5)_2Si\text{—}Si(C_6H_5)_2 \\ (C_6H_5)_2Si \quad\quad Si(C_6H_5)_2 \\ P \\ C_6H_5 \end{array} \tag{145}$$

Cleavage of a Si—CH_3 bond by *n*-butyllithium in THF/hexane also results in cyclization[249]:

$$\begin{array}{c} CH_3N \text{——————} NSi(CH_3)_3 \\ | \qquad\qquad\qquad | \\ Si(CH_3)_3 \qquad\quad H \end{array} \xrightarrow{n\text{-}C_4H_9Li} \begin{array}{c} CH_3N \text{——} NSi(CH_3)_3 \\ Si(CH_3)_2 \end{array} \tag{146}$$

The now well-known insertion reaction has been used to prepare the first tin(II)-nitrogen bond[250]:

$$\text{2,2'-biphenylylenedioxy-Sn} + 2C_6H_5NCO \xrightarrow{\text{toluene}} \begin{array}{c} O \quad\ C_6H_5 \\ \| \quad / \\ \text{—O—C—N} \\ \backslash \\ Sn \\ / \\ \text{—O—C—N} \\ \| \quad \backslash \\ O \quad\ C_6H_5 \end{array} \tag{147}$$

as well as a 1-sila-2,4,6-triazacyclohexane-3,5-dione[251,252]:

$$\begin{array}{c} (CH_3)_2 \\ Si \\ H \quad\quad\quad H \\ N \quad\quad N \\ | \quad\quad\ | \\ (CH_3)_2Si \quad Si(CH_3)_2 \\ N \\ H \end{array} + C_6H_5NCO \longrightarrow \begin{array}{c} (CH_3)_2 \\ Si \\ C_6H_5\text{—}N \quad N\text{—}C_6H_5 \\ | \quad\quad\quad | \\ C \quad\quad C \\ O \quad N \quad O \\ H \end{array} \tag{148}$$

Cyclosilithiazanes have been obtained by the redistribution of groups between hexamethylcyclotrisilthian and nonamethylcyclotrisilazane.[253,254]

C. Group IV–Group VI Bonds

1. Alcoholysis

The reaction of an alcohol or water with a group IV–halide is the most frequently used method for producing cyclic group IV–oxygen bonds. Although the extent of reaction varies considerably with the nature of the alcohol, in general the silicon-chlorine bond is reactive enough to allow displacement of the chlorine by the simple reflux of reactants, neat or in solution, although the reaction is facilitated by the addition of a base, such as pyridine or triethylamine. The base takes up the liberated HCl and may also modify the reaction by its complexing ability.[255] Displacement of chlorine from germanium usually requires a base, such as triethylamine, and a strong base, such as $NaNH_2$, is necessary with tin.

The number of known group IV sulfur heterocycles is rather limited, but in general the thioalcoholysis methods are similar to the alcoholysis methods except that triethylamine is used in almost every case.

The following examples of the application of the alcoholysis method to the preparation of silicon-oxygen and silicon-sulfur heterocycles are roughly in order of increasing complexity of the reagent ROH or RSH.

a. Use of H_2O or H_2S. Although the reaction of difunctional chlorosilanes or -germanes with water often results in polymerization, a number of cyclic structures have been obtained by this method:

$$R_2Si(Cl)\text{—}\!\!\frown\!\!\text{—}SiR_2(Cl) + H_2O \xrightarrow[\Delta]{256} R_2Si\langle\!\frown\!\rangle SiR_2 \;(\text{ring closed by O}) + 2HCl \quad (149)$$

$$RR'C[CH_2Si(CH_3)_2Cl][\text{—}Si(CH_3)(R'')Cl] + H_2O \xrightarrow{257,258} \text{cyclic } RR'C\text{—}Si(CH_3)_2\text{—}O\text{—}Si(H_3C)(R'') + 2HCl \quad (150)$$

$$2\,Si(CH_3)_2(Cl)\text{—}Si(CH_3)_2(Cl) + 2H_2X \longrightarrow \underset{X\text{—}Si(CH_3)_2\text{—}Si(CH_3)_2\text{—}X}{Si(CH_3)_2\text{—}Si(CH_3)_2} + 4HCl \qquad X = O^{259},\; X = S^{260} \quad (151)$$

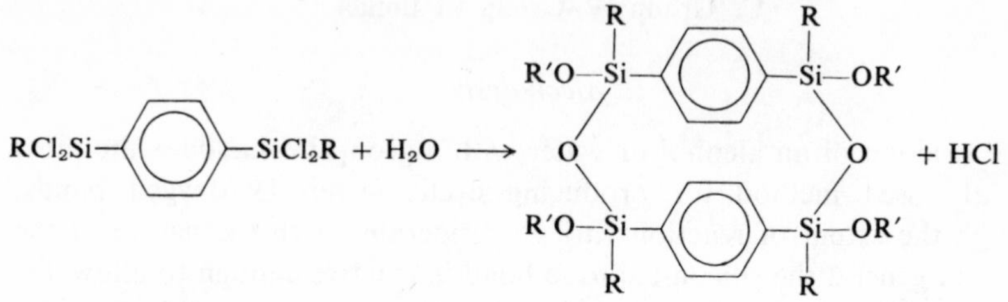

$$RCl_2Si-C_6H_4-SiCl_2R + H_2O \longrightarrow [R(R'O)Si-C_6H_4-Si(R)(OR')-O-]_2 \text{ (cyclic)} + HCl$$

$R = CH_3, C_6H_5, R' = H$[261]
$R = CH_3, R' = H$ and $Si(CH_3)_3$ (cohydrolysis with $(CH_3)_3SiCl$)[262] (152)

(see Ref. 263 for the napthalene analog)

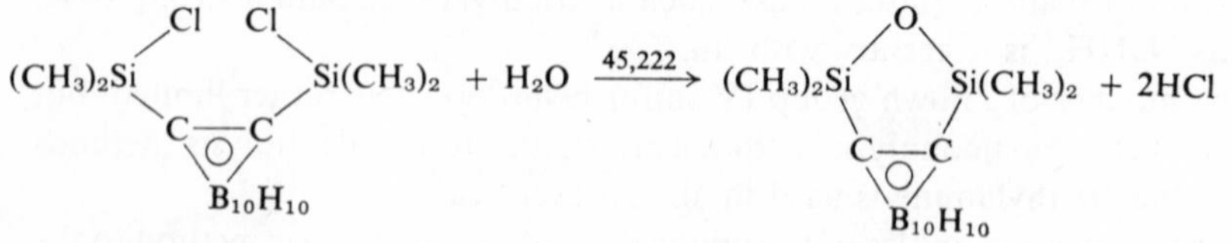

$$[(CH_3)_2SiCl]_2C_2B_{10}H_{10} + H_2O \xrightarrow{45,222} (CH_3)_2Si-O-Si(CH_3)_2 \text{ (bridged by } C_2B_{10}H_{10}) + 2HCl \quad (153)$$

$$(C_6H_5)_2M(I)-M(C_6H_5)_2-M(C_6H_5)_2-M(C_6H_5)_2(I) + H_2O \longrightarrow \text{cyclo-}[(C_6H_5)_2M]_4O + 2HI$$

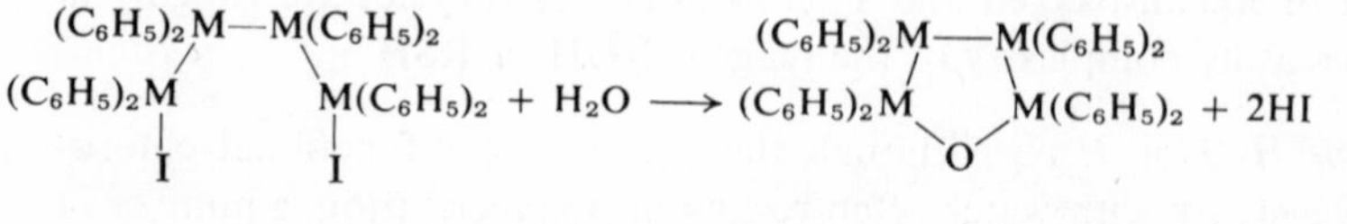

$M = Si$[264]
$M = Ge$[265,2...] (15...)

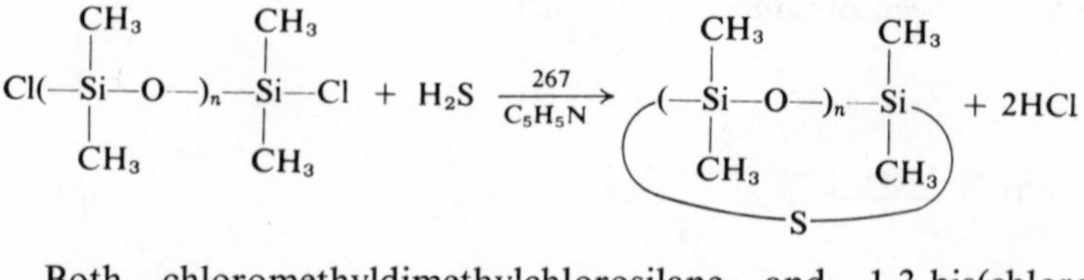

$$Cl(-Si(CH_3)_2-O-)_n-Si(CH_3)_2-Cl + H_2S \xrightarrow[C_5H_5N]{267} \text{cyclo-}[(-Si(CH_3)_2-O-)_n-Si(CH_3)_2-S-] + 2HCl \quad n = 1, 2 \quad (15...)$$

Both chloromethyldimethylchlorosilane and 1,3-bis(chloromethyl)tetramethyldisilazane react with H_2S in the presence of triethylamine to form the heterocycle[268]:

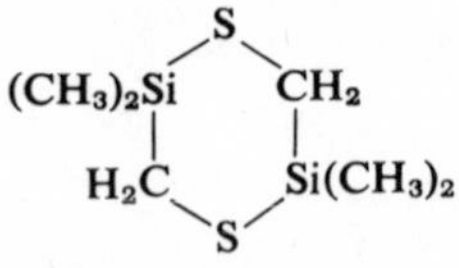

$$\text{cyclo-}[S-CH_2-Si(CH_3)_2-S-CH_2-Si(CH_3)_2-]$$

Germatetrahydrothiophenes have been obtained by reactions of the following type:

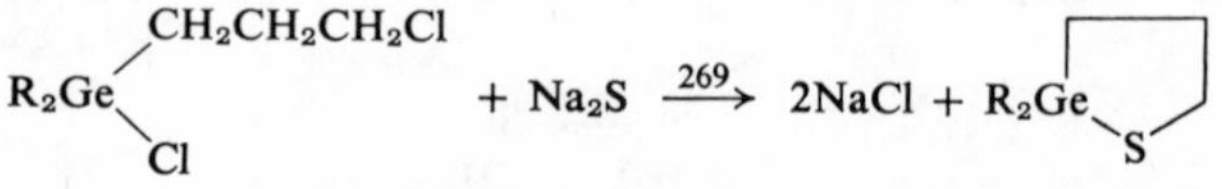

$$R_2Ge(CH_2CH_2CH_2Cl)Cl + Na_2S \xrightarrow{269} 2NaCl + R_2Ge(CH_2)_3S \text{ (ring)} \quad (156)$$

b. Use of diols or dithiols. The reaction of ethyleneglycol derivatives with dihalosilanes produces the dimeric 10-membered ring products[270] except when the glycol contains a tertiary carbon atom:

$$R_2SiCl_2 + HO\text{–}CR'_2\text{–}CR''_2\text{–}OH \longrightarrow R_2Si\begin{matrix} O\text{–}CR'_2 \\ O\text{–}CR''_2 \end{matrix} + 2HCl \qquad (157)$$

$R = R' = CH_3, R'' = H$; $R = R' = R'' = CH_3$[271]
$R = C_6H_{11}, R' = R'' = CH_3$[272]
$R = CH_3, R' = R'' = CF_3$[273]

The six- and seven-membered analogs[274,275], as well as compounds such as

R(OH)Si< ring –O–O–O– [276] and R_2Si< ring S–S [277]

can be prepared from the appropriate diols (or thiols) and halides in the presence of triethylamine or pyridine. Sulfuric acid reacts with dimethyldichlorosilane to produce[278,279]

$(CH_3)_2Si(OSO_2O)_2Si(CH_3)_2$

and chloral hydrate yields[280]

$(CH_3)_2Si[OCH(CCl_3)O]_2Si(CH_3)_2$

Spiro structures are obtained when $SiCl_4$ reacts with diols[281]:

$$2\ HO(CH_2)_nOH + SiCl_4 \longrightarrow Si[O(CH_2)_nO]_2 \qquad (158)$$

The reaction of glycolic acid with chloromethyldimethylchlorosilane produces a mixture 90 percent of which is polymeric, the other 10 percent a mixture of the isomers[282]

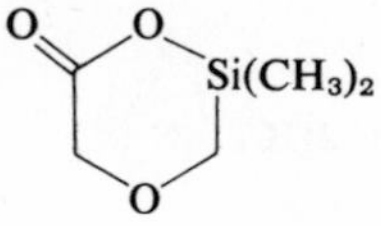

$Si(CH_3)_2$ and $Si(CH_3)_2$

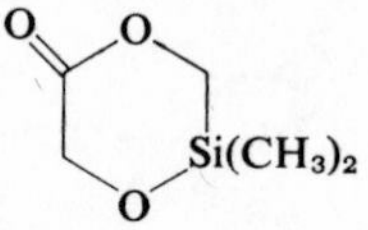

Thioglycolic acid produces the O—Si isomer in good yield[282]:

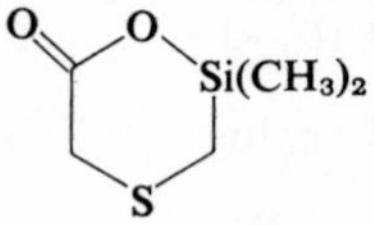

$Si(CH_3)_2$

Chloromethyldimethylchlorosilane reacts with  HS SH and  HS SH to give the compounds[283]

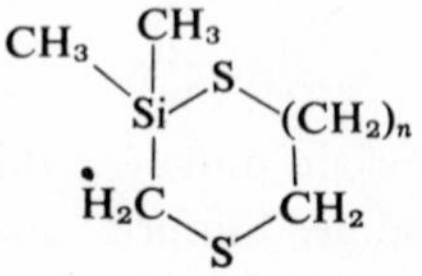

$n = 1, 2$

Various aromatic alcohols can also be used. The reaction of diorganodichlorosilanes with 2-hydroxyacetophenone or -propiophenone or salicylanilide in the presence of pyridine produces heterocycles of the type

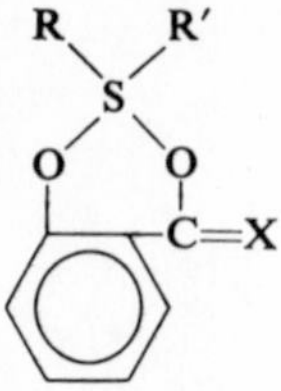

$R = R' = CH_3$; $X = N—C_6H_5$[284]

$R, R' = CH_3, CH = CH_2, C_6H_5$; $X = CH_2, CHCH_3$[285]

Catecohol, 1,8-dihydroxynaphthalene, and 2,2′-dihydroxybiphenyl react with diorganodichlorosilanes in the absence of base to form the five-, six- and seven-membered rings, respectively:

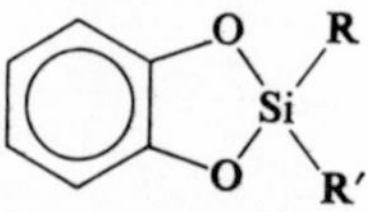

[285,286] 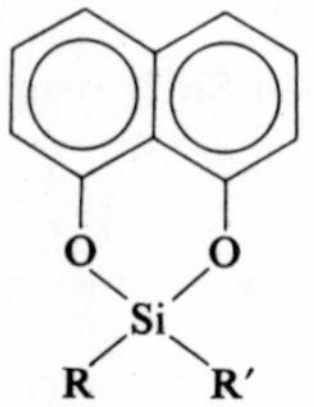 [287] 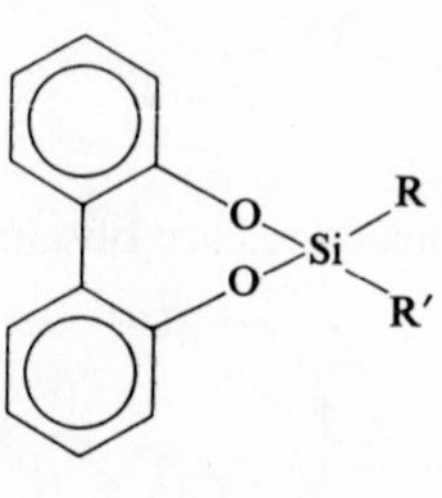 [285,286,288]

The analogous spiro compounds were obtained with $SiCl_4$[255,289] (pyridine and triethylamine complexes were found for the catechol derivatives[255]).

The treatment of *o*-hydroxybenzylalcohol with dimethyldichlorosilane in the presence of triethylamine affords the benzo-1,3-dioxane heterocycle[290]:

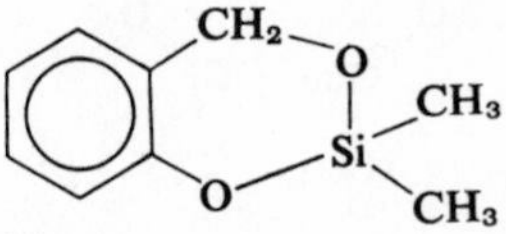

The germanium analog was obtained similarly.[290] A variety of isomeric benzo-1,4-dioxane systems and their sulfur and nitrogen analogs have been prepared by the general reaction

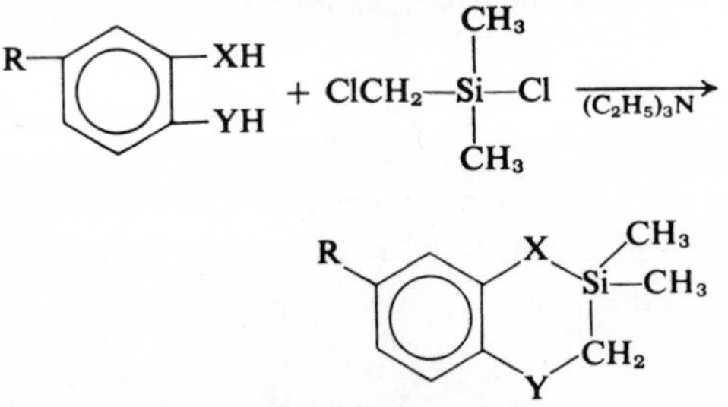

$R = H, X = Y = O$[220,221] (159)
$X = O, Y = NH$[220]
$X = NH, Y = S$[220]
$R = CH_3, X = Y = S$[291]
$X = Y = O$[291]

When $R = CH_3$, both isomers can be isolated:

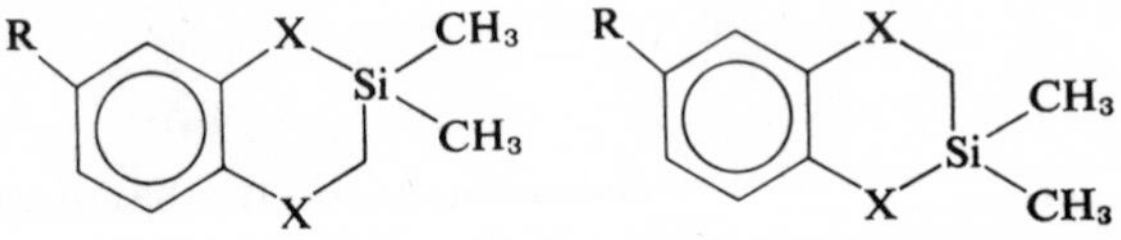

The germanium and tin analogs were prepared by the same method.[291] The silicon-oxygen derivative

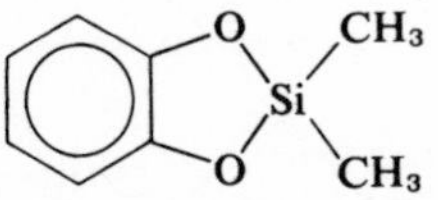

can also be obtained from the disodium salt of catechol.[276] (See also Ref. 273 for the use of lithium.)

As indicated above, the formation of germanium- and tin-oxygen and sulfur heterocycles proceeds quite analogously to the silicon derivatives except that a base, usually triethylamine—in a few cases, with tin, $NaNH_2$ or NaOH—is almost always used.

Hydrolysis of tetrabutyldichlorodigermane results in the six-membered ring[292]

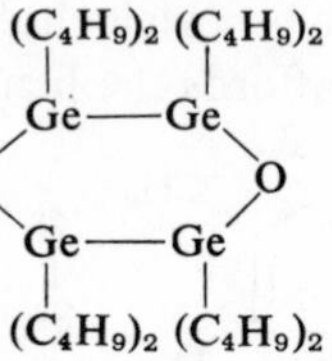

$$\begin{array}{c} (C_4H_9)_2 \; (C_4H_9)_2 \\ Ge\text{—}Ge \\ O \qquad\quad O \\ Ge\text{—}Ge \\ (C_4H_9)_2 \; (C_4H_9)_2 \end{array}$$

The reaction of ethylene glycol with di-*n*-butyltin dichloride in the presence of sodium hydroxide also results in the dimeric ring structure.[293] The use of dithioethane or -propane gives the structures

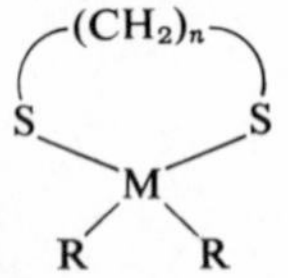

$(CH_2)_n$, S, S, M, R, R

$n = 2$, $R = C_6H_5$, $M =$ Ge, Sn, Pb[294,295]
$n = 2$, $R = CH_3$, $M =$ Sn[296]
$n = 3$, $R = CH_3$, $M =$ Ge, Sn[297]

and thioglycolic acid or 2-hydroxyethanethiol produces the mixed oxygen-sulfur ring.[294] The four-membered and unsaturated five-membered analogs have been made by way of the sodium salts:

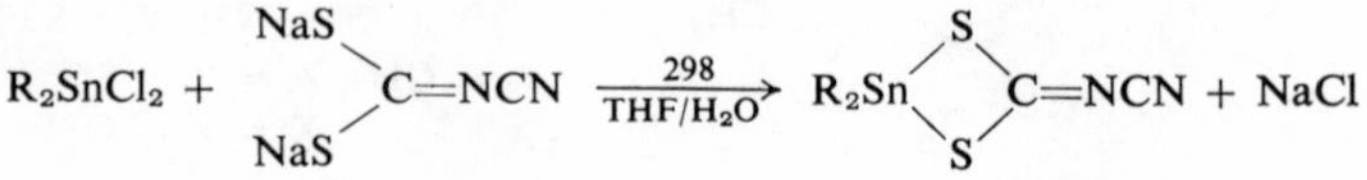

$$R_2SnCl_2 + (NaS)_2C{=}NCN \xrightarrow[THF/H_2O]{298} R_2Sn(S)_2C{=}NCN + NaCl \tag{160}$$

$$R_2SnCl_2 + (NaS)XC{=}CX(SNa) \xrightarrow{299} R_2Sn(S)_2C_2X_2 + 2NaCl \tag{161}$$

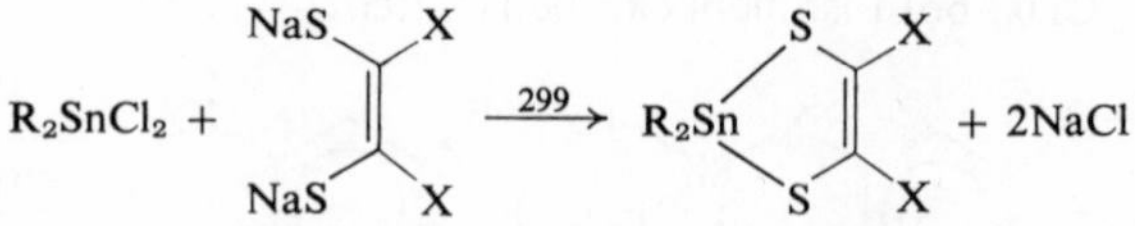

A number of germatranes have been obtained with triethanolamine[300,301]:

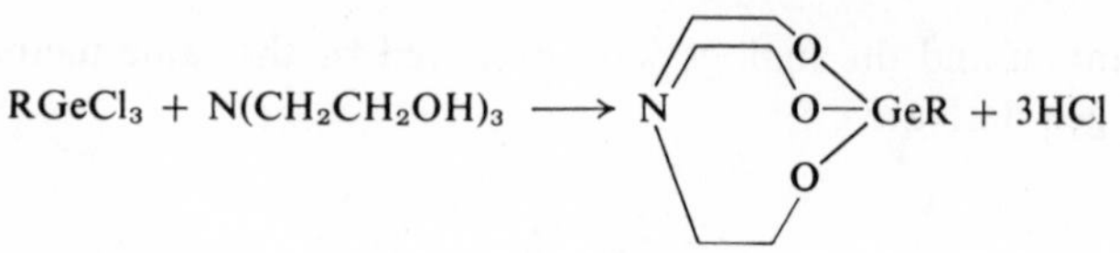

$$RGeCl_3 + N(CH_2CH_2OH)_3 \longrightarrow N(CH_2CH_2O)_3GeR + 3HCl \tag{162}$$

The spiro systems

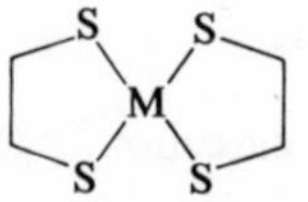

M = Si, Ge, Sn[302]

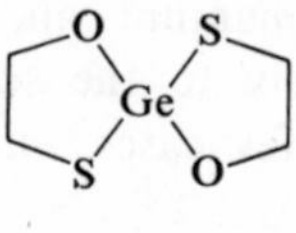

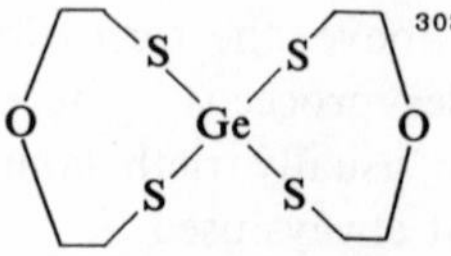

(spiro Ge with O, S / S, O rings) and (spiro Ge with S, S / S, S rings containing O)[303] result from the tetrachloride and the appropriate thiol in the presence of pyridine[294] or sodium ethoxide.[302]

The aromatic five- and six-membered analogs can be obtained by reactions such as

$$\text{CH}_3\text{C}_6\text{H}_3(\text{SH})_2 + \text{Cl}_2\text{M}(\text{CH}_3)_2 \xrightarrow[(\text{C}_2\text{H}_5)_3\text{N}]{290} \text{CH}_3\text{C}_6\text{H}_3\text{S}_2\text{M}(\text{CH}_3)_2 + 2\text{HCl} \quad (163)$$

$$2\,\text{CH}_3\text{C}_6\text{H}_3(\text{SH})_2 + \text{SiCl}_4 \xrightarrow{304} (\text{CH}_3\text{C}_6\text{H}_3\text{S}_2)_2\text{Si} + 4\text{HCl} \quad (164)$$

$$\text{R-C}_6\text{H}_3(\text{XH})(\text{YH}) + \text{ClCH}_2\text{GeCl}(\text{CH}_3)_2 \xrightarrow[(\text{C}_2\text{H}_5)_3\text{N}]{305} \text{R-C}_6\text{H}_3(\text{X})(\text{Y})\text{CH}_2\text{Ge}(\text{CH}_3)_2 + 2\text{HCl}$$

R = H, X = Y = O; X = O, Y = NH (165)
R = CH_3, X = Y = S

$$\text{C}_6\text{H}_4(\text{OH})_2 + \text{R}_2\text{SnCl}_2 \xrightarrow[\text{NaNH}_2]{306} \text{C}_6\text{H}_4\text{O}_2\text{SnR}_2 + 2\text{NaCl} + 2\text{NH}_3 \quad (166)$$

The tin(II) derivatives can also be prepared by this reaction (167).[306]

2. *Condensation Reactions*

All preparative methods involving reactions of the following types:

a. $\begin{cases} R_2MO + 2R'OH \longrightarrow R_2M(OR')_2 + H_2O \\ MO + 2R'OH \longrightarrow M(OR)_2 + H_2O \end{cases}$

b. $R_3MOH + R'OH \longrightarrow R_3MOR' + H_2O$

c. $R_3MOR'' + R'OH \longrightarrow R_3MOR' + HOR''$

where R′ and R″ can be alkyl, aryl, acetyl, and the like, are classified arbitrarily as condensation reactions. This category is then subdivided into oxide reactions (type *a*), group IV–ol reactions (type *b*), and ether reactions (type *c*).

a. Oxide reactions. A number of germanium and tin heterocycles have been prepared from metal oxides as shown below. The oxide and alcohol are refluxed in benzene and the water removed azeotropically; yields are generally quite satisfactory:

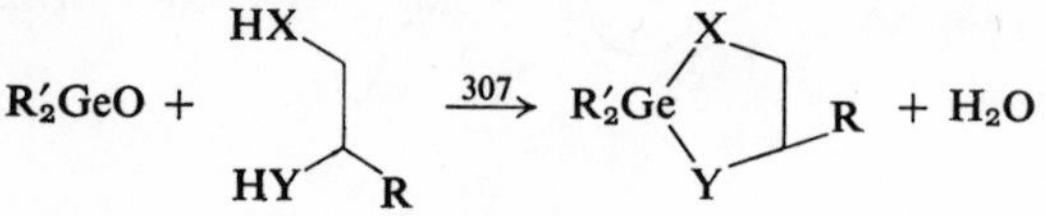

$$R'_2GeO + HX\text{—}CH_2\text{—}CH(R)\text{—}YH \xrightarrow{307} R'_2Ge\langle X\text{—}CH_2\text{—}CH(R)\text{—}Y\rangle + H_2O$$

$$\begin{aligned} R' &= C_4H_9,\ C_6H_5;\ X = Y = O;\ R = H \\ R' &= C_4H_9;\ X = Y = O;\ R = CH_3 \\ R' &= C_4H_9,\ C_6H_5;\ X = O;\ Y = NH \end{aligned} \qquad (167)$$

$$R_2GeO + (HOCH_2CH_2)_2NH \xrightarrow{307} R_2Ge\langle O\text{—}CH_2CH_2,\ O\text{—}CH_2CH_2\rangle NH + H_2O \qquad (168)$$

$$(C_2H_5)_2GeO + H_2SO_4 \xrightarrow{308} (C_2H_5)_2Ge\langle O\text{—}S(O_2)\text{—}O,\ O\text{—}S(O_2)\text{—}O\rangle Ge(C_2H_5)_2 + H_2O \qquad (169)$$

$$R_2SnO + HO\text{—}CH_2CH_2\text{—}OH \xrightarrow{309,293} R_2Sn\langle O\text{—}CH_2CH_2\text{—}O,\ O\text{—}CH_2CH_2\text{—}O\rangle SnR_2 + H_2O \qquad (170)$$

$$+ \ o\text{-}C_6H_4(OH)_2 \xrightarrow{309} R_2Sn\langle O_2C_6H_4\rangle + H_2O \qquad (171)$$

$$+ \ HO\text{—}CH(CH_3)\text{—}CH(CH_3)\text{—}OH \xrightarrow{309} R_2Sn\langle O\text{—}CH(CH_3)\text{—}CH(CH_3)\text{—}O\rangle + H_2O \qquad (172)$$

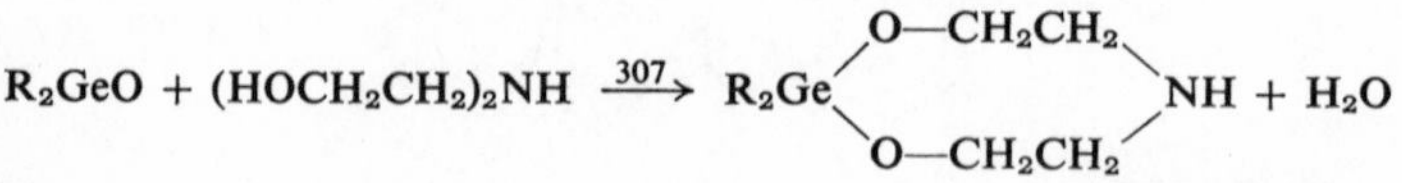

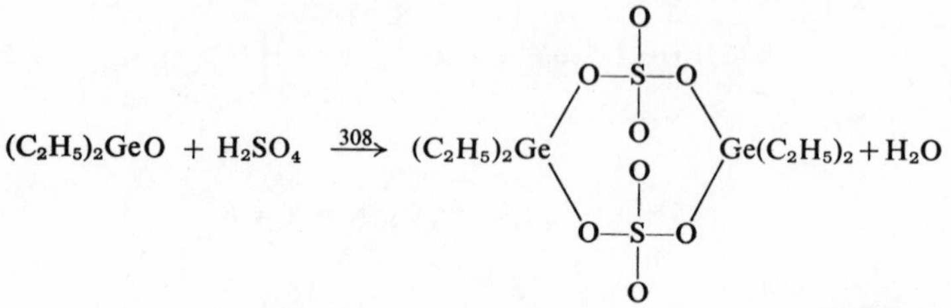

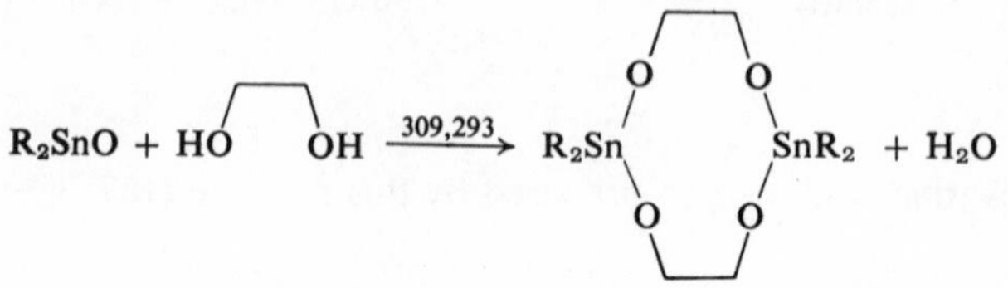

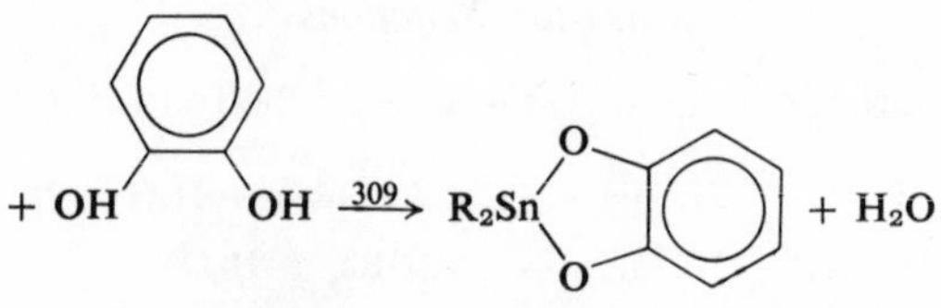

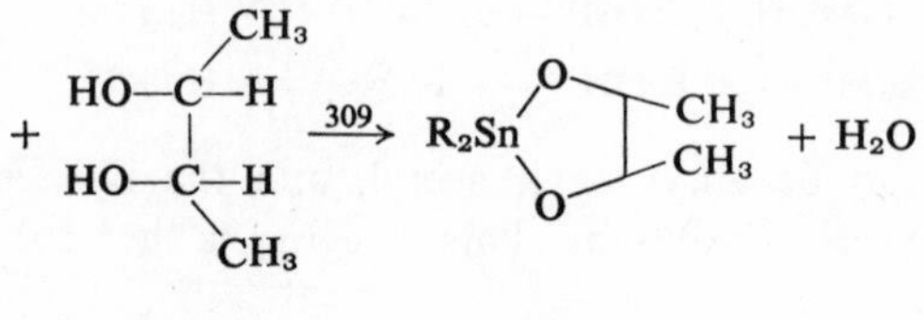

$$+ \ HSCH_2C(=O)\text{—}OH \xrightarrow{310} \ \text{S—}CH_2\text{—}C(=O)\text{—O—}SnR_2\text{—S (ring)} + H_2O \qquad (173)$$

$$+ \ [HOOC(CH_2)_nS]CR'R'' \xrightarrow{311} R_2Sn\left\langle\begin{matrix} O_2C(CH_2)_nS \\ O_2C(CH_2)_nS \end{matrix}\right\rangle C\left\langle\begin{matrix} R' \\ R'' \end{matrix}\right. + H_2O \qquad (174)$$

$$+ \ [HOOC(CH_2)_n\text{—}S]_2C{=}S \xrightarrow{312} R_2Sn\left\langle\begin{matrix} O_2C(CH_2)_n\text{—}S \\ O_2C(CH_2)_n\text{—}S \end{matrix}\right\rangle C{=}S + H_2O \qquad (175)$$

$$+ \ [HOOC(CH_2)_n]_2S \xrightarrow{312} R_2Sn\left\langle\begin{matrix} O_2C(CH_2)_n \\ O_2C(CH_2)_n \end{matrix}\right\rangle S + H_2O \qquad (176)$$

Stannous oxide reacts in the presence of copper with catechol, 2,2′-dihydroxydiphenyl, or 2,3-dihydroxynaphthalene to give the tin(II) heterocycles:

$$C_6H_4\left\langle\begin{matrix} O \\ O \end{matrix}\right\rangle Sn \qquad C_6H_4C_6H_4\left\langle\begin{matrix} O \\ O \end{matrix}\right\rangle Sn \qquad C_{10}H_6\left\langle\begin{matrix} O \\ O \end{matrix}\right\rangle Sn$$

The reaction can be carried out in solution or neat at a temperature high enough to melt the diol.[313]

b. Group IV–ol reactions. A number of heterocycles have been prepared by condensation of two SiOH groups:

$$2 \ HO\text{—}Si(C_6H_3)_2\text{—}Si(C_6H_5)_2\text{—}OH \xrightarrow[\text{formic acid}]{314,315} \begin{matrix} (C_6H_5)_2Si\text{—}Si(C_6H_5)_2 \\ O \qquad\qquad O \\ (C_6H_5)_2Si\text{—}Si(C_6H_5)_2 \end{matrix} + 2H_2O \qquad (177)$$

$$(CH_3)_2Si(CH_2)_6Si(CH_3)_2(OH)_2 \xrightarrow{316} [(CH_3)_2Si(CH_2)_6Si(CH_3)_2O]_2 + 2H_2O \quad (178)$$

$$HO-Si(CH_3)_2-C_6H_4-Si(CH_3)_2-OH \xrightarrow[HCl]{317} \left[-Si(CH_3)_2-C_6H_4-Si(CH_3)_2-O-\right]_n + nH_2O \quad n = 7, 8 \quad (179)$$

$$HO-Si(CH_3)_2-C_6X_4-Si(CH_3)_2-OH \xrightarrow{318} [(CH_3)_2Si-C_6X_4-Si(CH_3)_2-O-]_2 + 2H_2O \quad X = F, Cl \quad (180)$$

A patent describes the reaction between boron and silicon diols:

$$RB(OH)_2 + R_2Si(OH)_2 \xrightarrow[\text{acidic catalyst}]{319} R_2Si(OBR)_2O \quad (181)$$

The reaction of alcohols with alkoxysilanes is a cyclization method almost as common as the alcoholysis procedure. In fact many of the heterocycles mentioned there have also been prepared by this procedure; for example, the reaction of dimethyldiethoxysilane and ethylene glycol in the presence of *p*-toluenesulfonic acid monohydrate produces the dimeric ring[270,320]

$$[(CH_3)_2Si(OCH_2CH_2O)]_2$$

Similarly, with the appropriate glycols, the following were obtained:

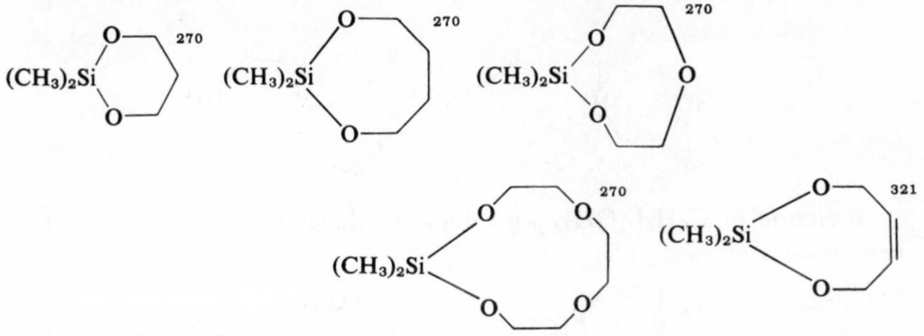

$(CH_3)_2Si$ [270] $(CH_3)_2Si$ [270] $(CH_3)_2Si$ [270]

$(CH_3)_2Si$ [270] $(CH_3)_2Si$ [321]

Hydrolysis of 3,6-bisdimethylmethoxysilyl-1,4-cyclohexadiene produces[322]

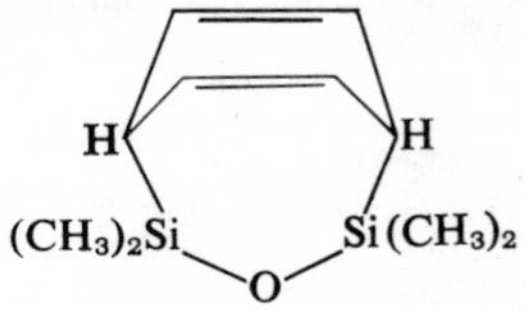

$(CH_3)_2Si$ — O — $Si(CH_3)_2$

and acid hydrolysis of 1-[*m*-(ethoxydimethylsilyl)phenyl]-3-[*p*-ethoxydimethylsilyl)phenyl]-hexafluoropropane produced[323]

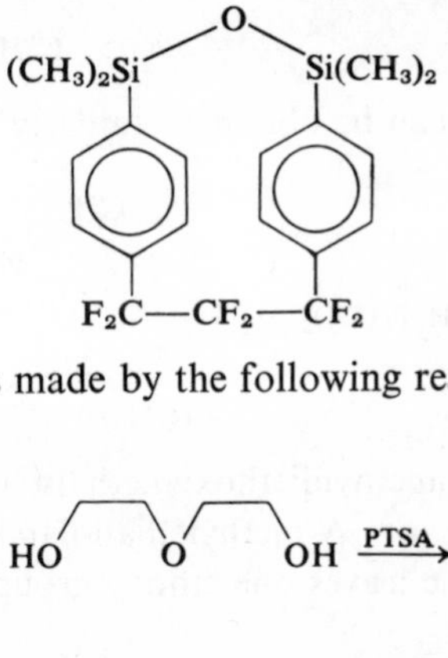

$(CH_3)_2Si$ — O — $Si(CH_3)_2$

$F_2C—CF_2—CF_2$

A 10-membered ring was made by the following reaction[270]:

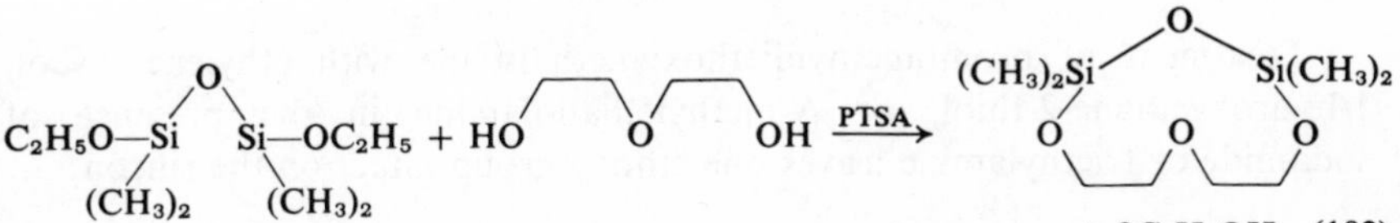

$$C_2H_5O—Si(CH_3)_2—O—Si(CH_3)_2—OC_2H_5 + HOCH_2CH_2OCH_2CH_2OH \xrightarrow{PTSA} (CH_3)_2Si(O)Si(CH_3)_2 \text{ ring} + 2C_2H_5OH \quad (182)$$

The silatranes have been made from $N(CH_2CH_2OH)_3$ and various alkoxy silanes:

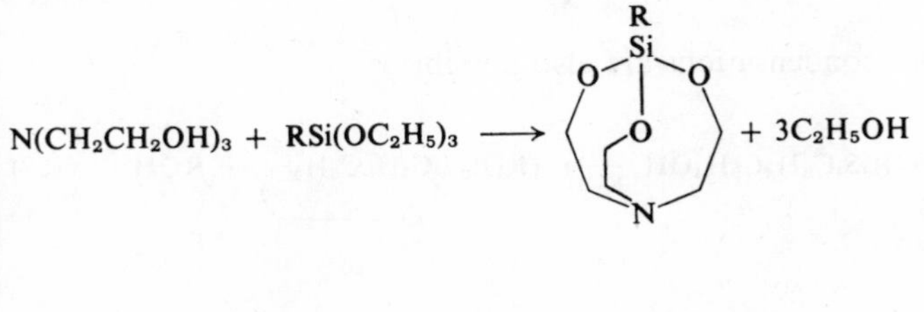

$$N(CH_2CH_2OH)_3 + RSi(OC_2H_5)_3 \longrightarrow RSi(OCH_2CH_2)_3N + 3C_2H_5OH \quad (183)$$

$R = C_6H_5$[324]
$R = CH{=}CH_2$[325]
$R = OC_2H_5$[325,326]

Use of tetraethoxysilane also results in the spiro structure[324]:

[spiro structure: HO–CH2CH2–N(CH2CH2O)2Si(OCH2CH2)2N–CH2CH2–OH]

Reaction of $C_6H_5B(OH)_2$ with dialkoxysilanes produces two B—O—Si heterocycles[327]:

$$C_6H_5B(OH)_2 + C_6H_5Si(CH_3)(OC_2H_5)_2 \longrightarrow [C_6H_5BOSi(CH_3)(C_6H_5)O]_2 \quad (184)$$

$$C_6H_5B(OH)_2 + [C_2H_5O(C_2H_5)_2Si]_2O \longrightarrow \text{cyclo-}[(C_2H_5)_2SiOB(C_6H_5)OSi(C_6H_5)_2O] \quad (185)$$

Various P—O—Si rings can be obtained similarly[328]:

$$C_2H_5O(Si(CH_3)_2{-}O)_3C_2H_5 + CH_3P(O)(OH)_2 \xrightarrow{100^\circ} \text{cyclo-}[CH_3P(O)OSi(CH_3)_2OSi(CH_3)_2O] + 2C_2H_5OH \quad (186)$$

Treatment of bromomethyldiethoxymethylsilane with ethylene glycol, 1-hydroxyethane-2-thiol, or *N*-methylethanolamine in the presence of sodamide or triethylamine leaves one ethoxy group intact on the silicon:

$$HXCH_2CH_2OH + BrCH_2Si(CH_3)(OC_2H_5)_2 \longrightarrow \text{cyclo-}[OSi(CH_3)(OC_2H_5)CH_2XCH_2CH_2] \quad (187)$$

$X = O, S$[329,221]
$= NCH_3$[330,221]

Intramolecular condensations are also possible:

$$(RO)_3SiC_3H_6C_6H_4OH \xrightarrow[KOH]{331} (RO)_2SiC_3H_6C_6H_4\text{–}O \text{ (ring)} + ROH \quad (188)$$

$$O[Si(CH_3)_2(CH_2)_4OH]_2 \xrightarrow{258} (CH_3)_2\overline{Si(CH_2)_4O} + H_2O \qquad (189)$$

$$O[Si(CH_3)_2CH_2OH]_2 \xrightarrow[CaO]{332,333} (CH_3)_2\overline{Si\text{–}CH_2\text{–}O\text{–}Si(CH_3)_2\text{–}CH_2\text{–}O} \qquad (190)$$

$$O[Si(CH_3)_2CH_2OH]_2 \xrightarrow[H_2SO_4]{332,333} (CH_3)_2\overline{Si\text{–}O\text{–}Si(CH_3)_2\text{–}CH_2\text{–}O\text{–}CH_2} + H_2O \qquad (191)$$

$$(HOCH_2)_2C(CH_2O)_2Si(CH_3)_2 \xrightarrow[H^+]{334} C(CH_2OH)_4 + (CH_3)_2Si(OCH_2)_2C(CH_2O)_2Si(CH_3)_2 \qquad (192)$$

$$\left[CH_3\overset{O}{\overset{\|}{C}}\text{–}O(CH_2)_3\text{–}\underset{H_5OC_2}{\overset{CH_3}{Si}}\right]_2 O \xrightarrow[NaOC_2H_5]{335} CH_3Si\text{–}O\text{–}SiCH_3 \text{ (bridged by } O\text{–}(CH_2)_3 \text{ and } (CH_2)_3\text{–}O) \qquad (193)$$

$$O[Si(CH_3)_2(CH_2)_3OH]_2 \xrightarrow[CaO]{332} Me_2\overline{Si(CH_2)_3O} + H_2O \qquad (194)$$

$$o\text{-}HOC_6H_4(CH_2)_3Si(OCH_3)_3 \xrightarrow{336} \overline{C_6H_4\text{–}O\text{–}Si(OCH_3)_2\text{–}(CH_2)_3} + CH_3OH \qquad (195)$$

An unusual spiro-titanate ester is formed in the reaction between diphenylsilanediol and *n*-butyltitanate[337]:

$$(C_6H_5)_2Si(OH)_2 + (n\text{-}C_4H_9O)_4Ti \longrightarrow$$

$$Ti[\text{–}O\text{–}Si(C_6H_5)_2\text{–}O\text{–}Si(C_6H_5)_2\text{–}O\text{–}Si(C_6H_5)_2\text{–}O\text{–}Si(C_6H_5)_2\text{–}O\text{–}]_2 + 4n\text{-}C_4H_9OH \qquad (196)$$

Acetoxysilane derivatives also undergo condensation. Spirosystems are formed in the reaction of glycols with tetracetoxysilane:

$$Si(OCOCH_3)_4 + HO(CR_2)_nOH \xrightarrow{338,339} (R_2C)_n\overset{O}{\underset{O}{\diagdown}}Si\overset{O}{\underset{O}{\diagup}}(CR_2)_n + 4CH_3\overset{O}{\overset{\|}{C}}OH \qquad (197)$$

Cyclic structures are also obtained in reactions of dialkyldiacetoxysilanes with glycerol and various other diols:

$$\text{(cyclic } -\text{O}-(\text{CH}_2)_n-\text{O}-\text{SiR}_2)\ ^{340} \qquad (\text{CH}_3)_2\text{Si}\langle\text{O}-\text{CH}_2-\text{X}-\text{CH}_2-\text{O}\rangle\ ^{341} \quad \text{X = S, NH, NC}_6\text{H}_5$$

Dimethyldiacetoxysilane reacts with pentaerythritol according to[298,342–344]

$$2(\text{CH}_3)_2\text{Si}(\text{OCOCH}_3)_2 + (\text{HOCH}_2)_4\text{C} \longrightarrow (\text{CH}_3)_2\text{Si}\langle\text{OCH}_2\rangle_2\text{C}\langle\text{CH}_2\text{O}\rangle_2\text{Si}(\text{CH}_3)_2 + 4\text{CH}_3\text{COOH} \quad (198)$$

Alkoxygermanes and -stannanes undergo similar condensations:

$$(\text{C}_4\text{H}_9)_2\text{Ge}(\text{OR})_2 + \text{R}'(\text{OH})_2 \xrightarrow[\text{PTSA}]{345} (\text{C}_4\text{H}_9)_2\text{Ge}\langle\text{O}_2\rangle\text{R}' + 2\text{ROH} \quad (199)$$

$$\text{R}' = -\text{CH}(\text{CH}_2)_n\text{CH}_2-,\ n = 0, 1, 2;\ -\text{CH}_2\text{CH}_2\text{OCH}_2\text{CH}_2-$$

$$\text{R}_2\text{Ge}(\text{OC}_2\text{H}_5)_2 + \text{HOR}'\text{COOH} \xrightarrow{346} \text{R}_2\text{Ge}\langle\text{O},\ \text{OOC}\rangle\text{R}' + 2\text{C}_2\text{H}_5\text{OH} \quad \text{R}' = \text{C}_6\text{H}_4, \text{C}_6\text{H}_5\text{CH} \quad (200)$$

$$(\text{C}_4\text{H}_9)_2\text{Sn}(\text{OH}_2)_2 + \text{HO}-(\text{CH}_2)_n-\text{HO} \xrightarrow{347} (\text{C}_4\text{H}_9)_2\text{Sn}\langle\text{O}_2\rangle(\text{CH}_2)_n + \text{C}_2\text{H}_5\text{OH} \quad (201)$$

$$n = 2, 3, 4, 5, 6$$

c. Ether reactions. Silanes containing several alkoxy, siloxy, or acetoxy groups can undergo intramolecular condensation:

$$(\text{CH}_3)_3\text{SiO}(\text{CH}_2)_3\text{SiR}(\text{OR}')_2 \xrightarrow{348} (\text{R}'\text{O})(\text{R})\text{Si}\langle(\text{CH}_2)_3,\ \text{O}\rangle + (\text{CH}_3)_3\text{SiOR} \quad (202)$$

$$(\text{CH}_3)_2\text{Si}[\text{O}(\text{CH}_2)_3\text{Si}(\text{OR})_3]_2 \xrightarrow{348} \begin{array}{l}(\text{RO})_2\text{Si}-(\text{CH}_2)_3-\text{O}\\ \ \ \ \ \ \ \ \ \ \ \ |\qquad\qquad\qquad\ \ |\\ \ \ \ \ \ \ \ \ \ \ \ \text{O}-(\text{CH}_2)_3-\text{Si}(\text{OR})_2\end{array} + (\text{CH}_3)_2\text{Si}(\text{OR})_2 \quad (203)$$

$$(\text{C}_2\text{H}_5\text{O})(\text{CH}_3)\text{Si}(\text{OC}_2\text{H}_5)(\text{CH}_2)_3\text{OCCH}_3(=\text{O}) \xrightarrow{349} (\text{C}_2\text{H}_5\text{O})(\text{CH}_3)\text{Si}\langle(\text{CH}_2)_3,\ \text{O}\rangle + \text{CH}_3\overset{\text{O}}{\overset{\|}{\text{C}}}\text{OC}_2\text{H}_5 \quad (204)$$

3. *Cleavage of Heterocycles with Metal Halides*

The reaction of organometallic oxygen and nitrogen bonds with metal-halide derivatives seems to be fairly general.[227] A 1953 patent assigns the structure

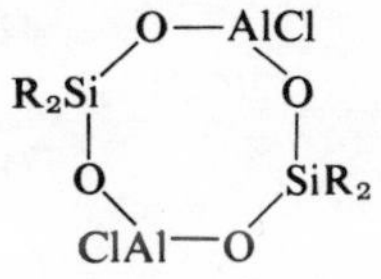

O—AlCl
R$_2$Si O
O SiR$_2$
ClAl—O

to the product of the reaction between $[R_2SiO]_4$ and $AlCl_3$.[350] This reaction was later reinvestigated, and one of the products was shown to be the following complex[351]:

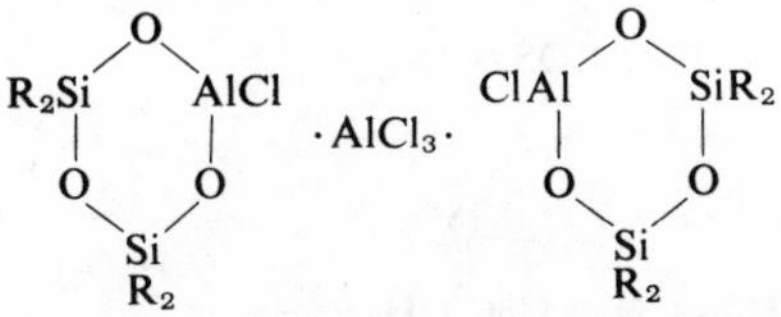

R$_2$Si(O)AlCl(O)(O)Si R$_2$ · $AlCl_3$ · ClAl(O)SiR$_2$(O)(O)Si R$_2$

Bis(2,2′-diphenylenedioxy)silane is the product of the reaction of $SiCl_4$ with 2,2′-diphenylenedioxydimethylsilane.[289] The germanium derivative is obtained by the same route:

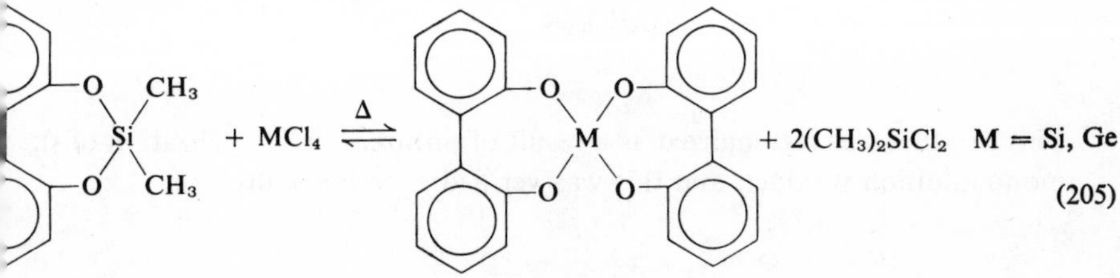

$$\text{(2,2'-diphenylenedioxy)Si(CH}_3)_2 + MCl_4 \underset{}{\overset{\Delta}{\rightleftharpoons}} \text{bis(2,2'-diphenylenedioxy)M} + 2(CH_3)_2SiCl_2 \quad M = Si, Ge \tag{205}$$

A 10-membered heterocycle is obtained from a 1,2-disiloxycyclohexene[352]:

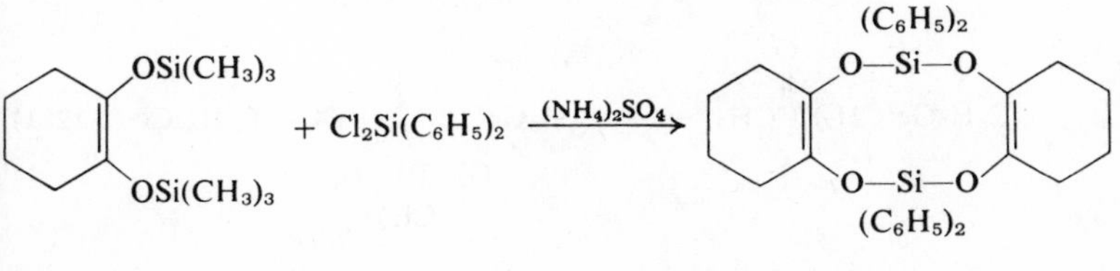

$$\text{1,2-bis(OSi(CH}_3)_3)\text{cyclohexene} + Cl_2Si(C_6H_5)_2 \xrightarrow{(NH_4)_2SO_4} \text{[O—Si(C}_6H_5)_2\text{—O] bridged bis(cyclohexene)} \tag{206}$$

Tin heterocycles can also be transformed to their silicon analogs[353]:

$$(C_4H_9)_2Sn\underset{O(CH_2)_3-O}{\overset{O(CH_2)_3-O}{\diagup\diagdown}}Sn(C_4H_9)_2 + 2(C_6H_5)_2SiCl_2 \longrightarrow$$

$$2(C_6H_5)_2Si\langle{}^{O-}_{O-}\rangle + 2(C_4H_9)_2SnCl_2 \qquad (207)$$

4. *Hydride Addition Reactions*

The well-known reaction of group IV hydrides with unsaturated substrates has been applied to the preparation of a number of heterocycles:

$$CH_2{=}CHCH_2\overset{O}{\overset{\|}{C}}O\overset{H}{\overset{|}{Si}}(CH_3)_2 \xrightarrow{354} \text{cyclic } O{=}C{-}O{-}Si(CH_3)_2 \qquad (208)$$

$$C_6H_5\overset{Cl}{\overset{|}{Ge}}H_2 + CH_2{=}CH\overset{O}{\overset{\|}{C}}CH_3 \xrightarrow{355} \text{cyclic } C_6H_5Ge(Cl){-}O{-}C(CH_3) \qquad (209)$$

The major product of the reaction between diphenylgermane and methyl vinyl ketone is the cyclic material[356]:

$$(C_6H_5)_2Ge\text{(ring: }{-}O{-}CH(CH_3){-}\text{)}$$

This product was rationalized as a result of intramolecular cyclization of the mono addition product, and this was verified experimentally[356]:

$$(C_6H_5)_2\overset{H}{\overset{|}{Ge}}(CH_2)_2\overset{O}{\overset{\|}{C}}CH_3 \xrightarrow[AIBN]{100°,\ 16\ hr} (C_6H_5)_2Ge\text{(ring: }{-}O{-}CH(CH_3){-}\text{)} \qquad (210)$$

$$C_6H_5\underset{X}{\underset{|}{\overset{H}{\overset{|}{Ge}}}}(CH_2)_4\overset{O}{\overset{\|}{C}}CH_3 \longrightarrow C_6H_5Ge(X)\text{(ring: }{-}O{-}CH(CH_3){-}\text{)} \qquad X = C_6H_5, Cl \qquad (211)$$

5. *Alcoholysis of Group IV Amines*

Group IV amines are generally amenable to attack by OH groups:

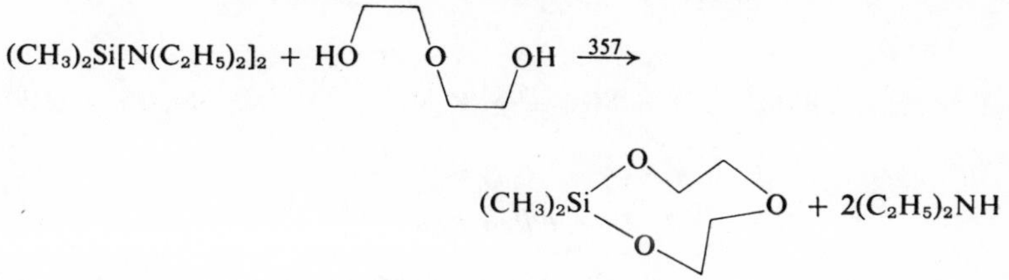

$$(CH_3)_2Si[N(C_2H_5)_2]_2 + HOCH_2CH_2OCH_2CH_2OH \xrightarrow{357} (CH_3)_2Si(OCH_2CH_2OCH_2CH_2O) + 2(C_2H_5)_2NH \quad (212)$$

$$2(CH_3)_2SiHN(C_2H_5)_2 + HOCH_2CH_2OH \xrightarrow{358} (CH_3)_2Si(OCH_2CH_2O)_2Si(CH_3)_2 + 4(C_2H_5)_2NH \quad (213)$$

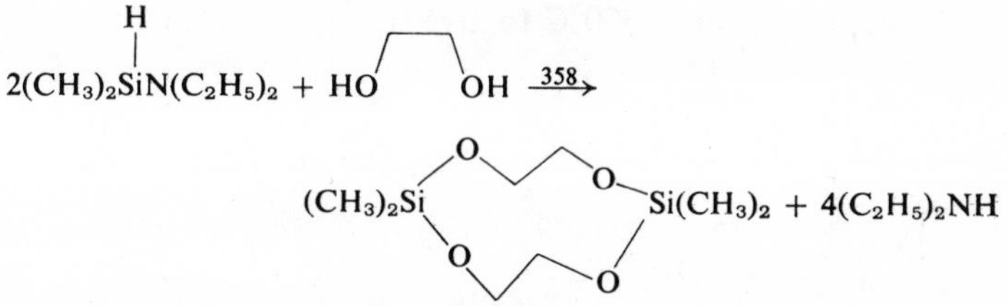

$$HN(R')CH(R)COOH + CH_3C(O)N(CH_3)-Si(C_6H_5)(CH_3)-N(CH_3)C(O)CH_3 \xrightarrow{359} R'-N\langle Si(CH_3)(C_6H_5)-O-C(=O)-CH(R)\rangle + 2CH_3C(O)NHCH_3 \quad (214)$$

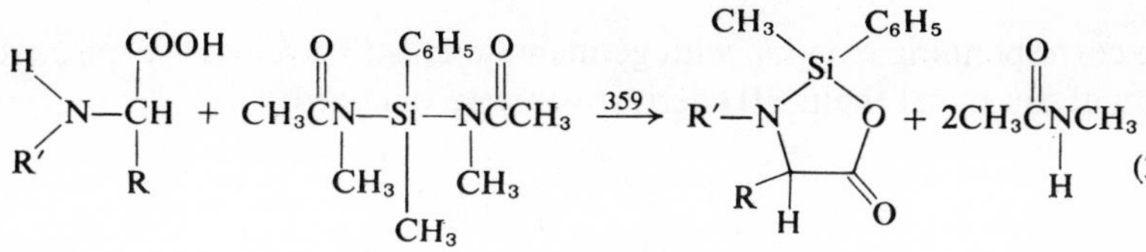

6. *Ring Expansions*

Sila- and germacyclobutanes have been found to undergo ring insertion reactions with elemental sulfur and selenium, sulfur dioxide, sulfur trioxide, and phenyl(bromodichloromethyl) mercury:

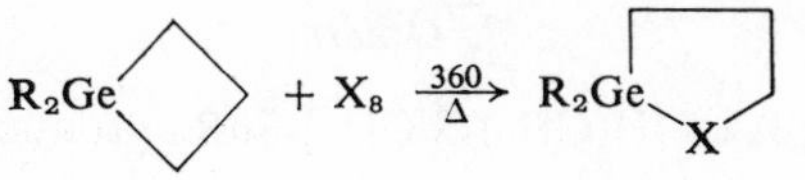

$$R_2Ge(CH_2)_3 + X_8 \xrightarrow[\Delta]{360} R_2Ge(CH_2)_3X \qquad X = S, Se \quad (215)$$

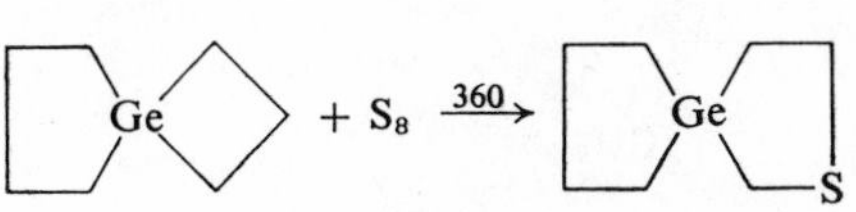

$$(CH_2)_4Ge(CH_2)_3 + S_8 \xrightarrow{360} (CH_2)_4Ge(CH_2)_3S \quad (216)$$

$$R_2Ge(CH_2)_3 + SO_2 \xrightarrow{361} R_2Ge(CH_2)_3S(=O)O \quad (217)$$

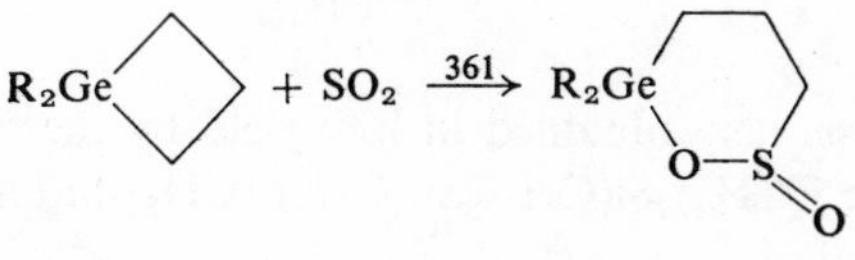

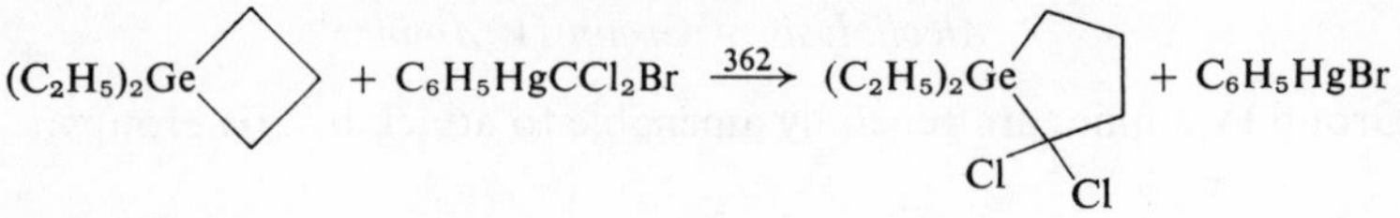

$$(C_2H_5)_2Ge\square + C_6H_5HgCCl_2Br \xrightarrow{362} (C_2H_5)_2Ge(\text{cyclo-}CH_2CH_2CH_2CCl_2) + C_6H_5HgBr \quad (218)$$

$$R_2M\square + SO_3 \xrightarrow{363} R_2M(CH_2CH_2CH_2SO_2O) \qquad M = Si, Ge \quad (219)$$

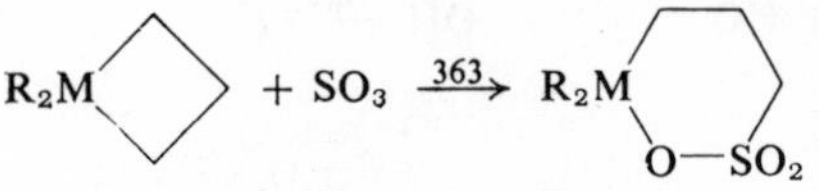

7. *Direct Reaction*

Elemental silicon reacts with catechol and 2,2′-dihydroxybiphenyl in the presence of copper at 200 to 300°C to give the spiro systems[364]

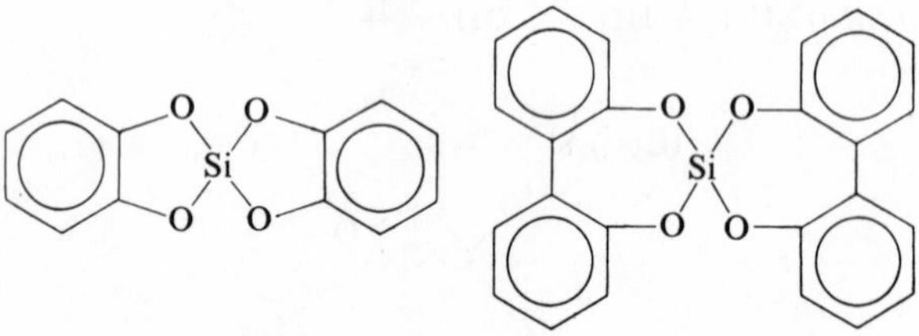

The corresponding reaction with germanium fails to produce the spiro esters, but with tin metal the tin(II) derivatives were obtained[365]:

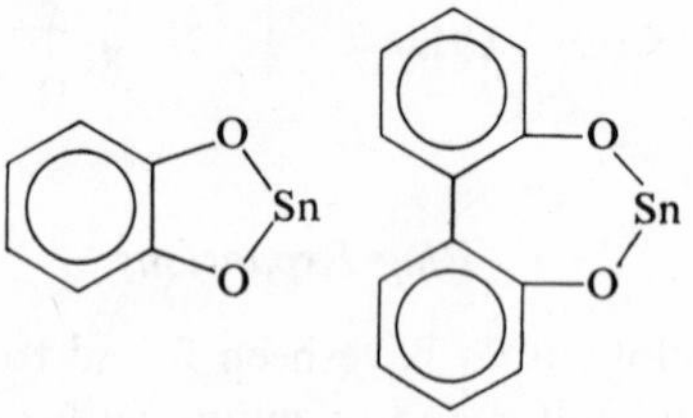

The method is limited by the thermal stability of the starting materials.

8. *Others*

Alkaline hydrolysis of $[H(CH_3)_2SiCH_2]_2SnR_2$ produces:

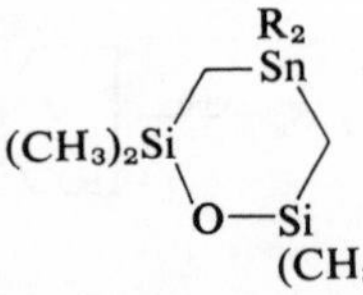

$$R_2Sn(CH_2Si(CH_3)_2OSi(CH_3)_2CH_2) \quad ^{366,367}$$

This compound was also obtained in low yield by thermal cracking of the linear polysiloxane $[(CH_3)_2Si(CH_2)_3Sn(CH_3)_2(CH_2)_3Si(CH_3)_2O—]_x$.[366]

Organosilicon lactones have been secured by intramolecular cyclization in the presence of metal ions[368] or acid[369]:

$$(CH_3)_3Si(CH_2)_2C(=O)-OH \longrightarrow (CH_3)_2Si\langle CH_2CH_2C(=O)O\rangle \qquad (220)$$

Three heterocycles

$$(CH_3)_2Si\langle CH_2 \rangle Si(CH_3)_2\langle O-Si(CH_3)_2-O\rangle \qquad (CH_3)_2Si\langle O\rangle Si(CH_3)_2\langle CH_2-Si(CH_3)_2-CH_2\rangle \qquad (CH_3)_2Si\langle CH_2, CH_2\rangle Si(CH_3)_2$$

were found in the cracking products of the compound

$$\text{cyclo-}[-Si(CH_3)_2-O-Si(CH_3)_2-CH_2-Si(CH_3)_2-O-Si(CH_3)_2-CH_2-] \quad 176$$

A 16-membered ring was obtained by sulfuric acid cleavage of an organometallic ketone[370]:

$$[(CH_3)_3Si(CH_2)_2]_2C{=}O \xrightarrow[H^+/H_2O]{} \text{cyclo-}[-Si(CH_3)_2-(CH_2)_2-C(=O)-(CH_2)_2-Si(CH_3)_2-O-Si(CH_3)_2-(CH_2)_2-C(=O)-(CH_2)_2-Si(CH_3)_2-O-] \qquad (221)$$

The reaction of benzophenone with silicon tetrachloride and magnesium in ether yields the spiro ester[371]

$$(C_6H_5)_2C-O \quad O-C(C_6H_5)_2$$
$$\quad\quad\quad\quad Si$$
$$(C_6H_5)_2C-O \quad O-C(C_6H_5)_2$$

Octaphenylcyclothiotetrasilane was prepared by the following route[372]:

$$LiSi(C_6H_5)_2-Si(C_6H_5)_2-Si(C_6H_5)_2-Si(C_6H_5)_2Li + SCl_2 \longrightarrow \text{cyclo-}[-Si(C_6H_5)_2-Si(C_6H_5)_2-Si(C_6H_5)_2-Si(C_6H_5)_2-S-] + 2LiCl \qquad (222)$$

Several silicon-oxygen heterocycles have been obtained by the reaction[373]

$$R_3SiO(CH_2)_nCl \xrightarrow{\text{Na/toluene}} R_2Si\underset{O}{\diagdown\!\!\diagup}(CH_2)_n \qquad n = 3, 4, 5 \tag{223}$$

D. Group IV–Transition-Metal Bonds

Recently there has been a great deal of interest in compounds containing main-group elements bonded to transition metals,[374] and ring structures have been shown to exist in several examples in which transition metals are bonded to the elements silicon, germanium, tin, and lead.[375] In this concluding section we describe the preparation of a selected number of these compounds.

Alkynyltin compounds react with tri-iron dodecacarbonyl in petroleum ether at about 110° to give various compounds containing iron-tin rings by loss of the alkynyl groups[376]:

$$R_2Sn(C{\equiv}CR')_2 + Fe_3(CO)_{12} \longrightarrow (OC)_4Fe\langle(SnR_2)_2\rangle Fe(CO)_4 \tag{224}$$

Tetravinylsilane reacts with dicobalt octacarbonyl to give nonacarbonyl(vinylsilicon)tricobalt[377]:

$$Si(CH{=}CH_2)_4 + Co_2(CO)_8 \longrightarrow CH_2{=}CH{-}Si[Co(CO)_3]_3 \tag{225}$$

A similar ring structure is obtained through aryl silicon bond cleavage[378]:

$$(C_6H_5)_4Si + Co_2(CO)_8 \longrightarrow [(OC)_3Co]_3Si{-}Si[Co(CO)_3]_3 \tag{226}$$

The product is a violet-colored, crystalline material stable in air if refrigerated. An analogous tin-containing ring compound has been synthesized by a route starting from *n*-butyltin trichloride[379]:

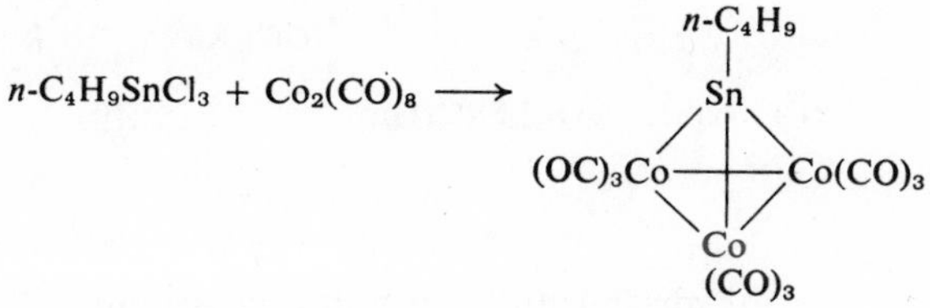

$$n\text{-}C_4H_9SnCl_3 + Co_2(CO)_8 \longrightarrow \text{cluster: } n\text{-}C_4H_9Sn \text{ bonded to three } Co(CO)_3 \text{ units } [(OC)_3Co,\ Co(CO)_3,\ Co(CO)_3] \quad (227)$$

Dimeric dialkyltin-iron tetracarbonyls that involve rings and clusters result from tributyltin chloride with iron pentacarbonyl[380]:

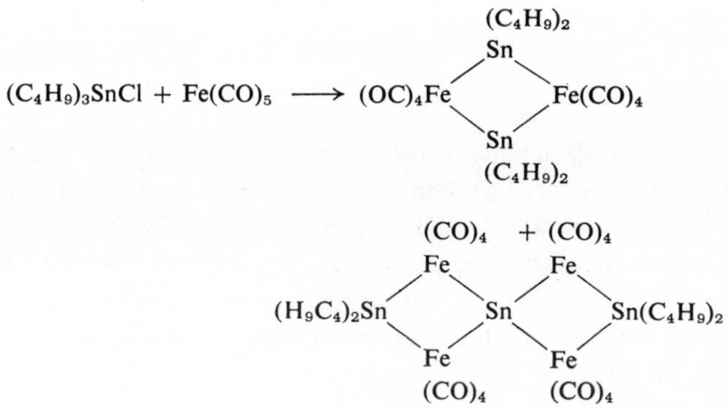

$$(C_4H_9)_3SnCl + Fe(CO)_5 \longrightarrow (OC)_4Fe[Sn(C_4H_9)_2]_2Fe(CO)_4 + (H_9C_4)_2Sn[Fe(CO)_4]_2Sn[Fe(CO)_4]_2Sn(C_4H_9)_2 \quad (228)$$

The spiro compound is a bright red color. An x-ray crystallographic study has confirmed the structure of the methyltin compounds.[381] A three-membered ring containing an iron-iron bond has been confirmed in the product of the reaction of iron pentacarbonyl with either hexabultylditin or with stannous chloride after long refluxing in THF[380,382,383]:

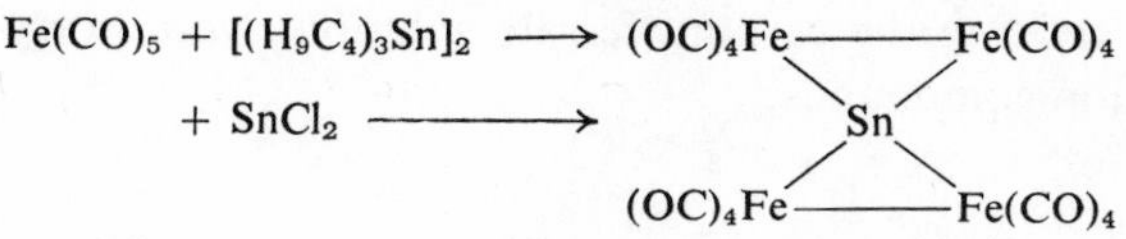

$$Fe(CO)_5 + [(H_9C_4)_3Sn]_2 \longrightarrow \qquad + SnCl_2 \longrightarrow \quad Sn[Fe_2(CO)_8]_2 \text{ (spiro: } (OC)_4Fe\text{—}Fe(CO)_4 \text{ pairs bridged by Sn)} \quad (229)$$

The resulting product is diamagnetic, suggesting covalent iron-iron interactions, and an x-ray crystallographic study has been carried out.[382] Tin-ruthenium heterocycles are also known.[384] The ring compound shown below was a by-product in the action of a triorganotin hydride on triruthenium dodecacarbonyl. The same compound results when the trimethylsilyl derivative of ruthenium carbonyl is treated with trimethyltin hydride[384]:

$$\left.\begin{array}{l} R_3SnH + Ru_3(CO)_{12} \searrow \\ (CH_3)_3SnH + (CH_3)_3Si[Ru(CO)_4]_2Si(CH_3)_3 \nearrow \end{array}\right. (OC)_3Ru\overset{SnR_2}{\underset{SnR_2}{\diamond}}Ru(CO)_3 \quad \begin{array}{l}\text{(with } R_3Sn \text{ on each Ru)}\end{array} \qquad (230)$$

A germanium-bridged, iron carbonyl complex has been obtained by reacting dimethylgermanium dihydride with tri-iron dodecacarbonyl[385]:

$$(CH_3)_2GeH_2 + Fe_3(CO)_{12} \longrightarrow (H_3C)_2Ge\begin{array}{l} \diagup Fe(CO)_3 - Ge(CH_3)_2 \\ \qquad\qquad\qquad\quad | \\ \diagdown Fe(CO)_3 - Ge(CH_3)_2 \end{array} \qquad (231)$$

Treatment of tetracarbonylferrate solutions with triaryl or trialkyltin and lead halides or hydroxides has been shown to lead to a variety of complexes of the general formula $(R_3M)_2Fe(CO)_4$, where M = Sn, Pb. Although generally much more stable than the analogous carbon complexes of the type $R_2Fe(CO)_4$, which have never been isolated, the straight-chain aliphatic lead compounds readily disproportionate to the lead tetralkyls and red dimeric complexes of the formula $[R_2PbFe(CO)_4]_2$[386]:

$$R_3PbX + [Fe(CO)_4]^- \longrightarrow R_3PbFe(CO)_4 \longrightarrow R_2Pb\overset{Fe(CO)_4}{\underset{Fe(CO)_4}{\diamond}}PbR_2 \qquad (232)$$

The analogous complexes of tin[386,387] and germanium[388,389] have also been prepared by treating the carbonylferrate solutions with diorganodihalo tin and germanium compounds:

$$R_2MCl_2 + [Fe(CO)_4]^- \longrightarrow (OC)_4Fe\overset{MR_2}{\underset{MR_2}{\diamond}}Fe(CO)_4 \qquad \begin{array}{l} M = Ge, Sn \\ R = CH_3, C_2H_5, n\text{-}C_4H_9 \end{array} \qquad (233)$$

The addition of tin(II) chloride to a solution of $PtCl_4^{2-}$ in acetone gives the complex anion $[Pt_3Sn_8Cl_{20}]^{4-}$, in which the platinum atoms are in the zero oxidation state[374,375]:

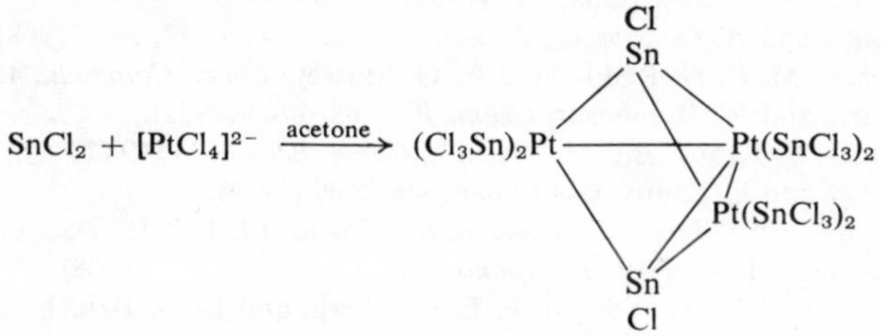

$$SnCl_2 + [PtCl_4]^{2-} \xrightarrow{\text{acetone}} (Cl_3Sn)_2Pt\text{—}Pt(SnCl_3)_2 \text{ / } Pt(SnCl_3)_2 \text{ (bridged by two ClSn groups)} \qquad (234)$$

A cobalt-cobalt bond supported by bridging tin and carbonyl groups is found in the product of the reaction of bis(acetylacetonato)dichlorotin with the tetracarbonylcobalt anion[390]:

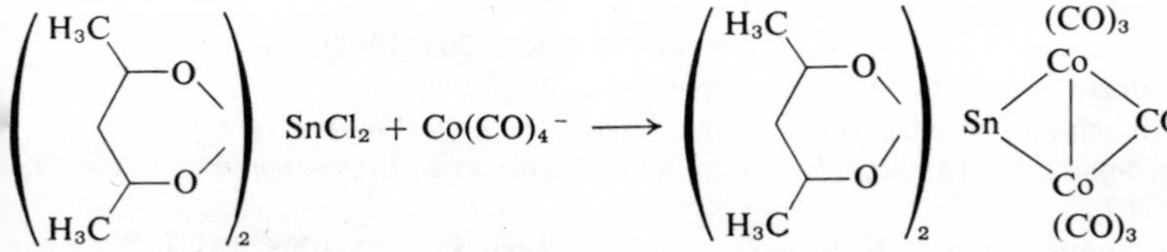

$$(acac)_2SnCl_2 + Co(CO)_4^- \longrightarrow (acac)_2Sn[Co(CO)_3]_2(\mu\text{-}CO) \qquad (235)$$

The red crystalline heterocycle is diamagnetic.

ACKNOWLEDGMENT

We express our appreciation to Mrs. T. S. R. Shirley for her help in the literature search for this article.

References

1. A. Bygden, *Chem. Ber.*, **48**, 1236 (1915).
2. G. Grüttner and M. Wiernik, *Chem. Ber.*, **48**, 1474 (1915).
3. K. A. Andrianov and L. M. Khananashvili, *Organometal. Chem. Rev.*, **2**, 141 (1967).

3a. I. Haiduc, *The Chemistry of Inorganic Ring Systems*, Wiley-Interscience, New York, 2 parts, 1970.

4. I. Haiduc, *J. Chem. Educ.*, **38**, 134 (1961).
5. J. J. Zuckerman, *Advan. Inorg. Chem. Radiochem.*, **6**, 383 (1964).
6. G. I. Nikishin, A. D. Petrov, and S. I. Sadykh-Zhade, *Khim. i Prakt. Primenenie Kremneorgan. Soedin. Tr. Konf. Leningrad*, **I**, 68 (1958); *Chem. Abstr.*, **53**, 17097 (1959).

7. A. D. Petrov, N. P. Smetankina, and G. I. Nikishin, *Izv. Akad. Nauk. SSSR, Otd. Khim. Nauk*, 1414 (1958).
8. W. English, A. Taurins, and R. Nicholls, *Can. J. Chem.*, **30**, 646 (1952).
9. A. D. Petrov, G. I. Nikishin, N. P. Smetankina, and Yu. P. Egorov, *Izv. Akad. Nauk SSSR, Otd. Khim. Nauk*, 947 (1955).
10. S. C. Cohen and A. G. Massey, *J. Organometal. Chem.*, **12**, 341 (1968).
11. S. C. Cohen, M. L. N. Reddy, and A. G. Massey, *Chem. Commun.*, 451 (1967).
12. R. Schwartz and W. Reinhardt, *Chem. Ber.*, **65**, 1743 (1932).
13. G. Grüttner, E. Kraus, and M. Wiernik, *Chem. Ber.*, **50**, 1549 (1917).
14. G. Grüttner and E. Kraus, *Chem. Ber.*, **49**, 2666 (1916).
15. E. C. Juenge and S. Gray, *J. Organometal. Chem.*, **10**, 465 (1967).
16. F. J. Bajer and H. W. Post, *J. Organometal. Chem.*, **11**, 187 (1968).
17. N. S. Nametkin, V. M. Vdovin, P. L. Grinberg, and E. D. Babieh, *Dokl. Akad. Nauk SSSR*, **161**, 358 (1965).
18. P. Mazerolles, *Bull. Soc. Chim. France*, 1907 (1962).
19. E. C. Juenge and H. E. Jack, *Abstr. Midwest Regional A.C.S. Meeting*, Manhattan, Kans., November, 1968.
20. K. C. Williams, *J. Organometal. Chem.*, **19**, 210 (1969).
21. R. West, *J. Amer. Chem. Soc.*, **76**, 6012 (1954).
22. H. Gilman and O. L. Marrs, *Chem. Ind.* (*London*), 208 (1961).
23. H. Gilman and O. L. Marrs, *J. Org. Chem.*, **30**, 325 (1965).
24. H. Gilman and W. H. Atwell, *J. Org. Chem.*, **28**, 2906 (1963).
25. M. Kumada, K. Tamao, T. Takubo, and M. Ishikawa, *J. Organometal. Chem.*, **9**, 43 (1967).
26. L. H. Sommer and O. F. Bennett, *J. Amer. Chem. Soc.*, **79**, 1008 (1957).
27. L. H. Sommer and O. F. Bennett, *J. Amer. Chem. Soc.*, **81**, 251 (1959).
28. H. A. Clark, U.S. Pat 2,563,004 (1949); *Chem. Abstr.*, **45**, 10676 (1951).
29. M. Kumada and M. Ishikawa, unpublished work quoted in *Organosilicon Compounds* by V. Bazant, V. Chvalovsky, and J. Rathousky, Academic, New York, 1965, vol. 2, part 22, p. 447.
30. L. H. Sommer and G. A. Baum, *J. Amer. Chem. Soc.*, **76**, 5002 (1954).
31. V. M. Vdovin, N. S. Nametkin, and P. L. Grinberg, *Dokl. Akad. Nauk SSSR*, **150**, 799 (1963).
32. H. Gilman and W. H. Atwell, *J. Amer. Chem. Soc.*, **86**, 2687 (1964).
33. W. H. Knoth and R. V. Lindsey, Jr., *J. Org. Chem.*, **23**, 1392 (1958).
34. H. Gilman and W. H. Atwell, *J. Organometal. Chem.*, **2**, 277 (1964).
35. W. J. Kriner, *J. Org. Chem.*, **29**, 1601 (1964).
36. N. S. Nametkin, V. M. Vdovin, and A. V. Zelenaia, *Dokl. Akad. Nauk SSSR*, **170**, 1088 (1966).
37. D. Seyferth and C. J. Attridge, *J. Organometal. Chem.*, **21**, 103 (1970).
38. V. F. Mironov, T. K. Gar, S. A. Mihayljants, and E. M. Berliner, *Abstr. Fourth Intern. Conf. Organometal. Chem.*, Bristol, England, July, 1969.
39. H. Gilman and W. H. Atwell, *J. Amer. Chem. Soc.*, **86**, 5589 (1964).
40. R. A. Benkeser, Y. Nagai, J. L. Noe, R. F. Cunico, and P. H. Gund, *J. Amer. Chem. Soc.*, **86**, 2446 (1964).
41. R. A. Benkeser, J. L. Noe, and Y. Nagai, *J. Org. Chem.*, **30**, 378 (1965).
42. R. A. Benkeser and R. F. Cunico, *J. Organometal. Chem.*, **4**, 284 (1965).
43. E. Rosenberg and J. J. Zuckerman, *Abstr. 158th Natl. Meeting A.C.S.*, New York, September, 1969.
44. R. West and E. G. Rochow, *Naturwissenschaften*, **40**, 142 (1953).

45. S. Papetti, B. B. Schaeffer, H. J. Troscianiec, and T. L. Heying, *Inorg. Chem.*, **3**, 1444 (1964).
46. S. Papetti, U.S. Pat. 3,137,719 (1964); *Chem. Abstr.*, **61**, 5692 (1964).
47. L. I. Zakharkin, V. I. Brgadze, and O. Yu. Okhlobystin, *J. Organometal. Chem.*, **4**, 211 (1965).
48. M. Kumada, H. Tsunemi, and S. Iwasaki, *J. Organometal. Chem.*, **10**, 111 (1967).
49. R. West and E. G. Rochow, *J. Org. Chem.*, **18**, 1739 (1953).
50. D. Wittenberg and H. Gilman, *J. Amer. Chem. Soc.*, **80**, 2677 (1958).
51. R. Fessenden and M. D. Coon, *J. Org. Chem.*, **26**, 2530 (1961).
52. C. Tamborski and H. Rosenberg, *J. Org. Chem.*, **25**, 246 (1960).
53. H. Gilman and E. A. Zeuch, *J. Amer. Chem. Soc.*, **82**, 3605 (1960).
54. I. Haiduc and H. Gilman, *J. Organometal. Chem.*, **11**, 55 (1968).
55. H. Gilman and R. D. Gorsich, *J. Amer. Chem. Soc.*, **77**, 6380 (1955).
56. I. M. Gverdtsiteli, T. P. Doksopulo, M. M. Menteshashvili, and I. I. Abkhazava, *Soobshch. Akad. Nauk Gruz. SSR*, **40**, 338 (1965); *Chem. Abstr.*, **64**, 11239 (1966).
57. H. Gilman and R. D. Gorsich, *J. Amer. Chem. Soc.*, **80**, 1883 (1957).
58. H. Gilman and R. D. Gorsich, *J. Amer. Chem. Soc.*, **80**, 3243 (1958).
59. I. M. Gverdtsiteli, T. P. Doksopulo, M. M. Menteshashvili, and I. I. Abkhazava, *Zh. Obshch. Khim.*, **36**, 114 (1966).
60. R. Gelius, *Angew. Chem.*, **72**, 322 (1960).
61. F. Johnson, U.S. Pat. 3,234,239 (1966); *Chem. Abstr.*, **64**, 11251 (1966).
62. S. C. Cohen and A. G. Massey, *Tetrahedron Letters*, 4393 (1966).
63. S. C. Cohen and A. G. Massey, *Chem. Commun.*, 457 (1966).
64. H. G. Kuivila and O. F. Beumel, Jr., *J. Amer. Chem. Soc.*, **80**, 3250 (1958).
65. L. I. Smith and H. H. Hoehn, *J. Amer. Chem. Soc.*, **63**, 1184 (1941).
66. F. C. Leavitt, T. A. Manuel, and F. Johnson, *J. Amer. Chem. Soc.*, **81**, 3163 (1969).
67. F. C. Leavitt, T. A. Manuel, F. Johnson, L. V. Matternas, and D. S. Lehman, *J. Amer. Chem. Soc.*, **82**, 5099 (1960).
68. E. H. Braye, W. Hübel, and I. Caplier, *J. Amer. Chem. Soc.*, **83**, 4406 (1961).
69. E. H. Braye, H. Hübel, and I. Caplier, U.S. Pat. 3,151,140 (1964); *Chem. Abstr.*, **61**, 16097 (1964).
70. H. Gilman, S. G. Cottis, and W. H. Atwell, *J. Amer. Chem. Soc.*, **86**, 1596 (1964).
71. H. Gilman, S. G. Cottis, and W. H. Atwell, *J. Amer. Chem. Soc.*, **86**, 5584 (1964).
72. H. Gilman and W. H. Atwell, *J. Organometal. Chem.*, **2**, 291 (1964).
73. K. Rühlmann, *Z. Chem.*, **5**, 354 (1965).
74. W. H. Atwell, D. R. Weyenberg, and H. Gilman, *J. Org. Chem.*, **32**, 885 (1967).
75. K. W. Hubel, E. H. Braye, and I. H. Capliers, U.S. Pat. 3,151,140 (1964); *Chem. Abstr.*, **61**, 16097 (1964).
76. J. G. Zavistoski and J. J. Zuckerman, *J. Org. Chem.*, **34**, 4197 (1969).
77. F. C. Leavitt and F. Johnson, U.S. Pat. 3,116,307 (1963); *Chem. Abstr.*, **60**, 6872 (1964).
78. M. D. Rauch and L. P. Klemann, *J. Amer. Chem. Soc.*, **89**, 5732 (1967).
79. H. Gilman and E. A. Zuech, *Chem. Ind. (London)*, 1227 (1958).
80. H. Gilman and E. A. Zuech, *J. Amer. Chem. Soc.*, **82**, 2522 (1960).
81. H. Gilman and E. A. Zuech, *J. Org. Chem.*, **26**, 2013 (1961).
82. H. Gilman and E. A. Zuech, *J. Org. Chem.*, **26**, 3481 (1961).
83. H. Gilman and E. A. Zuech, *J. Org. Chem.*, **27**, 2897 (1962).
84. D. Wasserman and R. E. Jones, Belg. Pat. 613,915 (1962); *Chem. Abstr.*, **58**, 1490 (1963).

85. D. Wasserman and R. E. Jones, Fr. Pat. 1,315,605 (1963); *Chem. Abstr.*, **58**, 11400 (1963).
86. D. Wasserman, R. E. Jones, S. A. Robinson, and J. D. Garber, *J. Org. Chem.*, **30**, 3248 (1965).
87. C. Tamborski and H. Gilman, U.S. Pat. 3,079,414 (1963); *Chem. Abstr.*, **59**, 5196 (1963).
88. E. J. Kupchik and V. A. Perciaccante, *J. Organometal. Chem.*, **10**, 181 (1967).
89. K. Oita and H. Gilman, *J. Amer. Chem. Soc.*, **79**, 339 (1957).
90. C. H. S. Hitchcock, F. G. Mann, and A. Vanterpool, *J. Chem. Soc.*, 4537 (1957).
91. E. J. Kupchik and J. A. Vrsino, *Chem. Ind. (London)*, 794 (1965).
92. E. Hengge and U. Brychcy, *Monatsh. Chem.*, **97**, 1309 (1966).
93. H. Gilman, *Trans. N.Y. Acad. Sci.*, **25**, 820 (1963).
94. H. Gilman and S. Inoue, *Chem. Ind. (London)*, 74 (1964).
95. Esso Research and Engineering Company, Fr. Pat. 1,467,549 (1968); *Chem. Abstr.*, **68**, 49769 (1968).
96. A. G. Brook and J. B. Pierce, *J. Org. Chem.*, **30**, 2566 (1965).
97. H. Gilman and D. L. Marrs, *J. Org. Chem.*, **29**, 3175 (1964).
98. H. Gilman and D. L. Marrs, *J. Org. Chem.*, **30**, 1942 (1965).
99. H. Gilman, E. A. Zuech, and W. Steudel, *J. Org. Chem.*, **27**, 1836 (1962).
100. K. A. Andrianov, L. M. Volkova, N. V. Delazari, and N. A. Chumaevskii, *Khim. Geterotsikl. Soedin., Akad. Nauk Latv. SSR*, 435 (1967).
101. C.-T. Wang, H.-C. Chou, and M.-S. Hung, *Hua Hsüeh Hsüeh Pao*, **30**, 91 (1964); *Chem. Abstr.*, **61**, 1888 (1964).
102. W. H. Knoth and R. V. Lindsey, *J. Amer. Chem. Soc.*, **80**, 4106 (1958).
103. W. H. Knoth, U.S. Pat. 2,983,744 (1961); *Chem. Abstr.*, **55**, 22132 (1961).
104. G. Rossmy and G. Koerner, *Makromol. Chem.*, **73**, 85 (1964).
105. K. A. Andrianov and S. E. Yakushkina, *Izv. Adad. Nauk SSSR, Otd. Khim. Nauk*, 1396 (1962).
106. T. Yu, L. Hsu, and S. Wu, *Hua Hsüeh Hsüeh Pao*, **24**, 170 (1958); *Chem. Abstr.*, **53**, 6233 (1959).
107. P. Mazerolles, M. Lesbre, and J. Dubac, *Compt. Rend.*, **260**, 2255 (1965).
108. P. Mazerolles, J. Dubac, and M. Lesbre, *J. Organometal. Chem.*, **5**, 35 (1966).
109. J. W. Curry, *J. Amer. Chem. Soc.*, **78**, 1686 (1956).
110. J. W. Curry and G. W. Harrison, Jr., *J. Org. Chem.*, **23**, 1219 (1958).
111. K. Kojima, *Bull. Chem. Soc. Jap.*, **31**, 663 (1950).
112. J. W. Curry, *J. Org. Chem.*, **26**, 1308 (1961).
113. D. L. Bailey, Ger. Pat. 1,115,929 (1961); *Chem. Abstr.*, **59**, 3957 (1963).
114. A. M. Polyakova, M. D. Suchkova, V. V. Korshak, and V. M. Vdovin, *Izv. Akad. Nauk SSSR, Ser. Khim.*, 1267 (1965).
115. M. F. Hawthorne, *J. Amer. Chem. Soc.*, **82**, 748 (1960).
116. G. S. Kolesnikov, S. L. Davydova, and T. I. Ermoleeva, *Vysokomolekul. Soedin.*, **1**, 591 (1951); *J. Polym. Sci.*, **43**, 593 (1960).
117. M. C. Henry and J. G. Noltes, *J. Amer. Chem. Soc.*, **82**, 561 (1960).
118. J. G. Noltes and G. J. M. Van der Kerk, *Chimia*, **16**, 122 (1962).
119. J. G. Noltes and G. J. M. Van der Kerk, *Rec. Trav. Chim.*, **81**, 41 (1962).
120. A. J. Neusink and J. G. Noltes, *J. Organometal. Chem.*, **16**, 91 (1969).
121. A. J. Leusink, J. G. Noltes, H. A. Budding, and G. J. M. van der Kerk, *Rec. Trav. Chim.*, **83**, 1036 (1964).
122. E. A. Cheruyshev and N. G. Tolstikova, *Izv. Akad. Nauk SSSR, Otd. Khim. Nauk*, 1146 (1963).

123. L. L. Muller and J. Hammer, *1,2-Cycloaddition Reactions*, Wiley-Interscience, New York, 1967, p. 44.
124. M. E. Volpin and D. N. Kursanov, *Izv. Akad. Nauk SSSR, Otd. Khim. Nauk*, 1903 (1960).
125. M. E. Volpin, Yu. D. Koreshkov, and D. N. Kursanov, *Izv. Akad. Nauk SSSR, Otd. Khim. Nauk*, 1355 (1961).
126. M. E. Volpin and D. N. Kursanov, *Zh. Obshch. Khim.*, **32**, 1137 (1962).
127. M. E. Volpin and D. N. Kursanov, *Zh. Obshch. Khim.*, **32**, 1142 (1962).
128. M. E. Volpin and D. N. Kursanov, *Zh. Obshch. Khim.*, **32**, 1455 (1962).
129. O. M. Nefedov, M. N. Manakov, and A. D. Petrov, *Izv. Akad. Nauk SSSR, Otd. Khim. Nauk*, 1717 (1961).
130. O. M. Nefedov, M. N. Manakov, and A. D. Petrov, *Dokl. Akad. Nauk SSSR*, **147**, 1376 (1962).
131. O. M. Nefedov and M. N. Manakov, *Angew. Chem. Intern. Ed.*, **5**, 1021 (1966).
132. W. H. Atwell and D. R. Weyenberg, *Angew. Chem. Intern. Ed.*, **8**, 469 (1969).
133. P. S. Skell and E. J. Goldstein, *J. Amer. Chem. Soc.*, **86**, 1442 (1964).
134. O. M. Nefedov, M. N. Manakov, and A. D. Petrov, *Izv. Akad. Nauk SSSR, Otd. Khim. Nauk*, 1228 (1962).
135. O. M. Nefedov, M. N. Manakov, and A. D. Petrov, U.S.S.R. Pat. 148,055 (1962); *Chem. Abstr.*, **58**, 9137 (1963).
136. O. M. Nefedov, M. N. Manakov, and A. D. Petrov, *Dokl. Akad. Nauk SSSR*, **154**, 395 (1964).
137. D. R. Weyenberg, U.S. Pat. 3,187,031 (1965); *Chem. Abstr.*, **63**, 9988 (1965).
138. D. R. Weyenberg, U.S. Pat. 3,187,032 (1965); *Chem. Abstr.*, **63**, 14905 (1965).
139. D. R. Weyenberg, L. H. Toporcer, and A. E. Bey, *J. Org. Chem.*, **30**, 4096 (1965).
140. O. M. Nefedov, M. N. Manakov, V. N. Medvedev, A. S. Khachaturov, and V. I. Shiryaev, *Khim. Geterotsikl. Soedin., Akad. Nauk Latv. SSR*, 299 (1966); *Chem. Abstr.*, **65**, 2287 (1966).
141. J. C. Thompson, P. L. Timms, and J. L. Margrave, *Chem. Commun.*, 566 (1966).
142. A. G. MacDiarmid and F. M. Rabel, unpublished results quoted in Ref. 131.
143. V. F. Mironov and T. K. Gar, *Izv. Akad. Nauk SSSR, Otd. Khim. Nauk*, 578 (1963).
144. V. F. Mironov and T. K. Gar, *Dokl. Akad. Nauk SSSR*, **152**, 1111 (1963).
145. O. M. Nefedov, C. P. Kolesnikov, A. S. Khachaturov, and A. D. Petrov, *Dokl. Acad. Nauk SSSR*, **154**, 1389 (1964).
146. V. F. Mironov and T. K. Gar, *Izv. Akad. Nauk SSSR, Ser. Khim.*, 755 (1965).
147. T. K. Gar and V. F. Mironov, *Izv. Akad. Nauk SSSR, Ser. Khim.*, 855 (1965).
148. G. Manuel and P. Mazerolles, *Bull. Soc. Chim. France*, 2447 (1965).
149. P. Mazerolles and G. Manuel, *Bull. Soc. Chim. France*, 327 (1966).
150. O. M. Nefedov and M. N. Manakov, *Izv. Akad. Nauk SSSR, Ser. Khim.*, 840 (1964).
151. W. H. Atwell and D. R. Weyenberg, *J. Organometal. Chem.*, **5**, 594 (1966).
152. W. H. Atwell and D. R. Weyenberg, *J. Amer. Chem. Soc.*, **90**, 3438 (1968).
153. P. L. Timms, D. D. Stump, R. A. Kent, and J. L. Margrave, *J. Amer. Chem. Soc.*, **88**, 940 (1966).
154. M. E. Volpin, Yu. D. Koreshkov, V. G. Dulova, and D. N. Kursanov, *Tetrahedron*, **18**, 107 (1962).
155. L. A. Leites, V. G. Dulova, and M. E. Volpin, *Izv. Akad. Nauk SSSR, Otd. Khim. Nauk*, 731 (1963).
156. M. E. Volpin, V. G. Dulova, and D. N. Kursanov, *Izv. Akad. Nauk SSSR, Otd. Khim. Nauk*, 727 (1963).
157. F. Johnson and R. S. Gohlke, *Tetrahedron Letters*, 1291 (1962).

158. R. West and R. E. Bailey, *J. Amer. Chem. Soc.*, **85**, 2871 (1963).
159. M. E. Volpin, Yu. T. Struchkov, L. V. Vilkov, V. S. Mastyukov, V. G. Dulova, and D. N. Kursanov, *Izv. Akad. Nauk SSSR, Ser. Khim.*, 2067 (1963).
160. F. Johnson, R. S. Gohlke, and W. A. Nasutavicus, *J. Organometal. Chem.*, **3**, 233 (1965).
161. N. G. Bokiy and Yu. T. Struchkov, *Zh. Strukt. Khim.*, **6**, 571 (1965).
162. N. G. Bokiy and Yu. T. Struchkov, *Zh. Strukt. Khim.*, **7**, 133 (1966).
163. M. E. Volpin, V. G. Dulova, Yu. T. Struchkov, N. K. Bokiy, and D. N. Kursanov, *J. Organometal. Chem.*, **8**, 87 (1967).
164. J. C. Thompson, J. L. Margrave, P. L. Timms, and C. S. Liu, *Abstr. Fourth Intern. Conf. Organometal. Chem.*, Bristol, England, July, 1969.
165. P. L. Timms, R. A. Kent, T. C. Ehlert, and J. L. Margrave, *J. Amer. Chem. Soc.*, **87**, 2824 (1965).
166. J. G. Zavistoski and J. J. Zuckerman, *J. Amer. Chem. Soc.*, **90**, 6612 (1968).
167. W. G. Woods, *J. Org. Chem.*, **23**, 110 (1958).
168. G. Fritz, D. Kummer, and G. Sonntag, *Z. Anorg. Allgem. Chem.*, **342**, 113 (1966).
169. G. Fritz and B. Raabe, *Z. Naturforsch.*, **11b**, 57 (1956).
170. G. Fritz and G. Teichmann, *Angew. Chem.*, **70**, 701 (1958).
171. G. Fritz, H. J. Buhl, and D. Kummer, *Z. Anorg. Allgem. Chem.*, **327**, 165 (1964).
172. N. S. Nametkin, L. E. Guselnikov, V. M. Vdovin, P. L. Grinberg, V. I. Zavyalov, and V. D. Oppenheim, *Dokl. Akad. Nauk. SSSR*, **171**, 630 (1966).
173. N. S. Nametkin, V. M. Vdovin, L. E. Guselnikov, and V. I. Zarialov, *Izv. Akad. Nauk SSSR, Ser. Khim.*, 584 (1966).
174. R. W. Coutant and A. Levy, *J. Organometal. Chem.*, **10**, 175 (1967).
175. F. Mares and V. Chvalovsky, *Collection Czech. Chem. Commun.*, **32**, 382 (1967).
176. N. S. Nametkin, L. E. Guselnikov, T. Kh. Islamov, M. V. Shishkina, and V. M. Vdovin, *Dokl. Akad. Nauk SSSR*, **175**, 136 (1967).
177. D. N. Andreev, *Izv. Akad. Nauk SSSR, Otd. Khim. Nauk*, 237 (1960).
178. A. L. Smith and H. A. Clark, *J. Amer. Chem. Soc.*, **83**, 3345 (1961).
179. V. M. Vdovin, K. S. Pushchevaya, and A. D. Petrov, *Izv. Akad. Nauk SSSR, Otd. Khim. Nauk*, 281 (1961).
180. G. Fritz and G. Teichmann, *Chem. Ber.*, **95**, 2361 (1962).
181. G. Fritz, H. Fröhlich, and D. Kummer, *Z. Anorg. Allgem. Chem.*, **353**, 34 (1967).
182. G. Fritz and N. Szczepanski, *Angew. Chem.*, **79**, 1067 (1967).
183. G. J. D. Peddle, D. N. Roark, A. M. Good, and S. G. McGeachin, *Abstr. Fourth Intern. Conf. Organometal. Chem.*, Bristol, England, July, 1969.
184. H. Gilman and D. Wittenberg, *J. Amer. Chem. Soc.*, **79**, 6339 (1957).
185. D. Wittenberg, H. A. McNinch, and H. Gilman, *J. Amer. Chem. Soc.*, **80**, 5418 (1959).
186. Yu. K. Yurev, *Chemistry and Technology of Organosilicon Compounds*, Ts. B.T.I., Leningrad, 1958, part I, p. 157, quoted in Ref. 3.
187. P. Mazerolles, M. Lesbre, and M. Joanny, *J. Organometal. Chem.*, **16**, 227 (1969).
188. E. H. Braye and H. Hubel, *Chem. Ind.* (*London*), 1250 (1959).
189. N. S. Nametkin, V. M. Vdovin, K. S. Pushchevaya, and V. I. Zavyalov, *Izv. Akad. Nauk SSSR, Ser. Khim.*, 1453 (1965).
190. K. S. Pushchevaya, cand. thesis, Univ. Moscow, 1965; quoted in Ref. 3.
191. V. M. Vdovin, N. S. Nametkin, K. S. Pushchevaya, and A. V. Topchiev, *Izv. Akad. Nauk. SSSR, Otd. Khim. Nauk*, 1127 (1962).
192. V. M. Vdovin, N. S. Nametkin, K. S. Pushchevaya, and A. V. Topchiev, *Izv. Akad. Nauk SSSR, Otd. Khim. Nauk*, 274 (1963).

193. N. S. Nametkin, V. M. Vdovin, and K. S. Pushchevaya, *Dokl. Akad. Nauk SSSR*, **150**, 562 (1963).
194. D. Seyferth, R. Damrauer, and S. S. Washburn, *J. Amer. Chem. Soc.*, **89**, 1538 (1967).
195. R. Polster, *Ann. Chem.*, **654**, 20 (1962).
196. R. Polster, Ger. Pat. 1,153,748 (1963); *Chem. Abstr.*, **60**, 551 (1964).
197. R. Polster, Ger. Pat. 1,156,807 (1963); *Chem. Abstr.*, **60**, 4182 (1964).
198. G. P. Mack and E. Parker, U.S. Pat. 2,604,483 (1952); *Chem. Abstr.*, **47**, 4358 (1952).
199. R. Müller and W. Müller, *Chem. Ber.*, **97**, 1111 (1964).
200. R. Müller and W. Müller, *Chem. Ber.*, **97**, 1115 (1964).
201. C. H. Yoder and J. J. Zuckerman, *Inorg. Chem.*, **4**, 116 (1965).
202. K. Lienhard and E. G. Rochow, *Angew. Chem.*, **75**, 980 (1963).
203. C. H. Yoder and J. J. Zuckerman, *Inorg. Chem.*, **3**, 1392 (1964).
204. M. Wieber and M. Schmidt, *Z. Naturforsch.*, **18b**, 849 (1963).
205. U. Wannagat and O. Brandstätter, *Monatsh. Chem.*, **97**, 1352 (1966).
206. U. Wannagat and O. Brandstätter, *Angew. Chem. Intern. Ed.*, **2**, 263 (1963).
207. F. A. Henglein and K. Lienhard, *Makromol. Chem.*, **32**, 218 (1959).
208. D. Kummer and E. G. Rochow, *Z. Anorg. Allgem. Chem.*, **321**, 21 (1963).
209. N. Derkach and N. Smetankina, *Zh. Obshch. Khim.*, **36**, 2009 (1966).
210. C. G. Pitt and K. R. Skillern, *Inorg. Nucl. Chem. Letters*, **2**, 237 (1966).
211. C. G. Pitt and K. R. Skillern, *Inorg. Chem.*, **6**, 865 (1967).
212. R. H. Baney and G. C. Haberland, *J. Organometal. Chem.*, **5**, 320 (1966).
213. U. Wannagat and H. Niederprüm, *Angew. Chem.*, **70**, 745 (1958).
214. U. Wannagat and H. Niederprüm, *Z. Anorg. Allgem. Chem.*, **311**, 270 (1961).
215. U. Wannagat and E. Bogusch, *Inorg. Nucl. Chem. Letters*, **1**, 13 (1965).
216. W. Fink, *Angew. Chem. Intern. Ed.*, **5**, 760 (1966).
217. Chi-Tao Wang, Hsiu-Chong Chon, and Mau-Shui Hung, *Acta Chim. Sinica* (*Hua Hsueh Hsueh Pao*), **30**, 91 (1964); *Chem. Abstr.*, **61**, 1888 (1964).
218. C. R. Krüger, *Inorg. Nucl. Chem. Letters*, **1**, 85 (1965).
219. D. Kummer and E. G. Rochow, *Angew. Chem.*, **75**, 207 (1963).
220. M. Wieber and M. Schmidt, *J. Organometal. Chem.*, **1**, 22 (1963).
221. W. Simmler, *Chem. Ber.*, **96**, 349 (1963).
222. S. Papetti and T. L. Heying, *Inorg. Chem.*, **2**, 1105 (1963).
223. C. H. Yoder and J. J. Zuckerman, *J. Amer. Chem. Soc.*, **88**, 4831 (1966).
224. O. Scherer and M. Schmidt, *Angew. Chem.*, **75**, 642 (1963).
225. K. Jones and M. F. Lappert, *Proc. Chem. Soc.*, 358 (1962).
226. K. Jones and M. F. Lappert, *J. Chem. Soc.*, 1944 (1965).
227. C. H. Yoder and J. J. Zuckerman, *J. Amer. Chem. Soc.*, **88**, 2170 (1966).
228. O. J. Scherer, J. Schmidt, J. Wokulat, and M. Schmidt, *Z. Naturforsch.*, **20b**, 183 (1965).
229. O. J. Scherer and D. Biller, *Z. Naturforsch.*, **22b**, 1079 (1967).
230. U. Wannagat, *Angew. Chem.*, **76**, 234 (1964).
231. M. V. Geroge, D. Wittenberg, and H. Gilman, *J. Amer. Chem. Soc.*, **81**, 361 (1959).
232. S. S. Dua and M. V. George, *J. Organometal. Chem.*, **10**, 219 (1967).
233. M. V. George, P. B. Telukdar, and H. Gilman, *J. Organometal. Chem.*, **5**, 397 (1966).
234. U. Wannagat, *Angew. Chem.*, **78**, 648 (1966).
235. U. Wannagat, E. Bogusch, and F. Höfler, *J. Organometal. Chem.*, **7**, 203 (1967).
236. O. J. Scherer and P. Hornig, *Angew. Chem.*, **79**, 60 (1967).

237. D. Kummer and E. G. Rochow, *Z. Anorg. Allgem. Chem.*, **321**, 21 (1963).
238. K. Lienhard and E. G. Rochow, *Angew. Chem. Intern. Ed.*, **2**, 325 (1963).
239. E. G. Rochow, *Bull. Soc. Chim. France*, 1360 (1963).
240. U. Wannagat, E. Bogusch, and R. Braun, *J. Organometal. Chem.*, **19**, 367 (1969).
241. I. Haiduc and H. Gilman, *J. Organometal. Chem.*, **18**, P5 (1969).
242. J. Ellermann and F. Paersch, *Angew. Chem. Intern. Ed.*, **6**, 355 (1967).
243. E. W. Abel and R. P. Bush, *J. Organometal. Chem.*, **3**, 245 (1965).
244. W. Fink, *Chem. Ber.*, **99**, 2267 (1966).
245. W. Fink, *Helv. Chim. Acta*, **49**, 1408 (1966).
246. J. M. Gaidis and R. West, *J. Amer. Chem. Soc.*, **86**, 5699 (1964).
247. W. Fink, *Helv. Chim. Acta*, **46**, 720 (1963).
248. E. Hengge, R. Petzold, and V. Brychcy, *Z. Naturforsch.*, **20b**, 397 (1965).
249. R. West and M. Ishikawa, *J. Amer. Chem. Soc.*, **89**, 5049 (1967).
250. P. G. Harrison and J. J. Zuckerman, *Inorg. Nucl. Chem. Letters*, **5**, 545 (1969).
251. W. Fink, *Chem. Ber.*, **97**, 1424 (1964).
252. J. J. Daly and W. Fink, *J. Chem. Soc.*, 4958 (1964).
253. K. Moedritzer and J. R. Van Wazer, *J. Phys. Chem.*, **70**, 2030 (1966).
254. K. Moedritzer and J. R. Van Wazer, *Inorg. Nucl. Chem. Letters*, **2**, 45 (1966).
255. C. M. S. Yoder and J. J. Zuckerman, *Inorg. Chem.*, **6**, 163 (1967).
256. W. A. Piccoli, G. C. Haberland, and R. L. Merker, *J. Amer. Chem. Soc.*, **82**, 1883 (1960).
257. K. A. Andrianov, N. V. Delazarin, L. M. Volkova, and N. A. Chumaevskii, *Dokl. Akad. Nauk SSSR*, **160**, 1307 (1965).
258. H. Niederprüm and W. Simmler, *Z. Anorg. Allgem. Chem.*, **345**, 53 (1966).
259. M. Kumada, M. Yamaguchi, Y. Yamamoto, J. Nakajima, and K. Shiina, *J. Org. Chem.*, **21**, 1264 (1956).
260. U. Wannagat and O. Brandstätter, *Monatsh. Chem.*, **94**, 1090 (1963).
261. K. A. Andrianov and V. E. Nikitenkov, *Izv. Akad. Nauk SSSR, Otd. Khim. Nauk*, 441 (1961).
262. K. A. Andrianov and N. V. Delazari, *Izv. Akad. Nauk SSSR, Otd. Khim. Nauk*, **7**, 1266 (1961).
263. R. S. Tkeshelashvili, K. A. Andrianov, and A. I. Nogaideli, *Izv. Akad. Nauk SSSR, Ser. Khim.*, **8**, 1396 (1965).
264. E. Hengge, H. Reuter, and R. Petzold, *Z. Naturforsch.*, **18b**, 425 (1963).
265. W. P. Neumann and K. Kühlein, *Tetrahedron Letters*, **23**, 1541 (1963).
266. W. P. Neumann and K. Kühlein, *Ann. Chem.*, **683**, 1 (1965).
267. K. A. Andrianov, I. Khaiduk, L. M. Khananashvili, and N. I. Nekhaeva, *Zh. Obshch. Khim.*, **32**, 3447 (1962).
268. M. Schmidt and M. Wieber, *Inorg. Chem.*, **1**, 909 (1962).
269. P. Mazerolles, J. Dubac, and M. Lesbre, *J. Organometal. Chem.*, **12**, 143 (1968).
270. R. H. Krieble and C. A. Burkhard, *J. Amer. Chem. Soc.*, **69**, 2689 (1947).
271. V. P. Davydova, M. G. Voronkov, and B. N. Dolgov, *Chemistry and Technology of Organosilicon Compounds*, Leningrad, 1958, part 1, p. 204.
272. H. Staudinger and W. Hahn, *Makromol. Chem.*, **11**, 24 (1953).
273. C. L. Frye, R. M. Salinger, and T. J. Patin, *J. Amer. Chem. Soc.*, **88**, 2343 (1966).
274. N. Spassky, *Comp. Rend.*, **251**, 2371 (1960).
275. A. Marchand and J. Valade, *J. Organometal. Chem.*, **12**, 305 (1968).
276. H. H. Ender, U.S. Pat. 3,078,293 (1963); *Chem. Abstr.*, **59**, 1681 (1963).
277. M. Schmidt and H. Schmidbaur, *Angew. Chem.*, **70**, 470 (1958).
278. F. P. Richter and B. A. Orkin, U.S. Pat. 2,590,039 (1952); *Chem. Abstr.*, **46**, 5892.

279. M. Schmidt and H. Schmidbaur, *Chem. Ber.*, **93**, 878 (1960).
280. F. A. Henglein and H. Niebergall, *Chem. Ztg.*, **80**, 611 (1956).
281. H. W. Kohlschütter and G. Jaekel, *Z. Anorg. Allgem. Chem.*, **271**, 185 (1953).
282. M. Wieber and M. Schmidt, *Chem. Ber.*, **96**, 2822 (1963).
283. M. Wieber and M. Schmidt, *Chem. Ber.*, **96**, 1019 (1963).
284. R. M. Ismail and E. Bessler, *J. Organometal. Chem.*, **13**, 253 (1968).
285. R. M. Ismail, *J. Organometal. Chem.*, **11**, 49 (1968).
286. H. J. Emeléus and J. J. Zuckerman, *J. Organometal. Chem.*, **1**, 328 (1964).
287. C. M. Silcox and J. J. Zuckerman, *J. Organometal. Chem.*, **5**, 483 (1966).
288. R. M. Ismail, *Z. Naturforsch.*, **18b**, 1124 (1963).
289. C. M. Silcox and J. J. Zuckerman, *J. Amer. Chem. Soc.*, **88**, 168 (1966).
290. M. Wieber and M. Schmidt, *J. Organometal. Chem.*, **1**, 93 (1963).
291. M. Wieber and M. Schmidt, *J. Organometal. Chem.*, **2**, 129 (1964).
292. E. J. Bulten and J. G. Noltes, *Tetrahedron Letters*, **29**, 3471 (1966).
293. J. Bornstein, B. R. LaLiberte, T. M. Andrews, and J. C. Montermoso, *J. Org. Chem.*, **24**, 886 (1959).
294. W. E. Davidson, K. Hills, and M. C. Henry, *J. Organometal. Chem.*, **3**, 285 (1965).
295. R. C. Poller, *Proc. Chem. Soc.*, 312 (1963).
296. M. Wieber and M. Schmidt, *Z. Naturforsch.*, **18b**, 847 (1963).
297. M. Wieber and M. Schmidt, *J. Organometal. Chem.*, **1**, 336 (1964).
298. R. Seltzer, *J. Org. Chem.*, **33**, 3896 (1968).
299. E. W. Abel and C. R. Jenkins, *J. Chem. Soc.* (*A*), 1344 (1967).
300. M. G. Voronkov, G. Zelcans, V. F. Mironov, J. Bleidelis, and A. Kemme, *Khim. Geterotsikl. Soedin.*, 227 (1968); *Chem. Abstr.*, **69**, 87129 (1968).
301. M. G. Voronkov, G. Zelcans, and V. F. Mironov, U.S.S.R. Pat. 190,897, *Chem. Abstr.*, **68**, 69128 (1968).
302. H. J. Backer and W. Drenth, *Rec. Trav. Chim.*, **70**, 559 (1951).
303. H. J. Backer and F. Stienstra, *Rec. Trav. Chim.*, **52**, 1038 (1933).
304. F. H. Fink, J. A. Turner, and D. A. Payne, *J. Amer. Chem. Soc.*, **88**, 1571 (1966).
305. M. Wieber and C. D. Frohning, *J. Organometal. Chem.*, **8**, 459 (1967).
306. H. J. Emeléus and J. J. Zuckerman, *J. Organometal. Chem.*, **1**, 328 (1964).
307. R. C. Mehrota and S. Mathur, *J. Organometal. Chem.*, **6**, 11 (1966).
308. H. H. Anderson, *J. Amer. Chem. Soc.*, **72**, 194 (1950).
309. W. J. Considine, *J. Organometal. Chem.*, **5**, 263 (1966).
310. W. A. Gregory, U.S. Pat. 2,636,891 (1953); *Chem. Abstr.*, **48**, 3397.
311. I. Hechenbleikner, R. E. Bresser, and O. A. Homberg, Belg. Pat. 630,459 (1963); *Chem. Abstr.*, **61**, 8339 (1964).
312. I. Hechenbleikner and O. A. Homberg, U.S. Pat. 3,209,017 (1965); *Chem. Abstr.*, **63**, 16382 (1965).
313. G. T. Cocks and J. J. Zuckerman, *Inorg. Chem.*, **4**, 592 (1965).
314. A. W. P. Jarvie, H. J. S. Winkler, and H. Gilman, *J. Org. Chem.*, **27**, 614 (1962).
315. H. J. S. Winkler and H. Gilman, *J. Org. Chem.*, **26**, 1265 (1961).
316. P. G. Campbell, Univ. Microfilms Publ. No. 24000; *Diss. Abstr.*, **17**, 2808 (1957).
317. V. E. Nikitenkov, *Zh. Obshch. Khim.*, **35**, 1666 (1965).
318. I. Haiduc and H. Gilman, *J. Organometal. Chem.*, **11**, 459 (1968).
319. S. J. Groszos, U.S. Pat. 2,957,900 (1960); *Chem. Abstr.*, **55**, 5424 (1961).
320. M. M. Sprung, *J. Org. Chem.*, **23**, 58 (1958).
321. G. B. Sterling and C. E. Pawloski, U.S. Pat. 3,256,308 (1966); *Chem. Abstr.*, **65**, 7217 (1966).
322. D. R. Weyenberg and L. H. Toporcer, *J. Amer. Chem. Soc.*, **84**, 2843 (1962).

323. S. A. Fuqua and R. M. Silverstein, *J. Org. Chem.*, **29**, 395 (1964).
324. A. B. Finestone, Ger. Pat. 1,131,681 (1962); *Chem. Abstr.*, **58**, 4598 (1963).
325. C. M. Samour, U.S. Pat. 3,118,921 (1964); *Chem. Abstr.*, **60**, 10715 (1964).
326. M. G. Voronkov, I. Mazeika, and G. Zelcans, *Khim. Geterotsikl. Soedin. Akad. Nauk Latv. SSR*, 58 (1965); *Chem. Abstr.*, **63**, 5506 (1965).
327. K. A. Andrianov, T. V. Vasilieva, and R. A. Romanova, *Dokl. Akad. Nauk SSSR*, **168**, 1057 (1966).
328. K. A. Andrianov, Z. V. Vasilieva, and L. M. Khananashvili, *Izv. Akad. Nauk SSSR, Otd. Khim. Nauk*, 1030 (1961).
329. K. Friederich, A. de Montigny, and W. Simmler, *Kolloid Z. Z. Polym.*, **218**, 7 (1967).
330. D. Gölitz, H. Sattleger, and W. Simmler, *Kolloid Z. Z. Polym.*, **218**, 1 (1967).
331. K. A. Andrianov, V. I. Pakhomov, and N. E. Lapteva, *Izv. Akad. Nauk SSSR, Otd. Khim. Nauk*, **11**, 2039 (1962).
332. J. L. Speier, M. P. David, and B. A. Eynou, *J. Org. Chem.*, **25**, 1637 (1960).
333. W. Simmler, H. Niederprüm, and H. Sattlegger, *Chem. Ber.*, **99**, 1368 (1966).
334. M. F. Shostakovskii, A. S. Atavin, B. A. Trofimov, and E. P. Vyalykh, *Zh. Obshch. Khim.*, **35**, 1759 (1965).
335. G. Koerner and G. Rossmy, *Makromol. Chem.*, **97**, 241 (1966).
336. K. A. Andrianov, V. I. Pakhomov, and N. E. Lapteva, *Izv. Akad. Nauk SSSR, Otd. Khim. Nauk*, **11**, 2040 (1962).
337. V. A. Zeitler and C. A. Brown, *J. Amer. Chem. Soc.*, **79**, 4618 (1957).
338. M. G. Voronkov, V. P. Davydova, and B. N. Dolgov, *Izv. Akad. Nauk SSSR, Otd. Khim. Nauk*, 698 (1958).
339. W. Hahn, *Makromol. Chem.*, **11**, 51 (1953).
340. M. G. Voronkov, V. P. Davydova, and B. N. Dolgov, *Izv. Akad. Nauk SSSR, Otd. Khim. Nauk*, 677 (1958).
341. V. P. Davydova, M. G. Voronkov, and B. N. Dolgov, *Proc. Conf. Chem. and Pract. Appl. Org.-Si Chem.*, Leningrad, **I**, 204 (1958).
342. V. P. Davydova and M. G. Voronkov, *Zh. Obshch. Khim.*, **28**, 1879 (1958).
343. F. C. Whitmore, F. W. Pietrusza, and L. H. Sommer, *J. Amer. Chem. Soc.*, **69**, 2108 (1947).
344. L. W. Breed, W. J. Haggerty, Jr., and J. Harvey, *J. Org. Chem.*, **25**, 1804 (1960).
345. M. G. Chandra, A. K. Rai, and R. C. Mehrotra, *J. Organometal. Chem.*, **4**, 371 (1965).
346. S. Mathur and R. C. Mehrotra, *J. Organometal. Chem.*, **7**, 227 (1967).
347. R. C. Mehrotra and V. D. Gupta, *J. Organometal. Chem.*, **4**, 145 (1965).
348. K. A. Andrianov, V. I. Pakhomov, and N. E. Lapteva, *Dokl. Akad. Nauk SSSR*, **151**, 849 (1963).
349. G. Koerner and G. Rossmy, *Angew. Chem.*, **75**, 1114 (1963).
350. J. F. Hyde, U.S. Pat. 2,645,654 (1953); *Chem. Abstr.*, **48**, 7050 (1954).
351. A. A. Zhdanov, K. A. Andrianov, and A. A. Bogdanova, *Izv. Akad. Nauk SSSSR, Otd. Khim. Nauk*, **7**, 1261 (1961).
352. K. Rühlmann, R. Volkmer, and C. Michael, *Ann. Chem.*, **706**, 18 (1967).
353. J. C. Pommier, M. Pereyne, and J. Valade, *Compt. Rend. Acad. Sci. Paris*, **260**, 6397 (1965).
354. V. F. Mironov and N. S. Fedotov, *Khim. Geterotsikl. Soedin., Akad. Nauk Latv. SSR*, 179 (1967).
355. J. Satgé, P. Riviere, and M. Lesbre, *Compt. Rend. (C)*, **265**, 494 (1967).
356. J. Satgé and P. Riviere, *J. Organometal. Chem.*, **16**, 71 (1969).

357. K. A. Andrianov, T. K. Dzhashiashvili, V. V. Astakhin, and G. N. Shumakova, *Zh. Obshch. Khim.*, **37**, 928 (1967).
358. K. A. Andrianov, T. K. Dzhashiashvili, V. V. Astakhin, and G. N. Shumakova, *Izv. Akad. Nauk SSSR, Ser. Khim.*, 2229 (1966).
359. J. F. Klebe and H. Finkbeiner, *J. Amer. Chem. Soc.*, **88**, 4740 (1966).
360. P. Mazerolles, J. Dubac, and M. Lesbre, *J. Organometal. Chem.*, **12**, 143 (1968).
361. J. Dubac and P. Mazerolles, *Compt. Rend. Acad. Sci. Paris, Ser. C.*, **267**, 411 (1968).
362. D. Seyferth, S. S. Washburne, T. F. Jula, P. Mazerolles, and J. Dubac, *J. Organometal. Chem.*, **16**, 503 (1969).
363. J. Dubac and P. Mazerolles, *J. Organometal. Chem.*, **20**, P5 (1969).
364. J. J. Zuckerman, *J. Chem. Soc.*, 873 (1962).
365. J. J. Zuckerman, *J. Chem. Soc.*, 1322 (1963).
366. R. L. Merker and M. J. Scott, *J. Amer. Chem. Soc.*, **81**, 975 (1959).
367. R. L. Merker, U.S. Pat. 2,956,045 (1960); *Chem. Abstr.*, **55**, 5552 (1961).
368. N. V. Komarov and N. V. Semenova, *Izv. Akad. Nauk SSSR, Ser. Khim.*, **10**, 1879 (1965).
369. L. H. Sommer, Brit. Pat. 685,533 (1953); *Chem. Abstr.*, **48**, 2760 (1954).
370. L. H. Sommer, R. P. Pioch, N. S. Maraus, G. M. Goldberg, J. Rockett, and J. Kerlin, *J. Amer. Chem. Soc.*, **75**, 2932 (1953).
371. F. S. Kipping and J. T. Abrams, *J. Chem. Soc.*, 81 (1944).
372. E. Hengge and U. Brychcy, *Monatsh. Chem.*, **97**, 84 (1966).
373. W. H. Knoth, Jr. and R. V. Lindsey, Jr., *J. Amer. Chem. Soc.*, **80**, 4106 (1958).
374. M. C. Baird, *Progr. Inorg. Chem.*, **9**, 1 (1968).
375. J. F. Young, *Adv. Inorg. Chem. Radiochem.*, **11**, 102 (1969).
376. R. B. King and F. G. A. Stone, *J. Amer. Chem. Soc.*, **82**, 3833 (1960).
377. S. F. A. Kettle and J. A. Khan, *Proc. Chem. Soc.*, 82 (1962).
378. S. F. A. Kettle and J. A. Khan, *J. Organometal. Chem.*, **5**, 588 (1966).
379. S. D. Ibekwe and M. J. Newlands, *Chem. Commun.*, 114 (1965).
380. J. D. Cotton, J. Duckworth, S. A. R. Knox, P. F. Lindley, I. Paul, F. G. A. Stone, and P. Woodward, *Chem. Commun.*, 253 (1966).
381. R. M. Sweet, C. J. Fritchie, and R. A. Schunn, *Inorg. Chem*, **6**, 749 (1967).
382. P. F. Lindley and P. Woodward, *J. Chem. Soc., A*, 383 (1967).
383. J. D. Cotton, S. A. R. Knox, I. Paul, and F. G. A. Stone, *J. Chem. Soc., A*, 264 (1967).
384. J. D. Cotton, S. A. R. Knox, and F. G. A. Stone, *Chem. Commun.*, 965 (1967).
385. E. H. Brooks, M. Elder, W. A. G. Graham, and D. Hall, *J. Amer. Chem. Soc.*, **90**, 3587 (1968).
386. F. Hein and W. Jehn, *Ann. Chem.*, **684**, 4 (1965).
387. W. Hieber and R. Breu, *Chem. Ber.*, **90**, 1270 (1957).
388. O. Kahn and M. Bigorgne, *Compt. Rend.*, **261C**, 2483 (1965).
389. O. Kahn and M. Bigorgne, *Compt. Rend.*, **292C**, 906 (1966).
390. D. J. Patmore and W. A. G. Graham, *Chem. Commun.*, 7 (1967).

Inorganic Derivatives of Germane and Digermane

CHARLES H. VAN DYKE

Department of Chemistry
Carnegie-Mellon University, Pittsburgh

I. INTRODUCTION

An aspect of preparative inorganic chemistry that has been of widespread interest for many years concerns the hydride derivatives of the lower group IVA elements.* Most of the work thus far has concerned the preparation and study of the silicon hydrides (silanes) and compounds derived from them, as described in several recent reviews of the subject by Stone,[1] Mackay,[2] MacDiarmid,[3,4] Ebsworth,[5] Aylett,[6] and Van Dyke.[7] Structural investigations in this field have been extremely important in the formulation and testing of theories concerning bonding in silicon compounds. Much of the fundamental structural information behind these bonding theories could not have been obtained readily from studies of analogous organosilicon compounds. Despite the obvious importance of extending the work to include the germanium series of hydrides and their derivatives in order to learn more about bonding in germanium compounds, investigations of the germanes and their derivatives have been, relatively speaking, rather sparse. The reason for this can be traced largely to the synthesis and stability problems that early investigators encountered with these compounds. Only recently have the more successful systematic preparative studies of the germanes and their derivatives been reported. Thermal-stability problems are often encountered with many of the compounds, although, as pointed out later, the problems are generally not insurmountable.

With convenient procedures at hand for the preparation of derivatives of the germanes, detailed structural studies of the compounds have now started

* For a general review of the hydrides of the lower group IVA elements the reader is referred to Ref. 1, literature up to about 1962, and Ref. 2, literature up to about 1966.

to appear in the literature. Important qualitative information about bonding in germanium compounds has been obtained from the studies; however, much more quantitative theoretical work needs to be done in this area before definite conclusions can be drawn.

A general review of the procedures used in the synthesis of inorganic derivatives of GeH_4 and Ge_2H_6 is presented in this chapter.* The coverage includes only those compounds which contain at least one Ge—H bond and is believed to be complete through May 1970. Derivatives that contain the GeH_3- and Ge_2H_5- groupings are generally referred to as *germyl* and *digermanyl* compounds, respectively, and can be considered the germanium analogs of methyl (CH_3-) and ethyl (C_2H_5-) or silyl (SiH_3-) and disilanyl (Si_2H_5-) compounds. An alternate nomenclature scheme for the derivatives uses the substituent as a prefix added to the base work "germane" or "digermane"; for example, GeH_3I is referred to as germyl iodide or iodogermane, and Ge_2H_5I as digermanyl iodide or iododigermane. The use of this alternate nomenclature scheme is preferred when the germane derivative contains more than one substituent. Thus the recommended name of GeH_2I_2 is diiodogermane.

Organogermanes and their derivatives have recently been reviewed by Hooton in Volume 4 of this series[10] and also by Glockling[8] and are not discussed in this chapter. The parent germanes, GeH_4 to Ge_9H_{20} and certain binary hydrides of germanium with silicon, tin, phosphorus, and arsenic likewise are not given much consideration, owing to their coverage by Jolly and Norman in Volume 4 of this series.[11]

In addition to having the reader become acquainted with the known derivatives of the germanium hydrides and methods for their synthesis, it is hoped that the material presented illustrates much of the current research

* Very little work has been done with derivatives of the higher germanes. Mackay and co-workers have shown that two highly unstable monosubstituted iodide derivatives of Ge_3H_8 are formed together with small amounts of polysubstituted products in the reaction of Ge_3H_8 with iodine at $-63°$.[9,9a] Although cleavage of Ge—Ge bonds was found to be negligible under these conditions attempts to isolate the iodides or convert them to the corresponding chlorides by their reaction with AgCl have been unsuccessful. Treatment of the unstable iodides with CH_3MgI at $-63°$ produces the more stable and volatile methyl derivatives of the corresponding iodides: $GeH_3GeH(CH_3)GeH_3$ (78.7%) and $CH_3GeH_2GeH_2GeH_3$ (12.5%), thus indicating that the iodine substitution is mainly on the central germanium atom under these conditions. This confirms the preliminary report made on the products of the Ge_3H_8-iodine reaction in which the unstable iodides were converted into what was thought to be mainly $(GeH_3)_2GeHD$ by their reaction with $LiAlD_4$.[9] A preliminary investigation of the reaction of iodine with n-Ge_4H_{10} at $-22°$ indicates that iodotetragermane is not produced in any significant amounts.[9b] The main reaction that does occur involves cleavage of Ge—Ge bonds, producing GeH_3I, Ge_3H_8, and several indirectly characterized iododi- and iodotri-germanes.[9b]

activity in this field, particularly the synthetic aspects. After some brief comments in Section II on the apparatus used in most of the experimental work with the compounds, a short review of the methods commonly used to prepare the parent germanes is presented in Section III. A very general account of the procedures used in the synthesis of inorganic derivatives of GeH_4 and Ge_2H_6 is then presented in Section IV. Preparative details for specific compounds are considered in Sections IV to XI. It is hoped that enough experimental information is given in these sections so that the reader can decide which procedure is most suitable to his preparative needs if the synthesis of a particular inorganic derivative is being contemplated.

No attempt is made in this chapter to cover all the physical properties known for inorganic derivatives of the germanes. In Section XII, page 219, selected physical properties of some of the characterized germanium hydrides and their derivatives are given. Key literature references to infrared and nuclear-magnetic-resonance (nmr) spectra and references to some recent structural parameters for the germanes and their derivatives also appear in this section. A detailed discussion of bonding in germanium compounds is not presented, although a brief section is included to illustrate the current activity in the field.

II. GENERAL EXPERIMENTAL CONSIDERATIONS

The use of high-vacuum systems is strongly recommended in all experimental work with the parent germanes and their inorganic derivatives, although the compounds generally are not spontaneously inflammable in the presence of air or oxygen under normal conditions.* The hydrolytic stability of the parent germanium hydrides is high; their inorganic derivatives, however, are much more susceptible to hydrolysis or especially decomposition in the presence of water. As a result, care should be taken to exclude moisture rigorously from any parts of the vacuum system in which these derivatives will be in contact. The apparatus and techniques of high-vacuum-system work, as well as the important advantages of this general experimental method in synthetic inorganic chemistry, have been adequately described elsewhere.[14–17]

Procedures for the synthesis of the germanes and their derivatives using vacuum-line techniques usually employ relatively small quantities of reactants. Extreme caution should be exercised if one finds it necessary to modify the vacuum-line procedures to larger-scale vacuum-line or bench experiments.

* Polymeric $(GeH)_x$, a yellow solid that is often formed as an undesirable side product in certain reactions, may decompose explosively into its elements on exposure to air.[12] The higher germanes are spontaneously inflammable in air and are presumed to be very toxic.[13]

In many instances the reactivity of germanium compounds with air or oxygen is unknown, especially at higher temperatures. In one instance known to the author, trace amounts of GeH_4 in a storage flask, presumed to be empty, ignited when glassblowing repairs were being made on the flask.

It is generally satisfactory to use Apiezon greases (N recommended) as a lubricant for the stopcocks and ground-glass joints of the vacuum line in work with germanium hydrides, although Apiezon M grease is reported to dissolve readily in chlorinated germanes.[18] Systems lubricated with KEL-F or Dow-Corning high-vacuum grease are also satisfactory, although the latter tends to become tough and rubbery after extended contact with chlorinated germanes, with striations occasionally appearing.[19] Teflon stopcocks and related joints can be used satisfactorily. Digermane is absorbed by stopcock grease and should be stored in an all-glass ampul or in a vessel equipped with a mercury-float valve.*[12] Mercury-free systems are not required in most synthetic work with the compounds, although the iodide derivatives should not be exposed for long periods to bulk quantities of mercury, such as found in manometers.

Gas-phase reactions of the germanes requiring static systems are usually carried out in round flasks of a suitable capacity. The vessels commonly have an attached finger tube into which the reactants can be condensed easily. Various gas-phase reactions have also been carried out in special flow systems, as described in this chapter. Liquid-phase reactions are usually performed in all-glass break-seal tubes or somewhat more conveniently in reactors that have teflon stopcocks. A particularly suitable vessel is described by Morrison and Hagen.[20] In certain instances reactions have also been carried out directly in nmr tubes, thus permitting one to follow conveniently the course of the reaction by nmr methods.

Reactants and products of the reactions encountered in this field can usually, but not always, be separated and purified by repeated fractional condensation and vaporization in evacuated U traps held at specific low temperatures on the vacuum line. In order to fractionate a mixture by this method, the boiling points of the components must be at least 20° apart. Several investigators have used gas-chromatographic techniques in purifying the germanium hydrides and their derivatives when the low-temperature fractionation techniques failed.

Derivatives of GeH_4 and Ge_2H_6 are normally prepared either by a direct reaction of the respective parent hydride with an appropriate reagent or by a particular conversion of one germyl or digermanyl compound to another. Other than the halogen (Cl, Br, I) and certain mixed hydride derivatives most germyl compounds are prepared from a derivative of germane or digermane

* See Ref. 14 for a description of a mercury-float valve.

(most frequently a halide) rather than from the parent hydride itself. As a result of the ease with which the germanes are halogenated, halogermanes are readily available for use as intermediates.

III. THE PARENT GERMANES

In order to synthesize a particular derivative of GeH_4 or Ge_2H_6, one almost invariably must first have the appropriate parent hydride available. Whether the parent hydride can be converted directly to the desired derivative or whether an intermediate must first be prepared depends, of course, on the particular compound sought. Germane is now commercially available,* although the ease with which it can be prepared hardly justifies the high cost involved in its purchase. At the present time there are no commercial sources of the higher germanes or any of the halogen intermediates (other than $HGeCl_3$) so often encountered in the synthesis of germyl and digermanyl compounds. This presents no problem, however, because the germanes and the halogen intermediates can be prepared from readily available reagents without much difficulty. A particularly convenient preparation of GeH_4 and Ge_2H_6 involves the hydroborate (BH_4^-) reduction of aqueous solutions of GeO_2.† The method described in detail by Jolly and Drake involves the addition of an alkaline solution of hydroborate and GeO_2 (as $HGeO_3^-$) to aqueous acetic acid.[12,21] Germane and Ge_2H_6 are obtained in 73 and 6.5 percent yields, respectively, when a BH_4^-/Ge(IV) ratio of 2.9 is used. The yield of Ge_2H_6 can be increased to 9.1 percent using by a BH_4^-/Ge(IV) ratio of 1.5. Several variations in experimental procedure for the hydroborate reduction have also been described. For example, Piper and Wilson describe the preparation of GeH_4 by the reduction of an aqueous acidic solution of GeO_2 with sodium borohydride at 0°.[22] Yields of 60 to 75 percent germane are reported; although, if the reduction is carried out at 35°, yields of 90 to 95 percent are obtained.[23] Procedures for preparing Ge_3H_8, as well as polymeric germanium subhydrides, GeH_x ($x \cong 1.0$), by this general method are also available.[12]‡

The classic procedure used to prepare GeH_4 and several higher members of the series involves the acidic solvolysis of Mg_2Ge in an aqueous or liquid ammonia medium. Dennis and coworkers used the aqueous procedure to prepare a mixture of GeH_4, Ge_2H_6, and Ge_3H_8 in about a 27 percent yield,[24] and increased yields ($\cong$70 percent) of the germanes, mostly GeH_4, are

* Available from Matheson Gas Products, East Rutherford, N.J. 07073.

† Available from Eagle-Pitcher Co., Philadelphia, Pa.

‡ The polymeric GeH_x subhydrides can also be prepared by the acidic solvolysis of various metal germanides.[23a]

reported in the ammonia solvent.[25] The acidic hydrolysis of Mg_2Ge has recently been used by Amberger to prepare Ge_4H_{10} and Ge_5H_{12}.[26] Gas-chromatographic investigations by Borer and Phillips have shown that at least seven volatile compounds, some of the higher hydrides tentatively identified, are produced in the acid hydrolysis of Mg_2Ge.[27] The solvolysis of Mg_2Ge with D_2O, in the presence of a deuterated acid, has been used to prepare GeD_4 and some of the higher deuterogermanes.[28,29] The best procedure for preparing the deuterogermanes by this method (^2H content $\geq$ 99 percent) seems to be the one in which D_3PO_4 is prepared in a suitable reaction vessel by the action of $POCl_3$ with D_2O. The DCl formed is pumped away, and Mg_2Ge is introduced into the vessel in the absence of air.[29] The mixture of deuterogermanes obtained yields GeD_4 (passes through a trap at $-120°$), Ge_2D_6 (just held in a trap at $-96°$), and Ge_3D_8 with traces of higher deuterogermanes (held at about -46 or $-63°$).[29]

Emeléus and Mackay also report that small quantities of Ge_2H_6 and Ge_3H_8 for experimental work can be prepared by the acidic hydrolysis of Mg_2Ge.[30] The mixture of germanes (27 percent net yield) obtained by adding Mg_2Ge* to 50 percent phosphoric acid contains GeH_4, Ge_2H_6, and Ge_3H_8 in the approximate mole ratio 9:4:1.[30] It is possible to purify the GeH_4, Ge_2H_6, and Ge_3H_8 formed by standard vacuum-line fractionations. (See the above fractionation temperatures for the deuterogermanes.) The solid-phase reaction of GeO_2 with a deficit of $LiAlH_4$ at 148 to 170° (no solvent) also produces a mixture of GeH_4, Ge_2H_6, and Ge_3H_8. Although the yields of hydrides produced are low—for example, GeH_4, 4.37 percent; Ge_2H_6, 2.72 percent; and Ge_3H_8, 0.6 percent—small amounts of the hydrides can be conveniently prepared relatively quickly by this procedure.[31]

Germane has also been prepared by the $LiAlH_4$[32,33] or $LiAl(t\text{-}C_4H_9O)_3H$[34] reduction of $GeCl_4$ in a nonaqueous medium. This general method is no longer of synthetic importance for GeH_4, although it is commonly used to prepare GeD_4; for example, GeD_4 can be prepared easily by the $LiAlD_4$ reduction of $GeCl_4$ in diethyl ether.†[35] The synthesis of Ge_2H_5D and Ge_2D_5H has been reported by the reaction of $LiAlD_4$ or $LiAlH_4$ on the corresponding monosubstituted chloro or iodo compound in 1,2-dimethoxyethane (monoglyme).[37] In a similar manner, GeH_3D can be prepared by a $LiAlH_4$ reduction of $GeDCl_3$, prepared from $GeCl_2$ and DCl.[28] This same deuterogermane can also be prepared by the $LiAlD_4$ reduction of GeH_3Br.[38] Sodium or potassium borohydride can also be used to reduce $GeCl_4$ to GeH_4.[39,39a]

* Prepared by heating stoichiometric amounts of the finely powdered elements in hydrogen at 750 to 800° for 24 hr.[30]

† Drake and Riddle report that in the reduction of $GeCl_4$ by $LiAlD_4$ in bis(2-methoxyethyl)ether(diglyme), extensive H–D exchange with the solvent protons occurs.[36]

The best method for preparing large quantities of the higher germanes is by the decomposition of GeH_4 in an ozonizer-type electric-discharge system.[13,40] Procedures have been described by which about 0.5 g of Ge_3H_8, about 0.1 g n-Ge_4H_{10}, about 0.02 g iso-Ge_4H_{10}, and smaller amounts of various isomers of the higher hydrides up to Ge_9H_{20} are obtained.[13,40] The normal and iso forms of both Ge_4H_{10} and Ge_5H_{12} have now been isolated from the reaction products and characterized by spectroscopic studies.[9b] A compound provisionally identified as neo-Ge_5H_{12} has also been isolated from the higher germanes produced in the GeH_4 discharge reaction.[9b] Perdeuterodigermane has been isolated from the decomposition of GeD_4 in a similar discharge system.[29] Gas-chromatographic techniques must be used in separating the germanes containing more than three germanium atoms.[13] Care must be taken in work with the higher germanes, because they are reported to be spontaneously inflammable in air and are possibly very toxic.[13]

IV. A SURVEY OF THE GENERAL METHODS USED IN THE SYNTHESIS OF INORGANIC DERIVATIVES OF GERMANE AND DIGERMANE

In this section a survey of general procedures used in the synthesis of inorganic derivatives of GeH_4 and Ge_2H_6 is presented. A more complete discussion concerning the details of these and other less common procedures actually used in the synthesis of specific derivatives is given in Sections V to XI.

A. By Reactions of the Germanium-Hydrogen Bond

1. With Halogens or Halides of Hydrogen, Silver or Boron

One of the most common procedures used to prepare halogermanes of the type GeH_3X, GeH_2X_2, and Ge_2H_5X, where X = Cl, Br, and I, is by the halogenation of the Ge—H bonds of the parent hydride with an appropriate halogenating agent. Of the elemental halogens, only bromine and iodine have been used successfully to halogenate the germanes partially; for example,

$$\rangle GeH + X_2 \longrightarrow \rangle GeX + HX \qquad X = Br, I$$

the reaction of bromine with GeH_4 at low temperatures produces GeH_3Br in up to 90 percent yields.[23] The reaction can also be used to prepare GeH_2Br_2 by using a larger mole ratio of bromine to GeH_4 in the reaction than that used to prepare GeH_3Br.[23,41] In order to prepare Ge_2H_5Br by this method, the bromine must first be diluted with nitrogen gas.[42]

Elemental iodine can be used to convert GeH_4 and Ge_2H_6 to GeH_3I and Ge_2H_5I, respectively.[43–45] If Ge_2H_5I is to be used as an intermediate in the synthesis of another digermanyl compound, it is usually prepared *in situ* at $-63°$, owing to its low thermal stability.

The gas-phase interaction of GeH_4 with anhydrous hydrogen chloride or hydrogen bromide in the presence of the corresponding aluminum halide catalyst can be used to prepare GeH_3Cl and GeH_3Br, respectively[46]:

$$\geq GeH + HX \xrightarrow{Al_2X_6} \geq GeX + H_2 \qquad X = Cl, Br$$

This procedure has not been used to prepare GeH_3I or any of the digermanyl halides.

A most convenient preparation of the monochloro and monobromo derivatives of GeH_4 and Ge_2H_6 involves passing the parent hydride over heated AgCl or AgBr[42]:

$$\geq GeH + AgX \longrightarrow \geq GeX + Ag + \tfrac{1}{2}H_2$$

This has been one of the chief methods used in the synthesis of Ge_2H_5Cl and Ge_2H_5Br.[42] It has also been used in the synthesis of GeH_2Cl_2 from GeH_4.[46a]

Barker and Drake have recently discovered that boron trichloride or boron tribromide can be used to halogenate GeH_4.[46b] For example, the interaction of GeH_4 with BBr_3 (3:1 molar ratio) produces equimolar amounts of GeH_3Br and GeH_2Br_2. Germyl chloride is produced in the analogous reaction of GeH_4 with BCl_3. The interaction of GeH_4 with a mixture of BBr_3 and BCl_3 (3:1:1 molar ratio) produces GeH_3Cl, GeH_3Br, GeH_2Cl_2, GeH_2Br_2 and GeH_2ClBr (in the ratio 1:10:1:15:4).

2. *With Transition-Metal Complexes*

One of the procedures used in the preparation of transition-metal complexes that contain germanium groupings attached directly to the metal involves the reaction of the Ge—H bonds of GeH_4 or one of its derivatives with certain transition-metal complexes. The starting metal complexes suitable thus far for this kind of reaction are those which contain either metal-hydrogen or metal-metal bonds. For example, the interaction of GeH_4 with $HMn(CO)_5$ or $Mn_2(CO)_{10}$ proceeds according to[47]

$$GeH_4 + HMn(CO)_5 \longrightarrow GeH_2[Mn(CO)_5]_2 + 2H_2$$

$$GeH_4 + Mn_2(CO)_{10} \longrightarrow 2GeH_2[Mn(CO)_5]_2 + H_2$$

The interaction of germyl halides with *trans*-X-Pt$[(C_2H_5)_3P]_2H$ in equimolar proportions likewise results in the cleavage of the Ge—H bond, producing the corresponding platinum-substituted germyl derivative[47a]:

$$\text{trans-X-Pt}[(C_2H_5)_3P]_2H + GeH_3X \longrightarrow \text{trans-X-Pt}[(C_2H_5)_3P]_2GeH_2X + H_2$$

$$X = Cl, Br, I$$

Germane reacts with the platinum hydride complexes in a similar manner, producing *trans*-X-Pt$[(C_2H_5)_3P]_2GeH_3$.[47a] The formation of six-co-ordinate platinum complexes has been noted in the reaction of excess GeH_3Cl with *trans*-X-Pt$[(C_2H_5)_3P]_2H$.[47b]

3. *With Alkali Metals*

The alkali-metal derivatives of GeH_4 of the type $MGeH_3$, where M = Li, Na, K, important synthetic intermediates in this field, are also prepared directly by the interaction of GeH_4 with an alkali metal in a suitable solvent, such as liquid ammonia,[25,48,49] organic ethers,[50,51] or hexamethylphosphortriamide[52]:

$$GeH_4 + M \longrightarrow MGeH_3 + \tfrac{1}{2}H_2 \qquad M = Li, Na, K$$

B. By Reactions of the Germanium-Halogen Bond

Most of the known inorganic derivatives of the germanes are prepared by an appropriate reaction of a halogermane intermediate. The kinds of reactions can be divided arbitrarily into the following general categories:

1. Reactions with silver, mercury, and lead salts
2. Exchange reactions involving hydrogen halides or silyl compounds
3. Reactions with organometallic reagents and other metal derivatives of the group IVA and transition elements
4. Reactions with $LiAl(PH_2)_4$ or $LiAl(AsH_2)_4$
5. Reactions with alkoxy, aryloxy, and the corresponding sulfur alkali-metal salts.

1. *With Silver, Mercury, or Lead Salts*

There have been numerous attempts to arrange a silver salt conversion series for germyl compounds analogous to the well-known conversion series for silyl compounds. Much more experimental work needs to be done in the germanium series, however, before definite conclusions can be drawn. A general lack of enthusiasm for setting up a silver salt conversion series can be traced to the fact that exchange reactions using derivatives other than silver

TABLE I
Reactions of Halogermanes with some Silver, Mercury, and Lead Salts

Starting halogermane	Conversion salt	Product	Reference
Ge_2H_5I	AgX (X = Cl, Br)	Ge_2H_5X (X = Cl, Br)	42
GeH_3Br	AgCl	GeH_3Cl	53
GeH_3Br	$HgCl_2$	GeH_3Cl	53, 54
GeH_3Br	AgF	GeH_3F, GeH_2F_2	55
GeH_3Cl	PbF_2	GeH_3F	56
Ge_2H_5I	PbF_2	Ge_2H_5F (?)	42
GeH_3Br (or GeH_3I)	AgCN	GeH_3CN	23, 57, 58
GeH_3Br	AgNCO	GeH_3NCO	23
GeH_3Br	AgNCS	GeH_3NCS	23
GeH_3Br	$AgOCOCH_3$	GeH_3OCOCH_3	23
GeH_3I[a]	HgO	$(GeH_3)_2O$	59
GeH_3I	HgS	$(GeH_3)_2S$[b]	59
GeH_2Br_2	PbF_2	GeH_2F_2	38
GeH_2Br_2	AgCl (or $HgCl_2$)	GeH_2Cl_2	54
GeH_3Cl	$Pb(OH)_2$	$(GeH_3)_2O$	59b

[a] For the synthesis of $(GeH_3)_2O$, it is better to react $(GeH_3)_2S$ with HgO.[59]

[b] The sulfide conversion can also be accomplished by treating GeH_3Br with lithium sulfide in dimethyl ether.[59a]

salts have been extremely successful for converting halogermanes to other derivatives. Some of these recently discovered exchange reactions using various covalent compounds are considered in Section IV-B-2.

Silver, mercury, and lead salts have been used to interconvert one halogermane to another or to convert a particular halogermane to another kind of germyl compound, such as a halogenoid or an acetate derivative. These reactions strongly resemble analogous conversions found in the chemistry of silyl compounds.[1–7] Some general examples of the conversions which use silver, mercury, and lead salts are summarized in Table I.

2. *With Hydrogen Halides or Silyl Compounds*

It has recently been discovered that certain germyl halides undergo a mutual exchange reaction with certain silyl derivatives producing the analogous germyl derivative and the silyl halide. The reactions have been particularly useful for preparing a variety of new germyl derivatives that would be difficult to prepare by other methods*:

$$GeH_3X + SiH_3Y \rightleftharpoons GeH_3Y + SiH_3X$$

X = halogen, Y = other atoms or groups (see Table II)

* For several of the halogen exchange reactions, Ebsworth and Cradock have used nmr methods to study the equilibrium positions and their variation with temperature.[60]

Certain exchange reactions involving GeH_3Cl or GeH_2Cl_2 with hydrogen halides likewise have been used for preparing GeH_3Br, GeH_3I, or GeH_2I_2[53]:

$$GeH_3X + HY \rightleftharpoons GeH_3Y + HX \qquad X = Cl;\ Y = Br, I$$

The results of most of the exchange processes of this kind used from a synthetic point of view are summarized in Table II.

The equilibrium constants for the exchanges seem to depend largely on the relative bond energies concerned. Thus, whatever the reason may be, it is clear that the Si—F bond is relatively much stronger than the Ge—F bond, whereas the strengths of the Si—I and Ge—I bonds are much closer. This has been generalized to argue that bond energies then favor the attachment of light and electronegative elements to silicon. Therefore, if one wants to synthesize a germyl compound with a Ge—N bond in it from, for example, a silicon analog, he must start with a germyl compound with germanium bonded to an element that is lighter (or at least not much heavier) and more

TABLE II

Exchange Reactions Involving Germyl Halides with Hydrogen Halides or Silyl Compounds

Starting halogermane	Hydrogen halide or silyl compound	Germyl compound Produced	Yield	Reference
GeH_3Cl	HBr	GeH_3Br	100	53
GeH_3Cl	HI	GeH_3I	88	53
GeH_3Br	HI	GeH_3I	70	53
GeH_2Cl_2	HI	GeH_2I_2	...	53
GeH_2Cl_2	HBr	GeH_2Br_2	...	46b
		GeH_2ClBr	...	
GeH_2Cl_2	HI	GeH_3I	...	46b
		GeH_2ClI	...	
		$GeHCl_2I$	...	
GeH_2Br_2	HI	GeH_2I_2	...	46b
		GeH_2BrI	...	
GeH_3F	$(CH_3)_3SiN_3$	GeH_3N_3	95	61
GeH_3Br	$(SiH_3)_2Se$	$(GeH_3)_2Se$	89	62
GeH_3Br	$(SiH_3)_2Te$	$(GeH_3)_2Te$	~100	62
GeH_3Br	$(SiH_3)_3P$	$(GeH_3)_3P$	~100	63, 64
GeH_3Br	$(SiH_3)_3As$	$(GeH_3)_3As$	...	65
GeH_3Br	$(SiH_3)_3Sb$	$(GeH_3)_3Sb$	~100	65
GeH_3F	SiH_3CN	GeH_3CN	100	66
GeH_3X (X = Cl, Br)	SiH_3PH_2	GeH_3PH_2	96	36, 67
GeH_3X (X = Cl, Br)	SiH_3AsH_2	GeH_3AsH_2	82	67
GeH_3F	$(CH_3)_3SiNCNSi(CH_3)_3$	$GeH_3NCNGeH_3$	95	68
GeH_2X_2(X= Cl,Br)	SiH_3PH_2	GeH_2XPH_2	...	69
GeH_3F	SiH_3NCO	GeH_3NCO	90	60
GeH_3F	SiH_3NCS	GeH_3NCS	90	60

electronegative than nitrogen. In this case GeH_3F would be appropriate. For preparing germyl derivatives of the second and lower-row elements by this method one can usually get away with using GeH_3Cl or GeH_3Br in the exchange (see Table II).

3. *With Organometallic Compounds and Other Metal Derivatives of the Group IVA and Transition Elements*

The synthesis of a number of important germyl compounds has been achieved by the interaction of various halogermanes with organometallic reagents and related kinds of metal derivatives of the group IV and transition elements. The method has been important in the preparation of several organogermanes not specifically covered in this review. Some examples are given below:

$$Ge_2H_5I + RMgBr \longrightarrow RGe_2H_5 + MgBrI \qquad R = CH_3, C_2H_5 \qquad \text{(Ref. 70)}$$

$$Ge_2H_6 + I_2 \longrightarrow \text{iododigermanes} \xrightarrow{CH_3MgI}$$

$$(CH_3)_2GeHGeH_3, CH_3GeH_2GeH_2CH_3, (CH_3)_2GeHGeH_2CH_3 \qquad \text{(Ref. 71)}$$

$$(CH_3)_3GeF + NaGeH_3 \longrightarrow (CH_3)_3GeGeH_3 + NaF \qquad \text{(Ref. 72)}$$

$$GeH_3Cl + KSiH_3 \longrightarrow GeH_3SiH_3 + KCl \qquad \text{(Ref. 73)}$$

$$GeH_3Br + NaMn(CO)_5 \longrightarrow GeH_3Mn(CO)_5 + NaBr \qquad \text{(Ref. 74)}$$

$$GeH_3Br + NaCo(CO)_4 \longrightarrow GeH_3Co(CO)_4 + NaBr \qquad \text{(Ref. 75)}$$

$$GeH_3X + NaRe(CO)_5 \longrightarrow GeH_3Re(CO)_5 + NaX \qquad \text{(Ref. 75a)}$$

$$Ge_2H_5I + NaMn(CO)_5 \longrightarrow Ge_2H_5Mn(CO)_5 + NaI \qquad \text{(Ref. 75b)}$$

4. *With* $LiAl(PH_2)_4$ *and* $LiAl(AsH_2)_4$

The synthesis of germyl phosphines and arsines of the type GeH_3EH_2, where E = P, As, has been investigated by a variety of methods. A particularly convenient method for their preparation is based on the reaction of germyl halides with $LiAl(PH_2)_4$ or $LiAl(AsH_2)_4$[67,76,77]:

$$4GeH_3Br + LiAl(EH_2)_4 \longrightarrow 4GeH_3EH_2 + LiBr + AlBr_3 \qquad E = P, As$$

5. *With Alkoxy, Aryloxy and the Corresponding Sulfur Alkali-metal Salts*

Mixed carbon-germanium ethers and sulfides have been conveniently prepared by the interaction of germyl halides with appropriate metal salts of the group VIA elements. The equations given below illustrate the general method:

$$GeH_3Cl + CH_3ONa \longrightarrow GeH_3OCH_3 + NaCl \qquad \text{(Ref. 78)}$$

$$GeH_3Cl + C_6H_5OK \longrightarrow GeH_3OC_6H_5 + KCl \qquad \text{(Ref. 79)}$$

$$GeH_3Cl \text{ (or I)} + CH_3SNa \longrightarrow GeH_3SCH_3 + NaCl \text{ (or I)} \qquad \text{(Refs. 80, 81)}$$

$$GeH_3Br + C_6H_5SK \longrightarrow GeH_3SC_6H_5 + KBr \qquad \text{(Ref. 79)}$$

C. By Hydrolysis, Alcoholysis, and Ammonolysis Reactions

The synthesis of germyl compounds by hydrolysis, alcoholysis, and ammonolysis reactions has been particularly important for the synthesis of $(GeH_3)_2O$, GeH_3OCH_3, and $(GeH_3)_3N$, respectively. No procedures have been reported by which $(GeH_3)_2O$ or GeH_3OCH_3 can be obtained by the direct hydrolysis or methanolysis of a germyl halide. Instead, in order to prepare the compounds by this method, one must use digermylcarbodiimide in the hydrolysis [for $(GeH_3)_2O$] or methanolysis (for GeH_3OCH_3) reaction.[68] This is probably the best general method for preparing the two germyl ethers.

Water has also been used in the procedure for preparing $(GeH_3)_2AsH$. This arsine derivative is prepared by the controlled condensation of GeH_3AsH_2, in the presence of water[81a]:

$$2GeH_3AsH_2 \xrightarrow{H_2O} (GeH_3)_2AsH + AsH_3$$

It has not been determined just how the water functions as a catalyst in the condensation.

The particularly elusive compound trigermylamine $(GeH_3)_3N$ has finally been prepared by the direct reaction of GeH_3Cl with ammonia[56,82]:

$$3GeH_3Cl + 4NH_3 \longrightarrow (GeH_3)_3N + 3NH_4Cl$$

Germyl fluoride does not undergo an analogous ammonolysis with ammonia but rather produces a solid material, which at one time was thought to be the ionic salt $[GeH_3NH_3]^+F^-$ on the basis of infrared measurements.[55] More recent work indicates that the reaction actually produces NH_4F and $(GeH_2)_x$, the latter undergoing further decomposition producing GeH_4 and germanium subhydrides.[82a]

D. By Reactions of Alkali-metal Derivatives of Germane

Although alkali-metal derivatives of GeH_4 of the type $MGeH_3$, where M = alkali metal, have been known and characterized for many years, their true value as synthetic intermediates has been realized only very recently. One of the chief methods for preparing organogermanes that contain GeH_3 groups involves the reactions of $MGeH_3$ derivatives with an appropriate organic halide, for example,

$$MGeH_3 + RX \longrightarrow RGeH_3 + MX \qquad \text{(Refs. 48, 52, 78, 83)}$$

$$M = \text{Na or K};\ R = CH_3,\ C_2H_5,\ n\text{-}C_3H_7,\ CH_3OCH_2,\ SiH_3CH_2,\ \text{etc.}$$

Reactions that involve the formation of organogermanes are not considered in this review; the reaction has been used in the synthesis of a number of inorganic derivatives, however, including some novel compounds, such as $K(GeH_3BH_3)$, $K(GeH_3As(CH_3)_2Cl)$, and $GeH_3As(CH_3)_2$:

$$NaGeH_3 + SiH_3Br \longrightarrow GeH_3SiH_3 + NaBr \quad \text{(Ref. 72)}$$

$$NaGeH_3 + (CH_3)_3SiCl \longrightarrow GeH_3Si(CH_3)_3 + NaCl \quad \text{(Ref. 72)}$$

$$NaGeH_3 + (CH_3)_3GeF \longrightarrow GeH_3Ge(CH_3)_3 + NaF \quad \text{(Ref. 72)}$$

$$KGeH_3 + \tfrac{1}{2}B_2H_6 \longrightarrow K(GeH_3BH_3) \quad \text{(Ref. 51)}$$

$$KGeH_3 + (CH_3)_2AsCl \longrightarrow K(GeH_3As(CH_3)_2Cl) + GeH_3As(CH_3)_2 + KCl \quad \text{(Ref. 84)}$$

E. By Reactions in Electrical-discharge and Photolysis Systems

Chemical reactions induced by ozonizer electrical-discharge methods have played an important role in the synthetic chemistry of various *mixed* hydrides of germanium.* The method has been used chiefly to prepare mixed hydrides containing the group VA elements—for example, GeH_3PH_2, GeH_3AsH_2[86]—and also mixed silicon-germanium hydrides of the type $Si_xGe_yH_{2x+2y+2}$.[87,88] It has also been applied toward the synthesis of reasonably large quantities of the higher germanes.[13,40] Other, more convenient procedures are now available for the preparation of the mixed hydrides containing the group VA elements; the discharge method, however, is still recommended for the synthesis of mixed silicon-germanium hydrides and the higher germanes.

More recently, the method has been used in the synthesis of various mixed hydrides containing the group VIA elements.[88a] For example, the formation of GeH_3SSiH_3 or GeH_3SeSiH_3 has been noted in the discharges of ternary mixtures composed of GeH_4, SiH_4 and either H_2S or H_2Se. These mixed hydrides have not yet been prepared by other procedures. The discharge of binary systems composed of GeH_4 and either H_2S, H_2Se, or CH_3SH produces hydrides of the type GeH_3EH, $(GeH_3)_2E$ (E=S, Se) or GeH_3SCH_3 as well as higher germanes.

Preparation of germyl compounds based on photolysis techniques has not been pursued much, although the method has shown promise of being extremely useful. The chief drawback of this method is the low yields of conversion, based in part on the formation of nontransparent films of polymeric germanium subhydrides on the surface of the reaction vessels. Development of a flow-system reactor would be most helpful in making this method a more general synthetic technique.

The photolysis of a mixture of CH_3OH and GeH_4 containing a mercury sensitizer with 2537 Å (Hg source) light was used in the first synthesis of GeH_3OCH_3.[89] The analogous mercury-sensitized photolysis of GeH_4

* A detailed discussion of the apparatus used and the theory involved in these syntheses is given in Refs. 11 and 85.

produces Ge_2H_6 and higher germanes,[90] and photolysis of SiH_4—GeH_4 and GeH_4—NO mixtures produces small amounts of GeH_3SiH_3 and $(GeH_3)_2O$, respectively.[89,91] Although all these compounds have now been synthesized by more conventional procedures, the method is still of importance, because it could be used conveniently to synthesize small amounts of a new compound that would be difficult to prepare by other methods. The general stability of the compound could be determined easily, along with its partial characterization by various spectral methods. This preliminary information about the compound would undoubtedly aid in the search for a more conventional synthesis if larger quantities of material were needed.

V. PREPARATION OF THE HALOGERMANES

A. Preparation of the Monohalides GeH_3X* and Ge_2H_5X

1. Direct Halogenation of Germane *or* Digermane (*applicable to* GeH_3Br, GeH_3I, Ge_2H_5Br, and Ge_2H_5I)

The preparation of the monobromo and monoiodo derivatives of GeH_4 and Ge_2H_6 has been satisfactorily achieved by the direct interaction of the parent hydride with the appropriate free halogen:

$$GeH_4 + X_2 \longrightarrow GeH_3X + HX \qquad X = Br, I$$

$$Ge_2H_6 + X_2 \longrightarrow Ge_2H_5X + HX \qquad X = Br, I$$

a. Germyl bromide GeH_3Br. A typical synthesis of GeH_3Br by this method, as reported by Onyszchuk and coworkers, involves the successive condensation of small amounts of bromine (7.9 mmoles) onto a slight excess of GeH_4 (8.1 mmoles) cooled to $-196°$.[23,92] It is best to carry out the reaction in a round bulb rather than a tubelike vessel in order to keep the surface area of the solid GeH_4 large. The mixture is warmed slowly from $-196°$ after each addition until the reddish color of bromine just disappears. The mixture is

* *Note added in proof:* Anderson and Drake have recently reported a particularly convenient synthesis of the important intermediate GeH_3Cl based on the reaction of GeH_4 with thionyl chloride.

$$2GeH_4 + 2SOCl_2 \longrightarrow 2GeH_3Cl + SO_2 + S + 2HCl$$

In a typical synthesis, thionyl chloride (4.0 mmole) and GeH_4 (4.0 mmole) are condensed into a thick-walled glass tube which is then sealed. The reaction is allowed to proceed for 22 hr. at room temperature. After opening the tube on the vacuum-line, GeH_3Cl can be recovered in yields exceeding 80% by its condensation in a trap at $-95°$. It is important to note that no GeH_2Cl_2, which is difficult to separate from GeH_3Cl, is formed in this reaction when the above procedure is followed. [J. W. Anderson and J. E. Drake, Synthesis in Inorganic and Metalorganic Chemistry; 1971 (in press)].

then cooled again to −196°, another portion of bromine is added, and the procedure is repeated. The pure GeH_3Br can be separated from the product mixture by its condensation in a −96° trap. Yields of up to 91 percent are reported.[23,92]

b. Germyl iodide GeH_3I. Several variations in experimental procedure have been described for the synthesis of GeH_3I by the interaction of GeH_4 with iodine. Van Dyke and Wang have prepared the compound by a simple 12-hr interaction of gaseous GeH_4 (in excess) with iodine (partly in the vapor phase) in a 1-liter Pyrex bulb.[44] Only moderate yields of GeH_3I are obtained by this procedure, and a considerable quantity of a nonvolatile yellow-orange oil and a solid material are formed. Sujishi and Ando have described various other experimental procedures for the synthesis.[45] For example, excess GeH_4 (2.3 mmoles) undergoes reaction with iodine (0.8 mmoles) after about 30 min in a sealed tube (≅1 atm pressure) at room temperature. The conversion of GeH_4 to GeH_3I by this procedure is reported to be 80 percent. An alternate procedure is to carry out the reaction in a hot-cold tube apparatus. The apparatus is designed so that the reaction zone containing sublimed iodine can be kept at room temperature while another part of the apparatus can be cooled to −78°. One of the vessels described consists of a glass tube (35 by 5 cm) on which a short (1.5 cm) side arm is attached. A ground-glass joint assembly is attached to the open end of the larger tube, so that the entire apparatus can be evacuated on the vacuum line. The procedure calls for introducing solid iodine into the small side arm followed by evacuating the entire vessel and subliming the iodine to the upper third of the larger tube. Germane is introduced into the assembly, and a −78° trap is placed around the lower part of the large tube. The reaction is allowed to proceed for about 20 hr, after which the conversion of GeH_4 to GeH_3I is about 93 percent. The rate of the iodination in the hot-cold tube assembly can be increased significantly by condensing the GeH_4 onto the iodine as a liquid and maintaining the iodine in a solution of a thin film of liquid product, presumably GeH_3I. Under these conditions the iodine is completely consumed in 1 to 3 hr, and the conversion of GeH_4 to GeH_3I is about 82 percent.[45]

c. Digermanyl bromide Ge_2H_5Br. In order to prepare Ge_2H_5Br from the interaction of Ge_2H_6 with bromine, Mackay, MacDiarmid, and coworkers report that the reaction must be carried out in the gas phase and the bromine vapor must be diluted with nitrogen.[42] Thus, on allowing bromine vapor (0.264 mmole) diluted to about 1 atm with nitrogen in a 500-ml flask to mix slowly with Ge_2H_6 (0.330 mmole) in a 300-ml flask, the characteristic color of the halogen is discharged instantly. Crude Ge_2H_5Br (0.625 mmole) obtained in the reaction condenses in a trap at −78° and was shown to contain traces of GeH_2Br_2. Traces of Ge_2H_5Br are obtained by the direct reaction of Ge_2H_6 with undiluted bromine vapor at low temperatures (−78°).

d. Digermanyl iodide Ge_2H_5I. The recommended procedure for preparing Ge_2H_5I involves the reaction of Ge_2H_6 with iodine at $-63°$, as described by Mackay and Roebuck.[43] The reaction proceeds smoothly under these conditions without any complicating factors, such as the halogen cleavage of the germanium-germanium bond. The yield of Ge_2H_5I based on the Ge_2H_6 consumed is 93 percent. The preparation cited calls for equivalent mole ratios of reactants—for example, 1.0 mmole—with a 1-hr contact time. Longer contact times or higher temperatures give lower yields of product.

Digermanyl iodide is an important intermediate in the synthesis of other compounds that contain the Ge_2H_5 grouping. The iodide decomposes rapidly on distillation in the vacuum-line and thus in all cases where it is used as an intermediate the compound is prepared by an *in situ* method. At 0°, Ge_2H_5I can be distilled with about 20% loss by decomposition.

2. *Halogenation of Germane by Trifluoromethyl Iodide*

Griffiths and Beach have reported a convenient preparation of GeH_3I from GeH_4 that takes advantage of the pseudohalogen character of CF_3I[93]:

$$GeH_4 + CF_3I \xrightarrow{25-235°} GeH_3I + CF_3H$$

Deuterogermane undergoes a reaction with CF_3I in an identical manner.[93] The reactions have been carried out either in sealed tubes or in flow systems. In the sealed-tube reactions the conversion at room temperature is very slow, taking about 2 months to form appreciable quantities of GeH_3I. At higher temperatures ($\simeq 130°$) the reaction proceeds much more readily, producing GeH_3I in good yields in a matter of hours. At higher temperatures ($>200°$) the reaction is very rapid; copious amounts of yellow-orange solids are produced, however, as well as the expected product. In a typical experiment that involves the reaction of GeD_4 with CF_3I at 135 to 140°, the yield of GeD_3I based on the amount of GeD_4 actually consumed in the reaction is 85 percent. Some breakdown of the germyl iodide inevitably occurs under the high-temperature conditions, producing, presumably, polyiodinated germanes but, surprisingly, no hydrogen iodide. A flow system in which the reactant gases are passed at pressures of 50 or 100 torr through a 1-liter glass bulb heated with a heating mantle (195 to 260°C) has been investigated and seems to offer some advantages over the static hot-tube method. The yield of GeH_3I produced by this method varies between 15 and 37 percent, the lower yields occurring in experiments with low flow rates.

3. *Halogenation of Germane by Hydrogen Halides (applicable to* GeH_3Cl *and* GeH_3Br*)*

Germyl chloride and bromide can be prepared conveniently by the gas-phase interaction of GeH_4 with anhydrous hydrogen chloride and hydrogen bromide, respectively, in the presence of the corresponding aluminum halide catalyst[46]:

$$GeH_4 + HX \xrightarrow{Al_2X_6} GeH_3X + H_2 \qquad X = Cl, Br$$

Both reactions are carried out at room temperature, and if higher temperatures are attempted, considerable decomposition occurs. The exact nature of these decompositions is not known; they do produce oily substances or solids, however, that deactivate the catalyst.[94] Germyl chloride is reported to undergo slow decomposition at ambient temperatures according to[46]

$$2GeH_3Cl \longrightarrow GeH_4 + 2HCl + Ge$$

Typical procedures call for combining the reactants in a relatively large ($\simeq$5-liter) glass bulb whose inside surface is lightly coated with the appropriate freshly sublimed aluminum halide catalyst. For the hydrogen chloride reaction, Dennis and Judy[46] and Steese[19] report that a 30-min contact time is sufficient in order to obtain reasonable yields of GeH_3Cl ($\simeq$48%). Longer reaction times, however, on the order of 12 hr, are generally recommended, especially as the catalyst becomes deactivated.[94] Yields of GeH_3Cl up to 80 percent based on the amount of GeH_4 consumed have been reported.[28] The catalyzed reaction of hydrogen bromide with GeH_4 in the presence of aluminum bromide is considerably more vigorous than the analogous hydrogen chloride reaction.[46]

There are numerous literature citations of the use of this method in the preparation of GeH_3Cl, an important intermediate in germyl chemistry. Citations also exist for preparing GeH_3Br by this method, although the fractionation of the bromogermanes formed is reported to be difficult as a result of the ease with which the compounds decompose, especially in the presence of the aluminum bromide catalyst.[46] Also the yield of GeH_3Br produced by this method is seldom greater than 10 percent.[23] In order to favor the formation of the monohalogenated product in each of these halogenation procedures, it is recommended that excess GeH_4 be used in the reaction.[94]

The iodination of GeH_4 using hydrogen iodide in the presence of aluminum bromide has been attempted, but the reaction leads to the formation of mixed halogermanes that cannot be separated by standard techniques.[46] The reaction may be more successful if an aluminum iodide catalyst is used.

Digermanyl halides have not been prepared by this method, although attempts have been made.[42]

4. *Halogenation of Germane and Digermane by Silver Halides* (*applicable to* GeH_3Cl, GeH_3Br, Ge_2H_5Cl, *and* Ge_2H_5Br)

Mackay, MacDiarmid, and coworkers have reported a relatively new method for preparing halogermanes that involves streaming GeH_4 or Ge_2H_6

over heated AgCl or AgBr.[42] The basic reactions can be represented by the following equations:

$$GeH_4 + AgX \longrightarrow GeH_3X + Ag + \tfrac{1}{2}H_2 \qquad X = Cl, Br$$

$$Ge_2H_6 + AgX \longrightarrow Ge_2H_5X + Ag + \tfrac{1}{2}H_2 \qquad X = Cl, Br$$

A suitable reaction vessel is a 40-cm heavy-walled Pyrex tube (OD 19 mm) with appropriate ground-glass joints at one end for connection to the vacuum line. A thermometer, inserted into a small glass tube, is placed next to the reaction vessel, and heating tape is wrapped about the entire assembly. The tube is partly filled with the appropriate silver halide (loosely packed with glass wool) near the center and under the heating tape, and evacuated on the vacuum line. Germane or Ge_2H_6 is condensed into the bottom of the tube, and the silver halide–glass-wool mixture is then heated to the appropriate temperature. The germanium hydride is pumped over the heated silver halide and condensed in at least two traps maintained at −196°. The heating is then discontinued, and the volatile materials passed through a cold trap to remove the monohalide formed. The unreacted hydride is recondensed into the reactor, and the procedure repeated. The heating need not be discontinued if the reaction tube contains a separate bypass return tube.[94] Optimum reaction temperatures and cold-bath temperatures that effectively remove the monohalide derivatives from the system are given in Table III. The halogermanes produced by this method are relatively easy to purify by vacuum fractionation techniques, except for the difficult separation of Ge_2H_5Br from GeH_2Br_2.

TABLE III

Some Details on the Synthesis of Germyl and Digermyl Halides by the Interaction of the Parent Hydrides with a Silver Halide

Halogermane	Yield (%)	Optimum reaction temperature (°C)	Recommended bath temperature to remove the halogermane (°C)
GeH_3Cl	71.2	260	−96
GeH_3Br[a]	30.3	255	−78
Ge_2H_5Cl[b]	25.8	90	−46
Ge_2H_5Br[b,c,d]	40	185	−64

[a] Small amounts of GeH_2Br_2 are also formed.

[b] The Ge_2H_6 must be diluted with about an equal quantity of n-C_5H_{12}.

[c] It is desirable to add a fresh charge of AgBr to the tube if the extent of reaction decreases after several passes.

[d] Hydrogen bromide is also formed. In this reaction the products were passed through a trap at −64° to remove crude Ge_2H_5Br and through a trap at −134° to remove the hydrogen bromide. The unreacted Ge_2H_6 remains in the trap at −134° and is recycled.

5. *Halogen Exchange Reactions*

Certain of the germyl and digermyl halides, notably GeH_3Cl (or Br) and Ge_2H_5I, can be synthesized more conveniently from GeH_4 or Ge_2H_6 than others can. As a result, procedures have been investigated by which one of the easily prepared halogermanes can be converted to another by a simple exchange reaction. Reagents effecting the exchange are either the silver halides (or less frequently mercury(II) halides), hydrogen halides, or in the case of fluorogermanes, silver or lead fluoride.

a. With silver or mercury(II) halides. The use of silver salts of halogen acids for interconverting halogermanes is not commonly employed for the preparation of the germyl halides, but the method has been very important for converting the readily available Ge_2H_5I (see Section V-A-1-d) to Ge_2H_5Br or Ge_2H_5Cl[42]:

$$Ge_2H_5I + AgX \longrightarrow Ge_2H_5X + AgI \qquad X = Cl, Br$$

The Ge_2H_5I used in the reaction need not be isolated but can be prepared *in situ* from Ge_2H_6 and I_2.[43] In the synthesis of Ge_2H_5Cl or Ge_2H_5Br by this method, as described by Mackay, MacDiarmid, and coworkers,[42] Ge_2H_6 and iodine are allowed to interact at $-64°$, and an excess quantity of the silver halide (10 g) is added *in vacuo* after the hydrogen iodide is removed. The mixture is warmed to 0° and allowed to pass through a 10-cm glass column that contains more of the silver halide (15 g). The yield of Ge_2H_5Cl isolated from this reaction varies considerably, but values up to 65.3 percent have been obtained. The yield of pure Ge_2H_5Br isolated from the reaction is reported to be 13.7 percent. In some of the experiments, higher yields of Ge_2H_5Br were obtained, but in these systems the product contained GeH_2Br_2 which is difficult to remove, although it can be done, from Ge_2H_5Br.

Little work has been reported on halide-halide interconversions using mercury(II) halides. Cradock and Ebsworth have shown that GeH_3Cl can be prepared by passing GeH_3Br vapor over mercury(II) chloride.[53,60] This bromide-chloride conversion can also be made by using AgCl.[53]

b. With hydrogen halides (applicable to GeH_3Br *and* GeH_3I*).* Cradock and Ebsworth have reported a convenient and efficient method for converting the readily accessible GeH_3Cl to GeH_3Br or GeH_3I based on the rapid exchange reaction that GeH_3Cl undergoes with either hydrogen bromide or hydrogen iodide.[53] The equilibrium positions for the reactions and their variation with temperature have also been reported.[60] In a similar exchange, GeH_3Br can be converted to GeH_3I in about 70 percent yield by its reaction with hydrogen iodide.[53]

Details for the preparation of GeH_3I illustrate the ease with which the

eaction can be performed.[53] A mixture of GeH_3Cl (1.30 mmoles) and hydro-
en iodide (1.45 mmoles) condensed in a trap (50-ml vol) is allowed to warm
o room temperature for 1 min. Fractional distillation of the products gives
ure GeH_3I (1.14 mmoles) in a trap at $-78°$.

c. With other covalent halides. The conversion of one germyl halide to
nother can be accomplished by using various other covalent halides,
lthough these reactions do not have preparative significance at present.
ermyl chloride is converted to GeH_3Br by its reaction with BBr_3, at room
emperature; the complex mixture of volatile products, however, consisting
ainly of GeH_3Br, BCl_3, GeH_4, B_2H_6, and a little GeH_2Br_2, could not be
eparated satisfactorily into its components by standard fractionation
echniques.[60] The interaction of GeH_3I with PCl_3 produces a mixture of
roducts containing both GeH_3Cl and GeH_3I in substantial amounts, but the
onversion is of no importance synthetically. An attempt to convert GeH_3Cl
o GeH_3F by using PF_3 at room temperature was unsuccessful.[60]

The formation of the mixed dihalide GeH_2BrI has been detected by
radock and Ebsworth in the ^1H nmr spectrum of an equimolar mixture of
eH$_2$I$_2$ and GeH_2Br_2[60]:

$$GeH_2I_2 + GeH_2Br_2 \rightleftharpoons 2GeH_2BrI$$

he solution shows three resonance signals, two of which are readily
ssignable to GeH_2Br_2 and GeH_2I_2. The third peak in the spectrum has been
ssigned to GeH_2BrI formed in an equilibrium with the two halides by
xchange.[60]

d. Conversion of halogermanes to fluorogermanes. There are at present no
irect methods of converting parent germanes to fluorogermanes. The only
osely related direct conversion of a hydride to a fluoro compound is the
uorination of $C_2H_5GeH_3$ by PF_5.[95]

The recommended procedure for preparing GeH_3F involves passing the
apor of GeH_3Br slowly through a column or trap loosely packed with AgF
r PbF_2 on glass wool or glass helices. The conversion has been carried out
25°,[55] 0°,[28] and what seems to be the most desirable temperature, $-22.9°$.[23]
ermyl fluoride disproportionates appreciably at room temperature (about
5 percent after 16 hr at 25°)[55]:

$$2GeH_3F \longrightarrow GeH_4 + GeH_2F_2$$

nd at the lower temperature the rate of disproportionation is lowered. The
onversion itself is strongly exothermic, and the yield of GeH_3F is sensitive
the purity of the AgF.*[28] The impurities normally found with the crude
eH$_3$F include GeH_4, SiF_4, GeH_2F_2 and unreacted GeH_3Br.

* Synthesis of AgF is described in Ref. 96.

The vapor-phase conversion of either GeH_3Br[60] or GeH_3Cl[56] to GeH_3F by using freshly precipitated PbF_2* has also been reported. Experimental details are the same as those described above.

Attempts have been made by Mackay, MacDiarmid, and coworkers to prepare Ge_2H_5F by the interaction of Ge_2H_5I with PbF_2 at room temperature.[42] The tentative identification of Ge_2H_5F in the reaction products was made by infrared methods, but the pure compound has not been isolated.

B. Preparation of Dihalogermanes GeH_2X_2

Dihalogermanes are prepared by extensions of several of the reactions discussed in Section V-A. No disubstituted halogen derivatives of Ge_2H_6 have been reported, although there is indirect evidence of the formation of polyiododigermanes in the reaction of Ge_2H_6 with iodine.[71]

1. Direct Halogenation of Germane (*applicable to* GeH_2Br_2)

Dibromogermane is formed, together with GeH_3Br, in the low-temperature reaction of GeH_4 with bromine.[23] The yield of GeH_2Br_2 can be increased by using larger quantities of bromine in the reaction than what are called for in the GeH_3Br synthesis.[41]

Two reports cite evidence that GeH_2I_2 cannot be prepared from GeH_4 and iodine.[97,98] The suggestion that GeH_2I_2 is too unstable at room temperature for isolation has been disproved, because the compound has now been prepared by an alternate reaction (see Section V-B-3).

2. Halogenation of Germane by Using Hydrogen Halides (*applicable to* GeH_2Cl_2 *and* GeH_2Br_2)

Variable amounts of GeH_2Cl_2 and GeH_2Br_2 are formed along with the monohalogermane in the reaction of GeH_4 with hydrogen chloride or hydrogen bromide in the presence of the appropriate aluminum halide catalyst.[46] Dennis and Judy report that from a typical GeH_4–hydrogen chloride reaction, GeH_3Cl and GeH_2Cl_2 are formed in 48 and 14 percent yields, respectively.[46] The yield of dihalogermane can be increased somewhat by using an excess of hydrogen halide in the reaction.

3. Halogen Exchange Reactions (*applicable to* GeH_2F_2, GeH_2Cl_2, *and* GeH_2I_2)

Exchange reactions are commonly used to prepare specific dihalogermanes. Silver chloride or $HgCl_2$ has been used to convert GeH_2Br_2 to GeH_2Cl_2,[5

* This reagent can be prepared by adding an aqueous solution of NaF to an aqueous solution of $PbNO_3$. The PbF_2 is washed with ethanol before drying.

PbF_2 has been used to convert GeH_2Br_2 to GeH_2F_2,[38] and the interaction of GeH_2Cl_2 with hydrogen iodide has been used to prepare pure GeH_2I_2.[53] Diiodogermane is a solid at room temperature, and so, in order to provide for an adequate mixing of the reagents, a solvent is used. The solvent that has been recommended is GeH_3Cl.[53] A typical procedure calls for allowing a mixture of about 60 percent GeH_3Cl and 40 percent GeH_2Cl_2, obtained directly in the reaction of GeH_4 with hydrogen chloride,[46] to react with a fivefold excess of hydrogen iodide at room temperature for 1 min in a tube attached to which is a side arm with a constriction. The products of the reaction which are volatile at room temperature—GeH_3I, hydrogen chloride, and the excess hydrogen iodide—are removed from the vessel, which is then sealed by a torch. The GeH_2I_2 is sublimed in the absence of grease from the solid products of the reaction into the constricted side arm, where it condenses as colorless crystals.

4. *Halogenation of Germane by Silver Chloride*

Guillory and Smith report the preparation of GeH_2Cl_2 by passing GeH_4 through a column packed with AgCl which is heated to about 215–220° with a silicone oil bath. Initially the major product formed is GeH_3Cl. This product must be recycled through the AgCl column in order to produce GeH_2Cl_2 in reasonable amounts. In the purification of the compound, GeH_2Cl_2 condenses in a $-58°$ bath (cyclopentanone slush), while GeH_4 and GeH_3Cl do not.[46a]

5. *The Synthesis of Difluorogermane* GeH_2F_2

The most widely quoted synthesis of GeH_2F_2 is based on the reaction of GeH_3Br with AgF.[55,97,98] The GeH_3F formed in the reaction easily undergoes disproportionation, providing a convenient source of GeH_2F_2. The compound can also be prepared by passing GeH_2Br_2 vapor over excess PbF_2.[38]

C. Preparation of Trihalogermanes $HGeX_3$

1. *Trichlorogermane* $HGeCl_3$*

Several reports of procedures for the synthesis of $HGeCl_3$ have proved to be faulty. Winkler first investigated the preparation of $HGeCl_3$ by passing hydrogen chloride over heated germanium or its sulfide.[99] The product

* Trichlorogermane is available commercially from Alfa Inorganic Inc. (Ventron). The reported commercial preparation of the compounds is by the direct reaction of germanium with hydrogen chloride.

obtained was thought to be $GeCl_2$, but in a subsequent paper, Winkler reports that the compound formed is $HGeCl_3$.[100] It was later shown that the Winkler method of preparing $HGeCl_3$ actually produces a mixture of $HGeCl_3$ and $GeCl_4$, approximately 70:30 $HGeCl_3$:$GeCl_4$, that cannot be fractionated by standard vacuum-trap distillations.[101] Dennis and coworkers report the preparation of the compound by the action of hydrogen chloride on $GeCl_2$[101]; Moulton and Miller, however, have found that the Dennis procedure is not completely satisfactory.[102] The main objection to the method is that in the conversion of $GeCl_4$ to $GeCl_2$ using germanium metal, the required high temperatures cause extensive decomposition of the $GeCl_2$.[102] Products of the decomposition are not only $GeCl_4$ and germanium but also subchlorides—chlorine-deficient $GeCl_2$, for example, $GeCl_x$—which do not react with hydrogen chloride.[102]

One of the most efficient methods of preparing pure $HGeCl_3$ seems to be the interaction of gaseous hydrogen chloride with GeS. The reaction was first studied by Dennis and Hulse,[103] although not as a synthetic procedure. Details for the synthesis of $HGeCl_3$ by this method have been given by Moulton and Miller.[102] The reaction of dry, gaseous hydrogen chloride with GeS* takes place at room temperature, although the rate is enhanced considerably at higher temperatures. The procedure calls for passing the hydrogen chloride over GeS at room temperature, collecting the volatile products in a trap maintained at −196°. The reaction is exothermic and the temperature soon increases. A satisfactory temperature and reaction rate are reached by maintaining the hydrogen chloride pressure in the range of 650 to 750 torr. At the end of the run the −196° bath around the trapped products is replaced by a −78° bath, allowing most of the excess hydrogen chloride to be expelled normally through a mercury blowoff tube. A −46° bath is then placed around the trap, allowing hydrogen sulfide, and some hydrogen chloride, to be expelled. All material volatile at this temperature in the trap is then pumped away. After 1.5 to 2 hr of pumping, the purified $HGeCl_3$ in the trap can be condensed into a storage bulb or other suitable vessel by trapping its vapors using a −196° bath. For increased yields the remaining GeS can be treated with additional hydrogen chloride, and the procedure repeated. The overall reported yield of pure $HGeCl_3$ by this procedure is 40 percent.[102]

Trichlorogermane easily decomposes in the vacuum line to give $GeCl_2$ and hydrogen chloride[102]:

$$HGeCl_3 \rightleftharpoons GeCl_2 + HCl$$

At normal temperatures and under conditions by which the equilibrium is not disturbed, the dissociation is slight. On distilling the compound, however

* Details of the synthesis of GeS have been given by Foster.[104]

the dissociation may be enhanced, owing to the more rapid diffusion of hydrogen chloride than $HGeCl_3$. For example, by pumping on a trap containing $HGeCl_3$ at -24 to $-33°$, hydrogen chloride is removed easily, allowing $GeCl_2$ to accumulate in the system.*[102] The dissociation is even more rapid at higher temperatures.

An additional problem is encountered in the systems, because $GeCl_2$ also decomposes, even at $-20°$, to form the germanium subhalide ($GeCl_x$) and $GeCl_4$[102]:

$$GeCl_2 \longrightarrow GeCl_x + GeCl_4$$

Telltale evidence of the decomposition of $HGeCl_3$ is the formation of white nonvolatile deposits in appropriate parts of the vacuum line that have been in contact with the sample. The deposits formed in distillations at low temperatures (-24 to $-33°$) are white, with only occasional traces of yellow. The residues of room-temperature distillations undergo rapid color changes from white to yellow to orange and orange-red, eliminating $GeCl_4$ throughout the decomposition.[102]

Although an azeotropic mixture of $HGeCl_3$ and $GeCl_4$, such as that obtained in Winkler's method, cannot be separated by distillation techniques, it has been possible to separate the two components by a chemical method based on the formation of a $HGeCl_3$ etherate.[105,106] On treating one volume of an $HGeCl_3$–$GeCl_4$ mixture with two volumes of absolute diethyl ether, two immiscible layers form. The yellow oily liquid layer is the $HGeCl_3$-etherate of the composition $HGeCl_3 \cdot 2(C_2H_5)_2O$), which is essentially insoluble in the excess ether. The tetrachloride is soluble in ether, and thus the pure $HGeCl_4$ as an etherate is readily separable from the mixture. It is important to keep moisture out of the system, owing to the increased solubility of the $HGeCl_3$ etherate in the ether as the amount of hydrogen chloride (from hydrolysis) in the system increases.[106] As a result of this convenient separation a number of reactions of $HGeCl_3$ have been reported recently using the etherate of $HGeCl_3$ rather than the pure compound itself.†

The ^1H nmr spectrum of the etherate shows the CH_3 and CH_2 absorptions of the ether and a singlet at 14.6 ppm downfield from $(CH_3)_4Si$. Noting that the ^1H resonance of pure $HGeCl_3$ is at 7.6 ppm downfield, one can conclude that in the etherate, the proton is highly polarized, possibly having an ionic structure such as:[107a]

$$[(C_2H_5)_2O \rightarrow H \leftarrow O(C_2H_5)_2]^+GeCl_3^-$$

Di-n-butyl ether and tetrahydrofuran also form etherates with $HGeCl_3$.[107a]

* Gar and Mironov report that they have been unable to repeat this experiment.[106]
† The chemistry of $HGeCl_3$ has recently been reviewed by Mironov and Gar.[107]

2. *Tribromogermane* $HGeBr_3$

The synthesis of $HGeBr_3$ has been achieved by several methods. Brewer and Dennis report the preparation of the compound either by the reaction of hydrogen bromide with $GeBr_2$ or the reaction of hydrogen bromide with germanium metal.[108] The latter reaction actually produces an azeotropic mixture of $HGeBr_3$ and $GeBr_4$ that cannot be separated by distillation techniques. More recently Mironov and Gar have shown that the yield of $HGeBr_3$ in the mixture produced by this reaction can be increased to about 30 percent by including copper powder with the germanium.[109] Thus, after passing a stream of dry hydrogen bromide through a mixture of 40 g of finely ground germanium and 12 g of copper, 200 g of a colorless turbid condensate resulted. The condensate itself was not analyzed directly but rather was allowed to reflux with cyclohexene for 2 hr. The cyclohexyltribromogermane produced in the reflux stage was used as proof of the presence of $HGeBr_3$ in the condensate:

$$C_6H_{10} + HGeBr_3 \longrightarrow C_6H_{11}\text{—}GeBr_3$$

The amount of the organotribromogermane produced was used to show that the condensate contained at least 30 percent $HGeBr_3$.[109]

Although the azeotropic mixture of $HGeBr_3$ and $GeBr_4$ cannot be separated by distillation techniques, the $HGeBr_3$ can be recovered from the mixture as an etherate in a manner identical to the chemical separation of $HGeCl_3$ from $GeCl_4$.[109] Tribromogermane readily forms an orange, oily etherate that is insoluble in excess ether. The tetrabromide is readily soluble in the ether and only partially soluble in the etherate. As in the case of $HGeCl_3$, many reactions of $HGeBr_3$ have been conducted with the $HGeBr_3$ etherate.

Another reported preparation of $HGeBr_3$ involves the dissolution of $Ge(OH)_2$ in aqueous hydrogen bromide.[106] The $Ge(OH)_2$ is prepared by the hydrolysis of the $HGeCl_3$ etherate:

$$HGeCl_3 \xrightarrow{H_2O} Ge(OH)_2 \xrightarrow{46\%\ HBr} HGeBr_3$$

3. *Triiodogermane* $HGeI_3$ *and Chlorodiiodogermane* $HGeI_2Cl$

All attempts to synthesize $HGeI_3$ by procedures discussed for the synthesis of $HGeCl_3$ and $HGeBr_3$ have been unsuccessful.[106,109] The only germanium compound that results in the attempted preparations is GeI_2

$$HGeCl_3 \xrightarrow{H_2O} Ge(OH)_2 \xrightarrow{HI} [HGeI_3]$$

$$HGeCl_3 \xrightarrow{HI} [HGeI_3] \longrightarrow GeI_2 + HI$$

It has been presumed that $HGeI_3$ is indeed formed in the reactions, although it is unstable and undergoes spontaneous decomposition.

Mazerolles and Manuel have found that a solution of GeI_2 with hydrogen iodide in benzene behaves as if $HGeI_3$ were present.[110] For example, the interaction of a mixture of GeI_2 with hydrogen iodide in benzene with methylvinylketone produces methyl (2-triiodogermylethyl) ketone in 62 percent yields:

$$GeI_2 + HI \underset{}{\overset{\text{ether}}{\rightleftharpoons}} HGeI_3 \xrightarrow{CH_2{=}CH\overset{O}{\overset{\|}{C}}CH_3} GeI_3CH_2CH_2\overset{O}{\overset{\|}{C}}{-}CH_3$$

Various other reactions of this kind have been investigated.[110] The interaction of dry GeI_2 with an excess of hydrogen chloride etherate is reported to produce an etherate of composition $HGeI_2Cl \cdot 2(C_2H_5)_2O$.[111]

VI. PREPARATION OF THE HALOGENOID DERIVATIVES OF GERMANE

Characterized halogenoid derivatives of the germanes are of the type GeH_3X, where X = CN, NCO, NCS, and N_3. All the compounds except GeH_3N_3 have been prepared most frequently by an exchange reaction involving a germyl halide and the appropriate silver halogenoid salt:

$$GeH_3Br + AgX \longrightarrow GeH_3X + AgBr \qquad X = CN, NCO, NCS$$

A more recently described procedure for the synthesis of the halogenoid derivatives in high yields is based on the exchange reaction that GeH_3F undergoes with halogenoid derivatives of SiH_4:

$$GeH_3F + SiH_3X' \longrightarrow GeH_3X' + SiH_3F \qquad X' = CN, CNO, CNS, N_3$$

No disubstituted derivatives of GeH_4 or monosubstituted halogenoid derivatives of Ge_2H_6 have been reported to date, although extensions of the reactions to be discussed should be applicable toward their preparation. The synthesis and characterization of halogenoid derivatives of Ge_2H_6 will undoubtedly be complicated by the low thermal stability of the compounds. Mackay and Roebuck report an unsuccessful attempt to prepare Ge_2H_5NCO by the interaction of Ge_2H_5I with AgNCO.[43] The only volatile products of the reaction are GeH_4 and cyanic acid.

A. Germyl Cyanide GeH_3CN

Germyl cyanide has been prepared by passing either GeH_3I[57] or GeH_3Br[23,58] over excess powdered AgCN.* The compound is also formed

* Germyl cyanide prepared by this method has been shown by infrared[57,112] and microwave[113] studies to be predominantly, if not exclusively, of the normal form, GeH_3CN as opposed to the possible iso form, GeH_3NC.

in the reaction of GeH_3Cl with AgCN; extensive decomposition reactions, however, apparently catalyzed by GeH_3Cl or hydrogen cyanide, accompanies the conversion at room temperature and prevents the use of this halide for synthetic purposes.[23] The decomposition is given by

$$x\text{GeH}_3\text{CN} \longrightarrow (\text{GeH}_2)_x + x\text{HCN}$$

The recommended procedure for the iodide or bromide conversion by this method involves passing the germyl halide vapor slowly through a column loosely packed with a mixture of AgCN and glass wool at room temperature.

Varma and Buckton have described the synthesis of the isotopically enriched compound $GeH_3C^{15}N$ by the 15-min reaction of GeH_3I with solid $AgC^{15}N$.[113] The $GeH_3C^{15}N$ obtained apparently needed no further purification for their microwave studies.

A convenient preparation of GeH_3CN by the interaction of GeH_3F with SiH_3CN has been described recently by Ebsworth, Lee, and Sheldrick.[66] In a typical experiment, GeH_3F (1 mmole) and SiH_3CN (0.9 mmole) are condensed into a 250-ml bulb that is then allowed to warm to room temperature. After a 5-min reaction, the bulb is cooled to $-64°$, and the products that are volatile at this temperature are removed by pumping. Germyl cyanide (0.9 mmole) is the sole product remaining in the bulb.

B. Germyl Isocyanate, GeH_3NCO

This compound has been prepared by the interaction of GeH_3I[114] or GeH_3Br[23] with AgNCO.* The procedure as described by Onyszchuk and co-workers normally calls for streaming the vapor of the halide over dry AgNCO mixed with glass wool at room temperature.[23] The conversion of GeH_3Br to GeH_3NCO is almost quantitative (93% yields reported).[23] Cradock and Ebsworth also report the preparation of the compound in yields up to 90% by the exchange reaction that GeH_3F undergoes with SiH_3NCO.[60]

$$\text{GeH}_3\text{F} + \text{SiH}_3\text{NCO} \longrightarrow \text{GeH}_3\text{NCO} + \text{SiH}_3\text{F}$$

Germyl isocyanate is stable at ambient temperatures and a sample withstood heating at 110° for 40 hrs. However after 12 hrs at 200–220° the sample decomposed to metallic germanium, hydrogen and cyanic acid.[23]

$$\text{GeH}_3\text{NCO} \longrightarrow \text{Ge} + \text{H}_2 + \text{HNCO}$$

* Germyl isocyanate prepared by this reaction has been shown to be exclusively of the "iso" structure, GeH_3NCO as opposed to the possible normal structure, GeH_3OCN, by microwave[114] and both ^1H and ^{14}N magnetic resonance studies.[115,115a] The suggestion[116] that infrared evidence detects the presence of some GeH_3OCN in the sample prepared in this manner has been disputed.[115]

C. Germyl Isothiocyanate GeH_3NCS

Germyl isothiocyanate and GeD_3NCS have been prepared by the room-temperature interaction of GeH_3Br (or GeD_3Br) with AgNCS.* Experimental procedures are the same as described above for the AgNCO conversions. The compound has also been prepared in yields up to 90 percent by the exchange reaction that GeH_3F undergoes with SiH_3NCS.[60]

A sample of GeH_3NCS completely decomposed to GeH_4 and a yellow, non-volatile residue after being heated for 20 hr. at 55°.[23] The residue is believed to be polymeric thiocyanic acid and solid germanium hydrides produced according to the following scheme

$$GeH_3NCS \longrightarrow GeH_2 + HNCS$$
$$3GeH_2 \longrightarrow GeH_4 + 2GeH$$
$$xHNCS \longrightarrow (HNCS)_x$$

D. Germyl Azide GeH_3N_3

Cradock and Ebsworth report the synthesis of GeH_3N_3 in high yield by the interaction of $(CH_3)_3SiN_3$† with GeH_3F[61]:

$$(CH_3)_3SiN_3 + GeH_3F \longrightarrow GeH_3N_3 + (CH_3)_3SiF$$

In a typical experiment, $(CH_3)_3SiN_3$ (10.0 mmoles) and GeH_3F (10.6 mmoles) are allowed to interact for 3 min at room temperature. The GeH_3N_3 is formed almost quantitatively (9.5 mmoles) and can be recovered from the products by repeated condensation in a trap at −78° from a trap at −46°. The liquid compound is unstable at room temperature but is stable below −20°.

VII. PREPARATION OF GERMYL COMPOUNDS CONTAINING GERMANIUM BOUND TO THE GROUP VIA ELEMENTS

Isolated and characterized germanes containing the group VIA elements are limited to derivatives of GeH_4 containing oxygen, sulfur, selenium, and tellurium. Compounds to be considered are $(GeH_3)_2O$, GeH_3OCH_3, $GeH_3OC_6H_5$, GeH_3OCOCH_3, $(GeH_3)_2E$, where E = S, Se, Te, GeH_3SCH_3, and $GeH_3SC_6H_5$. There has been convincing evidence of the existence of compounds of the type GeH_3EH, where E = S, Se, and Te, from proton nuclear-magnetic-resonance studies of the following systems.[36,119,120]

$$(GeH_3)_2E + H_2E \rightleftharpoons 2GeH_3EH \qquad E = S, Se, Te$$

* The GeH_3NCS and GeD_3NCS prepared in this manner have been shown to be of the iso structure by infrared studies.[117]

† $(CH_3)_3SiN_3$ can be prepared readily from the reaction of $(CH_3)_3SiCl$ with NaN_3.[118] The synthesis of GeH_3N_3 can also be achieved by the exchange of GeH_3F with SiH_3N_3.[60]

It has not been possible, however, to isolate these mixed Group IVA–Group VIA hydrides. NMR evidence has also been used to identify $(GeH_3Se)_2GeH_2$ and possibly $(GeH_3Se)_3GeH$ and $(GeH_3Se)_4Ge$ formed in the reaction of GeH_3PH_2 or GeH_3AsH_2 with hydrogen selenide.[120]

Drake and Riddle have recently shown that the action of a silent electric discharge on equimolar mixtures of GeH_4 with hydrogen sulfide or hydrogen selenide produces hydrides of the type GeH_3EH and $(GeH_3)_2E$, where E is S or Se, as well as higher germanes.[88a] The discharge of ternary mixtures composed of GeH_4, SiH_4, and either hydrogen sulfide or hydrogen selenide produces the mixed hydrides GeH_3SSiH_3 or GeH_3SeSiH_3 respectively. Compound identification in this study was mainly by ^1H nmr spectroscopy.

A. Digermyl Ether $(GeH_3)_2O$

The synthesis and isolation of $(GeH_3)_2O$ has been a challenge to inorganic chemists. The application of procedures used in the synthesis of $(SiH_3)_2O$ have, in general, met with failure; for example, $(GeH_3)_2O$ cannot by synthesized by the reaction of GeH_3Cl with excess water.[121] The only reported product of this reaction is an uncharacterized white solid. Likewise $(GeH_3)_2O$ cannot be isolated from the reaction of GeH_3Br with Ag_2O or Ag_2CO_3, although it may be an unstable intermediate[23]:*

$$2GeH_3Br + Ag_2O \longrightarrow [(GeH_3)_2O] + 2AgBr$$

$$\downarrow$$

$$2(GeH_2)_x + H_2O$$

$$\uparrow$$

$$2GeH_3Br + Ag_2CO_3 \longrightarrow [(GeH_3)_2O] + 2AgBr + CO_2$$

Sujishi and Goldfarb have detected $(GeH_3)_2O$ by infrared methods in the hydrolysis of GeH_3CN, but the pure compound could not be isolated.[122]

Goldfarb and Sujishi first reported the preparation of pure $(GeH_3)_2O$ by the interaction of $(GeH_3)_2S$ with red HgO at $-40°$.[59] Germyl iodide is converted to $(GeH_3)_2O$ in a similar manner, although the authors point out that the higher volatility of GeH_3I (vapor pressure at 0°, 20 torr) compared with that of $(GeH_3)_2S$ (vapor pressure at 0°, 5 torr) makes the separation of unconsumed GeH_3I from the $(GeH_3)_2O$ (vapor pressure at 0°, 66 torr) much more difficult than separating any unconsumed $(GeH_3)_2S$ from $(GeH_3)_2O$.[59] In a typical preparation a mixture of about 1 mmole of $(GeH_3)_2S$ and 32 g of granular red HgO is held at $-40°$ with occasional shaking for 40 min. Yellow and powdery red HgO are not suitable for the conversion. The

* Cradock has recently prepared pure $(GeH_3)_2O$ by streaming GeH_3Cl over lead hydroxide.[59b]

volatile products are distilled out of the reaction flask maintained at −40° through traps at −78° and −112°. The −78° trap stops the unreacted $(GeH_3)_2S$, and the pure $(GeH_3)_2O$ condenses in the trap at −112°. The unreacted $(GeH_3)_2S$ is returned to a vessel containing a fresh batch of HgO, and the procedure is repeated. After three such treatments of $(GeH_3)_2S$ with HgO, 0.35 mmole of pure $(GeH_3)_2O$ can be obtained. Digermyl ether slowly decomposes at 0° in an unknown manner, although it has been established that GeH_4 is not a decomposition product. The decomposition seems to be catalyzed by trace amounts of water.[59]

Cradock and Ebsworth have reported a somewhat more convenient and efficient procedure for preparing $(GeH_3)_2O$, based on the hydrolysis of digermylcarbodiimide $GeH_3NCNGeH_3$.*[68] A typical synthesis calls for allowing $GeH_3NCNGeH_3$ (2.36 mmoles) to react with water (1.95 mmoles) for 10 min at room temperature. The $(GeH_3)_2O$ (1.91 mmoles, 98 percent yield) can be purified in the vacuum line as described above.

B. Germyl Methyl Ether GeH_3OCH_3

Van Dyke and coworkers have prepared GeH_3OCH_3 in a 55 percent yield by the reaction of GeH_3Cl with freshly prepared $NaOCH_3$ in diglyme.[78] The vessel for the preparation is a 50-ml two-necked Claissen flask. Sodium methoxide is first prepared in the flask by allowing sodium (3.9 mg atoms) to react with excess CH_3OH (3 ml). The flask is then attached to the vacuum line by way of one neck of the flask and a 100-ml pressure-equalizing addition funnel containing approximately 30 ml of diglyme is inserted into the second neck. Hydrogen, air, and excess CH_3OH are pumped out of the flask, and the diglyme is then introduced onto the $NaOCH_3$. Germyl chloride (3.9 mmoles) is condensed into the reaction flask, and the contents are held at −78° for 15 hr. The products are allowed to distill out of the flask, still maintained at −78°, through successive traps at −64°, −96°, and −196°. A small amount of CH_3OH and some diglyme condense in the −64° trap, GeH_3OCH_3 condenses in the −96° trap, and small amounts of GeH_4 (and Ge_1H_6) condense in the −196° trap. The compound is reasonably stable thermally and can be transferred through the vacuum line in normal operations without any decomposition. Some decomposition of GeH_3OCH_3 can be noted after keeping GeH_3OCH_3 at 0° for 17.5 hr. Volatile products of the decomposition are GeH_4, CH_3OH, and $(GeH_3)_2O$.[78]

Cradock and Ebsworth report the preparation of GeH_3OCH_3 and GeH_3OCD_3 based on the reaction of $GeH_3NCNGeH_3$ with CH_3OH or CD_3OD, respectively.[68] Thus GeH_3OCH_3 (2.09 mmoles, 64 percent yield) is obtained on allowing a mixture of $GeH_3NCNGeH_3$ (1.69 mmoles) and

* For the preparation of $GeH_3NCNGeH_3$ see Section VIII-F.

CH_3OH (3.28 mmoles) to react at room temperature for 5 min. The compound is purified as described above.

C. Germyl Phenyl Ether $GeH_3OC_6H_5$

Glidewell and Rankin have prepared this compound by the interaction of potassium phenoxide with GeH_3Br in diethyl ether[79]:

$$GeH_3Br + C_6H_5OK \longrightarrow GeH_3OC_5H_5 + KBr$$

Potassium phenoxide is prepared by adding phenol (433 mmoles) to potassium (10.28 mmoles) in diethyl ether ($\cong$5 ml) in a glass break-seal ampul. The mixture is shaken, allowed to stand at room temperature, and then degassed at $-196°$. Germyl bromide (4.0 mmoles) is condensed into the tube, and the frozen mixture allowed to warm to room temperature rapidly with vigorous shaking. After an additional room-temperature reaction period of 1 hr, the $GeH_3OC_6H_5$ is recovered from the mixture by its condensation in a trap at $-22°$. The reaction is essentially quantitative.

D. Germyl Acetate GeH_3OCOCH_3

Onyszchuk and coworkers report the preparation of GeH_3OCOCH_3 by the reaction of GeH_3Br with silver acetate $AgOCOCH_3$[23]:

$$GeH_3Br + AgOCOCH_3 \longrightarrow GeH_3OCOCH_3 + AgBr$$

The general procedure followed is that of a typical silver salt conversion. Thus, vaporized GeH_3Br (3.20 mmoles) is allowed to pass through a tube loosely packed with a mixture of powdered glass wool and dry silver acetate. The method produces GeH_3OCOCH_3 in 96 percent yield.

Germyl acetate is reasonably stable thermally. About 60 percent of a sample of pure material decomposed after it was heated at 100 to 110° for 16 hr. The probable decomposition is given as

$$GeH_3C_2H_3O_2 \longrightarrow GeH_2 + HC_2H_3O_2$$

$$xGeH_2 \longrightarrow (GeH)_x + \tfrac{1}{2}xH_2$$

E. Digermyl Sulfide $(GeH_3)_2S$

The synthesis of $(GeH_3)_2S$ has been achieved by the reaction of GeH_3I with an excess amount of red HgS[45,59,123]; for example, GeH_3I (1 mmole) is condensed into a flask or tube containing red HgS (5 to 10 mmoles) and allowed to react for 15 min. Initially the solid HgS becomes rather caked, but it is dry after the reaction period. The $(GeH_3)_2S$ is separated from the more volatile products of the reaction (H_2S, GeH_4, etc.) by its condensation in a trap at $-36°$. Yields of pure $(GeH_3)_2S$ from 55 to 75 percent based on the amount of GeH_3I used have been obtained.[45*]

F. Germyl Methyl Sulfide GeH_3SCH_3

Wang and Van Dyke report the preparation of GeH_3SCH_3 by the interaction of GeH_3I or GeH_3Cl with CH_3SNa.[†80,81] A typical preparation of the compound involves the condensation of GeH_3I (3.5 mmoles) into a 250-ml round-bottomed flask that contained approximately 1 g of CH_3SNa. The reaction flask is allowed to warm to room temperature open to a trap maintained at $-196°$. The material that collects in the $-196°$ trap is condensed back to the reaction flask, and the procedure is repeated. After this is done four times the volatile products are passed through traps at $-46°$ and $-64°$ into a trap at $-196°$. Slightly impure GeH_3SCH_3 collects in the trap at $-64°$. For further purification the impure sample of GeH_3SCH_3 must be passed through a gas-chromatography unit. A 5 ft by ¼ in. glass column packed with benzyl ether (20 percent) on Chromosorb W support is satisfactory for the purification, although undoubtedly other column materials would also be suitable.

G. Germyl Phenyl Sulfide $GeH_3SC_6H_5$

Glidewell and Rankin have synthesized this compound by the interaction of GeH_3Br with C_6H_5SK in diethyl ether[79]:

$$GeH_3Br + C_6H_5SK \longrightarrow GeH_3SC_6H_5 + KBr$$

The reaction of thiophenol (4.39 mmoles) with potassium (3.86 mmoles) in diethyl ether ($\cong 5$ ml) at room temperature in a break-seal tube is the source of C_6H_5SK. Germyl phenyl sulfide is prepared by condensing GeH_3Br onto a frozen and degassed mixture of C_6H_5SK in diethyl ether, followed by warming the reaction mixture to room temperature rapidly with vigorous shaking. The reaction is allowed to proceed at room temperature for an additional 20 min, after which the $GeH_3SC_6H_5$ is recovered by its condensation in a trap at $-23°$. The reaction is essentially quantitative.

H. Digermyl Selenide $(GeH_3)_2Se$

Drake and Riddle have reported the preparation of $(GeH_3)_2Se$ by the interaction of either GeH_3PH_2 or GeH_3AsH_2 with hydrogen selenide[120]:

$$2GeH_3EH_2 + H_2Se \longrightarrow (GeH_3)_2Se + 2EH_3 \qquad E = P, As$$

* Rankin has recently prepared $(GeH_3)_2S$ by the reaction of GeH_3Br with lithium sulfide in dimethyl ether.[59a]

† CH_3SNa is prepared from the reaction of CH_3ONa with CH_3SH.[124]

Evidence of the formation of GeH_3SeH has been obtained by following the ^1H nmr spectrum of the reaction mixture[120]:

$$2GeH_3EH_2 + 2H_2Se \longrightarrow 2GeH_3SeH + 2EH_3 \qquad E = P \text{ or } As$$
$$2GeH_3SeH \longrightarrow (GeH_3)_2Se + H_2Se$$

In a typical preparation, GeH_3PH_2 (0.3 mmole) and hydrogen selenide (0.15 mmole) are condensed into an nmr tube that is then sealed and allowed to warm to room temperature. The ^1H nmr spectrum of the initial reaction mixture shows GeH_3PH_2 and hydrogen selenide plus a widely separated doublet assigned to PH_3, a doublet and associated quartet assigned to GeH_3SeH, and a singlet assigned to $(GeH_3)_2Se$. It is observed that the intensities of the new peaks grow at the expense of the peaks of the starting materials, and after about 1 hr the single peak due to $(GeH_3)_2Se$ becomes the main absorption. The behavior of the GeH_3AsH_2–H_2Se reaction also followed by ^1H nmr spectroscopy is essentially identical, although the rate of the reaction is much slower. For example, after 20 hr, the only additional peaks in the spectrum are those due to GeH_3SeH and AsH_3 but not $(GeH_3)_2Se$.

After the nmr spectra indicate that the reactions are essentially complete, the $(GeH_3)_2Se$ can be isolated in about 67 percent yields by opening the tubes under vacuum and fractionating the products.[120] Digermyl selenide has been recovered from the GeH_3PH_2 reaction by its condensation in a $-78°$ trap or from the GeH_3AsH_2 reaction by its condensation in a $-46°$ trap. In each case the PH_3 or AsH_3, plus traces of GeH_4, pass through traps maintained at the above temperatures.

Further absorptions can be observed in the nmr spectra of the GeH_3PH_2–H_2Se system after 10 hr and in the GeH_3AsH_2–H_2Se system after 100 hr, as shown in Figure 1. The new peaks can be rationalized as arising from the products of the condensation of $(GeH_3)_2Se$ according to:[120]

$$(GeH_3)_2Se + (GeH_3)_2Se \longrightarrow (GeH_3Se)_2GeH_2 + GeH_4$$
$$(GeH_3)_2Se + (GeH_3Se)_2GeH_2 \longrightarrow (GeH_3Se)_3GeH + GeH_4$$
$$(GeH_3)_2Se + (GeH_3Se)_3GeH \longrightarrow (GeH_3Se)_4Ge + GeH_4$$

One grouping of the new peaks can be assigned unambiguously to $(GeH_3Se)_2GeH_2$. Two additional peaks can be assigned only tentatively to the other condensed species. None of the condensed compounds has been isolated.

The cause of the condensation is not clear, because $(GeH_3)_2Se$ itself shows no sign of condensing after several days, either alone or in systems containing PH_3 or H_2Se.[120]

Cradock, Ebsworth, and Rankin have reported the preparation of

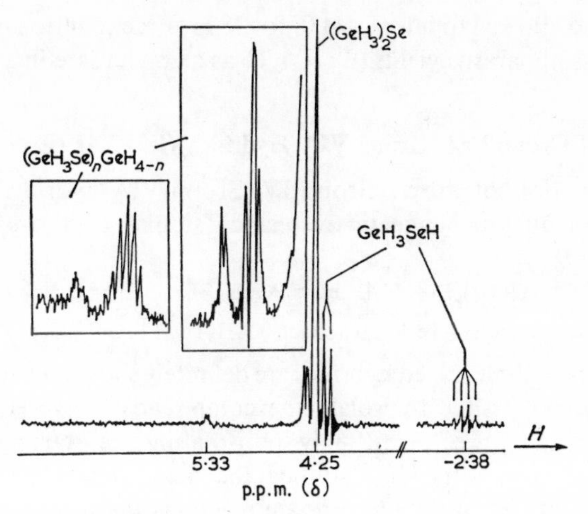

Figure 1. The ^{1}H nmr spectrum recorded at 60 MHz of the Ge—H and Se—H resonance regions at the conclusion of the reaction of H_2Se with GeH_3PH_2 (and GeH_3AsH_2). Inserts recorded at 100 MHz depict the condensed species. Chemical shifts given as positive to low field of $(CH_3)_4Si$. Reproduced with permission from reference 120.

$(GeH_3)_2Se$ by the exchange interaction of $(SiH_3)_2Se$ with GeH_3Br or by the interaction of $GeH_3NCNGeH_3$ with hydrogen selenide[62]:

$$2GeH_3Br + (SiH_3)_2Se \longrightarrow (GeH_3)_2Se + 2SiH_3Br$$

$$GeH_3NCNGeH_3 + H_2Se \longrightarrow (GeH_3)_2Se + H_2N_2C$$

In a typical experiment, $(SiH_3)_2Se$* (0.354 mmole) and GeH_3Br (0.727 mmole) are combined and allowed to react at 20° for 30 min. The $(GeH_3)_2Se$ can be separated from the products (0.315 mmole, 89 percent yield) by its condensation in a trap at −45°.

The most convenient synthesis of $(GeH_3)_2Se$ is based on the reaction of $GeH_3NCNGeH_3$ with hydrogen selenide. Treatment of $GeH_3NCNGeH_3$ (1.25 mmoles) with hydrogen selenide (≅3 mmoles) in a 100-ml ampul for 10 min at room temperature produces $(GeH_3)_2Se$ (1.05 mmoles, 84 percent yield).[62]

I. Digermyl Telluride $(GeH_3)_2Te$

Cradock, Ebsworth, and Rankin[62] report the quantitative formation of $(GeH_3)_2Te$ by the exchange reaction that $(SiH_3)_2Te$ undergoes with GeH_3Br. As an illustration, $(SiH_3)_2Te$† (1.32 mmoles) and GeH_3Br (4.1 mmoles) are

* Prepared by the reaction of Li_2Se with SiH_3Br at −96°.
† Prepared by the reaction of Li_2Te with SiH_3Br at −96°.

combined and allowed to interact at 0° for 18 hr. Fractionation of the products in an all-glass apparatus yields $(GeH_3)_2Te$ as a condensate in a trap at −23°.

J. Mixed Group IVA–Group VIA Hydrides GeH_3EH (E = S, Se, Te)

The formation of mixed Group IV–Group VIA hydrides has been detected by ^1H nuclear magnetic resonance studies of the systems shown below[81a,119,120]:

$$(GeH_3)_2E + H_2E \rightleftharpoons 2GeH_3EH \qquad E = S, Se$$

$$(GeH_3)_2Te + HCl \rightleftharpoons GeH_3TeH + GeH_3Cl$$

The structures of these mixed hydrides are definitely shown by the nmr studies; it has not been possible to isolate the compounds, however. Attempts to fractionate the mixtures result only in isolating the starting compounds. Sheldrick and coworkers have studied the mixed hydrides by combining equimolar quantities of the appropriate reactants in an nmr tube.[119] Drake and Riddle have studied the $(GeH_3)_2S$–H_2S system using an excess amount of hydrogen sulfide.[81a]

A description of the $(GeH_3)_2S$–H_2S system serves to illustrate the method.[81a] Digermyl sulfide (0.2 mmole) and hydrogen sulfide (1.0 mmole) are condensed into a semimicro nmr tube that is then sealed. The nmr spectrum of the sample, taken just after the mixture is warmed to room temperature, shows the signals expected for $(GeH_3)_2S$ and hydrogen sulfide plus a doublet-quartet pattern readily assignable to GeH_3SH. The spectrum is shown in Figure 2. After about $\frac{1}{2}$ hr the conversion is approximately 60 percent complete. Attempts to fractionate the mixture have been unsuccessful; the only products obtained are $(GeH_3)_2S$ and H_2S.[81a]

VIII. PREPARATION OF GERMYL COMPOUNDS CONTAINING GERMANIUM BOUND TO THE GROUP VA ELEMENTS

Characterized compounds of GeH_4 that contain the Group VA elements N, P, As, Sb are of the following types: $(GeH_3)_3E$, where E = N, P, As, Sb; $(GeH_3)_2EH$, where E = P, As; GeH_3EH_2, where E = P, As; GeH_2ClPH_2; $GeH_3PH_2BH_3$; and $GeH_3NCNGeH_3$.

Mass-spectrometric identification of several other interesting compounds containing the Group VA elements has been made. These include GeH_3NH_2 and $GeH_2(NH_2)_2$ produced in the acidic hydrolysis of a mixture of Mg_2Ge and Mg_3N_2,[125,126] and compounds of the type Ge_2PH_7, Ge_3PH_9, and GeP_2H_6 produced by the decomposition of a mixture of GeH_4 and PH_3 in an ozonizer silent electric discharge.[86] These compounds have not been characterized and so are not considered further here.

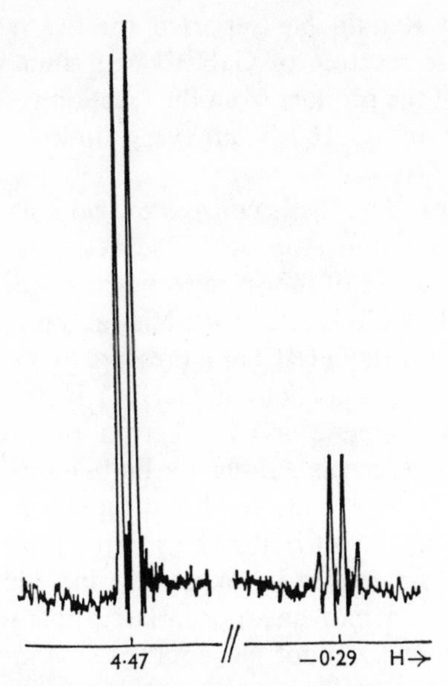

Figure 2. The ^{1}H nmr spectrum of germanethiol [p.p.m. (+) to low field of $(CH_3)_4Si$]. Reproduced with permission from reference 81a.

A. Compounds of the Type $(GeH_3)_3E$ (E = N, P, As, Sb)

1. *Trigermylamine* $(GeH_3)_3N$

Early attempts to prepare $(GeH_3)_3N$ by analogous reactions used in the synthesis of the silicon analog $(SiH_3)_3N$ met with failure. For example, Dennis and Work report that the interaction of GeH_3Cl with ammonia in the liquid phase at −78 to −50° for 2 hr produces GeH_4, a solid polymer $(GeH)_x$, and NH_4Cl[121]:

$$3xGeH_3Cl + 3xNH_3 \longrightarrow 3xNH_4Cl + xGeH_4 + 2(GeH)_x$$

Likewise, Sujishi and Keith report that $(GeH_3)_3N$ cannot be recovered from the products of the reaction of GeH_3CN with ammonia.[58]

Trigermylamine was first detected by Cradock and Ebsworth as a product of the exchange reaction that takes place between GeH_3F and $(SiH_3)_3N$.[60] The reactants, held at room temperature on the vacuum line or at −70° in TMS solution in nmr tubes, produced GeH_4 and a solid that was insoluble in TMS. The authors report the formation of an unidentified compound in the TMS solution that had a proton resonance at 5.09τ. The compound was not isolated, but this is the same chemical shift reported for pure $(GeH_3)_3N$ independently prepared by Rankin.[56,82]

The method that Rankin has reported for the synthesis of $(GeH_3)_3N$ involves the gas-phase reaction of GeH_3Cl with ammonia, followed by the immediate removal of the product from the reaction vessel.[56,82] The synthesis and characterization of $(GeH_3)_3N$ are very difficult, owing to the compound's great instability.

The apparatus for the synthesis consists of two bulbs, of 300- and 100-ml capacity, joined by a 6-mm stopcock. The vessel is first pretreated with SiH_3Cl to remove all traces of moisture from the inside surface. In a typical preparation the smaller bulb is filled with NH_3 at a pressure of 160 torr, and the larger bulb is filled with GeH_3Cl at a pressure of 40 torr. If higher overall pressures are used, the proportion of GeH_3Cl in the system must be increased to allow for the generation of NH_3 from the accelerated decomposition of $(GeH_3)_3N$. The stopcock between the bulbs is opened to allow the gases to mix. When this is done, a white solid that rapidly acquires a yellow color on standing forms in the larger bulb. The volatile products are immediately removed from the vessel either by expansion into an adjacent evacuated bulb or by condensation into an appropriate apparatus cooled by liquid nitrogen. The compound cannot be condensed and revolatilized in the vacuum line, owing to its rapid, almost complete decomposition. As a result a fresh sample of $(GeH_3)_3N$ must be prepared each time it is needed. The stability of the compound in the gas phase can be increased by adding argon ($\simeq 300$ torr) to the sample.

2. *Trigermylphosphine* $(GeH_3)_3P$

Ebsworth and coworkers have prepared $(GeH_3)_3P$ in high yield by the exchange reaction that occurs on mixing $(SiH_3)_3P$* with GeH_3Br[63,64]:

$$(SiH_3)_3P + 3GeH_3Br \longrightarrow (GeH_3)_3P + 3SiH_3Br$$

In a typical preparation a mixture of $(SiH_3)_3P$ (4.6 mmoles) and GeH_3Br (18.0 mmoles) is allowed to react in an evacuated tube for 4 hr at 0°. Trigermylphosphine is formed almost quantitatively and is purified by repeated condensation at $-46°$. There is no evidence of the formation of any mixed silylgermylphosphines. At room temperature, $(GeH_3)_3P$ slowly decomposes, with the formation of GeH_4 and a colorless liquid that is thought to contain compounds with GeH_3—P—GeH_2 chains.[64] At 55° the compound decomposes rapidly.[63,64]

Unsuccessful attempts to prepare $(GeH_3)_3P$ include the interaction of GeH_3Br with PH_3 in the presence of trimethylamine and the interaction of GeH_3Br with KPH_2 in dimethyl ether at $-78°$.[64]

* Trisilylphosphine can be prepared by the reaction of SiH_3Br with KPH_2 in dimethyl ether at $-90°$.[127,128]

3. *Trigermylarsine* $(GeH_3)_3As$

Ebsworth, Rankin, and Sheldrick report the preparation of $(GeH_3)_3As$ by the interaction of excess GeH_3Br with $(SiH_3)_3As$*[65]:

$$(SiH_3)_3As + 3GeH_3Br \longrightarrow (GeH_3)_3As + 3SiH_3Br$$

A typical experiment involves allowing a mixture of $(SiH_3)_3As$ and excess GeH_3Br to remain at 0° overnight in an all-glass break-seal ampul. The $(GeH_3)_3As$ is recovered by its condensation in a −46° bath; both SiH_3Br and GeH_3Br are volatile at this temperature. If unreacted $(SiH_3)_3As$ remains in the −46° bath, additional GeH_3Br should be added and the reaction allowed to continue. If necessary, the SiH_3Br formed can be removed from the system to aid in shifting the equilibrium toward the desired products.

The procedure as described in the literature actually involves allowing $(SiH_3)_3As$ (1.13 mmoles) and GeH_3Br (5.5 mmoles) to react at 0° for 5 hr.[65] The tube was surrounded by a − 46° bath, and the compounds volatile at this temperature were removed. Some of the SiH_3Br formed (2.3 mmoles) was recovered by fractionation, and the remaining SiH_3Br–GeH_3Br mixture, together with additional GeH_3Br (2.5 mmoles), was returned to the ampul, whose entire contents were then allowed to react overnight at 0°. After passing the products through a −46° bath, the less volatile fraction was found to be completely free of silyl compounds.

The reaction of SiH_3Br with $KAsH_2$ at low temperatures in dimethyl ether is reported to give $(SiH_3)_3As$.[129] Ebsworth and coworkers found that the analogous reaction between GeH_3Br and $KAsH_2$ under comparable conditions produces a mixture of GeH_3AsH_2, $(GeH_3)_2AsH$, and $(GeH_3)_3As$.[65] This is not the recommended method for preparing these compounds.

4. *Trigermylstibine* $(GeH_3)_3Sb$

An analogous exchange reaction used in the synthesis of $(GeH_3)_3As$ described above has been applied by Ebsworth and coworkers toward the synthesis of $(GeH_3)_3Sb$[65]:

$$(SiH_3)_3Sb + 3GeH_3Br \longrightarrow (GeH_3)_3Sb + + 3SiH_3Br$$

Excess GeH_3Br is combined with $(SiH_3)_3Sb$† in an all-glass break-seal tube and held at −23° until the $(SiH_3)_3Sb$ is completely consumed. If necessary, the SiH_3Br formed can be removed from the system to aid in displacing the equilibrium toward the desired products.

In the procedure followed in the literature, $(SiH_3)_3Sb$ ($\cong$1.0 mmole) and

* This compound is prepared by the reaction of SiH_3Br with $KAsH_2$ in dimethyl ether at low temperatures.[129]

† This compound can be prepared by the low-temperature reaction of SiH_3Br with Li_3Sb in dimethyl ether.[129]

GeH_3Br (4.8 mmoles) were allowed to react at $-23°$ for 4 hr. After allowing the products to pass through a trap at $-46°$, it was noted by infrared spectroscopy that the less volatile fraction (in the $-46°$ trap) still contained $(SiH_3)_3Sb$. Further fractionation enabled the unreacted GeH_3Br to be separated from the SiH_3Br formed. The GeH_3Br was then combined with the $(SiH_3)_3Sb$–$(GeH_3)_3Sb$ mixture, and the reaction was allowed to continue for an additional $3\frac{1}{2}$ hr at $-23°$. The infrared spectrum of the sample collecting in the $-46°$ trap after this second reaction period showed the presence of $(GeH_3)_3Sb$ only. The reaction as described above is essentially quantitative.

The reaction of Li_3Sb with GeH_3Br in dimethyl ether at $-78°$ also leads to the formation of $(GeH_3)_3Sb$.[65] This reaction produces only small amounts of $(GeH_3)_3Sb$; the yield can be increased significantly, however, by adding a catalytic amount of SiH_3Br to the reactants. The yield is still basically small, and, as a result, this is not a recommended procedure for preparing $(GeH_3)_3Sb$.

B. Compounds of the Type GeH_3EH_2 (E = P, As)

The synthesis of GeH_3PH_2 and GeH_3AsH_2 has been achieved by four different methods. The procedures for preparing the compounds based on the ozonizer-discharge-induced reaction of GeH_4 with PH_3[86] or GeH_4 with AsH_3[86] and the acid hydrolysis of germanide-phosphide or germanide-arsenide alloys[125,126]

$$GeH_4 + EH_3 \xrightarrow{\text{electric discharge}} GeH_3EH_2 + \text{other products} \qquad E = P, As$$

$$CaGe\text{—}Ca_3E_2 + H^+ \longrightarrow GeH_3EH_2 + \text{other products} \qquad E = P, As$$

have been adequately discussed by Jolly and Norman[11] and are not considered here. These methods, however, are somewhat more tedious than the two more recently reported procedures given below:

$$4GeH_3X + LiAl(EH_2)_4 \longrightarrow 4GeH_3EH_2 + LiX + AlX_3 \qquad X = \text{halogen}; E = P, As$$

$$GeH_3X + SiH_3EH_2 \longrightarrow GeH_3EH_2 + SiH_3X \qquad X = \text{halogen}; E = P, As$$

1. *Germylphosphine* GeH_3PH_2

a. By the reaction of germyl bromide with lithium tetrakis(dihydrogenphosphinido) aluminate. Wingleth and Norman have described a very convenient procedure for preparing GeH_3PH_2 in an 88 percent yield by the low-temperature interaction of GeH_3Br with $LiAl(PH_2)_4$[77]:

$$4GeH_3Br + LiAl(PH_2)_4 \longrightarrow 4GeH_3PH_2 + LiBr + AlBr_3$$

In a typical preparation, GeH_3Br (3.6 mmoles) and $LiAl(PH_2)_4$* (5.4 mmoles)

* Prepared by the 48 to 96 hr reaction of PH_3 with $LiAlH_4$ in triglyme[130]:

$$4PH_3 + LiAlH_4 \longrightarrow LiAl(PH_2)_4 + 4H_2$$

The actual compound used in this reaction has not been directly characterized and may be a complex mixture of $LiAlH_m(PH_2)_{4-m}$ species.

in an ether solvent are allowed to react for 1 hr at −45° followed by slow warming of the mixture to room temperature. The GeH_3PH_2 can be removed from the reaction products by its condensation in a −96° trap.* Small quantities of uncharacterized higher molecular weight germyl phosphines are obtained in the reaction if the ratio of $GeH_3Br:LiAl(PH_2)_4$ is greater than 4:1. Monogermyl phosphine or other germyl phosphines could not be obtained by the analogous reaction of potassium dihydrogenphosphinide (KPH_2) with GeH_3Br.[64]

b. By the interaction of silylphosphine with germyl halides. A particularly convenient preparation of GeH_3PH_2 is based on the almost quantitative exchange that occurs on allowing GeH_3Cl or GeH_3Br to interact with SiH_3PH_2[67]:

$$GeH_3X + SiH_3PH_2 \longrightarrow GeH_3PH_2 + SiH_3X \qquad X = Cl, Br$$

The deuterated derivatives GeD_3PH_2 and GeH_3PD_2 are prepared in an analogous manner using GeD_3Cl or SiH_3PD_2 in the reaction.[36] As an illustration, interaction of SiH_3PH_2 (0.34 mmole) with GeH_3Cl (0.28 mmole) at −63° produces pure GeH_3PH_2 (0.27 mmole) after a 2-hr reaction period.[67] The GeH_3Cl must be allowed to react completely because it is very difficult to separate from GeH_3PH_2.

2. *Germylarsine* GeH_3AsH_2

Extensions of the aluminate and exchange reactions discussed in the previous section for the preparation of GeH_3PH_2 have been used by Drake and coworkers to prepare GeH_3AsH_2 conveniently in good yields.[67,76] Typical procedures are given below.

a. By the reaction of germyl halides with lithium tetrakis(dihydrogenarsinido) aluminate. The $LiAl(AsH_2)_4$ slurry† is made by keeping an atmosphere of AsH_3 (≅10 mmoles) over a stirred mixture of $LiAlH_4$ (≅2.1 mmoles) in diglyme (≅30 ml).[76] Periodically the vessel is cooled to −196°, hydrogen is pumped away, and the AsH_3 atmosphere is renewed. The reaction is allowed to continue for about 2 days, after which approximately 80 percent of the AsH_3 has undergone conversion. The desired GeH_3AsH_2 can be obtained by condensing a germyl halide onto the $LiAl(AsH_2)_4$ slurry.[76] Germyl bromide gives the best yield of product, and the procedure calls for condensing this halide onto the $LiAl(AsH_2)_4$ slurry at −196° and allowing the mixture to warm to −45°. After about ½ hr the reaction solution turns to a green-brown

* GeH_3PH_2 and Ge_2H_6 (if any is formed in the reaction) both condense in a −96° trap. These two hydrides can be separated from one another by gas-chromatographic techniques.[36]

† The actual compound used in this reaction has not been directly characterized and may be a complex mixture of $LiAlH_m(AsH_2)_{4-m}$ species.

color. The system is then allowed to warm to room temperature, and the volatile products are condensed out of the reaction vessel. The GeH_3AsH_2 can be separated from the reaction products by its condensation in a trap cooled to $-78°$. The yields of GeH_3AsH_2, based on the $LiAl(AsH_2)_4$ used, obtained from GeH_3I, GeH_3Br, and GeH_3Cl are 59, 62, and 9 percent, respectively. A considerable portion of the germyl halide used in the reaction is converted to Ge_2H_6 (in amounts varying from *ca* 10–40 percent). The addition of H_2S does not significantly improve the yield of GeH_3AsH_2.[76]

b. By the reaction of silylarsine with germyl halides

$$SiH_3AsH_2 + GeH_3X \rightleftharpoons GeH_3AsH_2 + SiH_3X \qquad X = Cl, Br$$

Silylarsine (0.48 mmole) and GeH_3Cl (0.38 mmole) are condensed together without a solvent and held at $-45°$ for 12 hr.[67] The yield of GeH_3AsH_2 can be improved significantly by cooling the reaction vessel to $-96°$ every 2 hr or so and allowing the SiH_3Cl formed to distill into another trap at $-196°$. The GeH_3AsH_2 (0.31 mmole) can be purified by its condensation in a trap held at $-96°$. Silyl chloride and unreacted SiH_3AsH_2 in the products pass through the $-96°$ trap. The use of GeH_3Br is not recommended in this synthesis. Silyl bromide and SiH_3AsH_2 have similar volatilities, and so the reaction equilibrium cannot be forced toward the desired products by the successive removal of the silyl halide in this case.

C. Compounds of the Type $(GeH_3)_2EH$ (E = P, As)

Drake and Riddle have reported the convenient preparation of $(GeH_3)_2PH$ and $(GeH_3)_2AsH$ by the induced condensation of GeH_3PH_2 and GeH_3AsH_2, respectively[81a]:

$$2GeH_3EH_2 \longrightarrow 3(GeH_3)_2EH + EH_3 \qquad E = P, As$$

The synthesis of $(GeH_3)_2PH$ is based on the self-condensation of GeH_3PH_2 at room temperature, and the addition of a small amount of water is recommended to assist the condensation of GeH_3AsH_2. It is difficult to obtain and especially keep the compounds in a pure state, owing to their tendency to condense further to the corresponding trigermyl derivative:

$$3(GeH_3)_2EH \longrightarrow 2(GeH_3)_3E + EH_3$$

Most of the synthetic work reported to date has been carried out in standard nmr tubes. An advantage of using nmr tubes is that the extent of the condensations can be followed conveniently by nmr spectroscopy. It has been possible to break open the tubes and isolate the compounds in the vacuum line.

1. *Digermylphosphine* $(GeH_3)_2PH$

In a typical preparation as described in the literature, GeH_3PH_2 ($\cong$1 mmole) was condensed into a semimicro tube that was then sealed and allowed to warm to room temperature.[81a] The ^1H spectrum of the sample initially showed the presence of GeH_3PH_2; after several hours at room temperature, however, the intensity of the signals for GeH_3PH_2 began to diminish, and new signals simultaneously appeared for PH_3, $(GeH_3)_2PH$, and $(GeH_3)_3P$. The extent of conversion after a certain number of hours is given in Table IV. A condensation time of about 60 hr is recommended for the synthesis of $(GeH_3)_2PH$. After this time period, the nmr tube can be cooled to $-196°$, broken open under vacuum, and the products fractionated in the vacuum line. The $(GeH_3)_2PH$ and $(GeH_3)_3P$ condense in a trap at $-63°$, while GeH_4, PH_3, and unreacted GeH_3PH_2 pass through at this temperature. The $(GeH_3)_3P$ is removed from the $(GeH_3)_2PH$ by its condensation in a trap at $-45°$. The $(GeH_3)_2PH$ prepared in this manner rapidly condenses further to give $(GeH_3)_3P$ and PH_3. The formation of GeH_4 has also been noted in the condensations of germylphosphines.

2. *Digermylarsine* $(GeH_3)_2AsH$

It has been observed that GeH_3AsH_2 undergoes self-condensation at room temperature, forming $(GeH_3)_2AsH$ and $(GeH_3)_3As$ in a manner comparable with the self-condensation of GeH_3PH_2 discussed in the preceding section.[81a] As before, the reaction can be followed conveniently by nmr spectroscopy. Much longer time intervals are involved in the arsenic system, however, before appreciable self-condensation occurs; for example, the products of the fastest condensation were 54 percent mono-, 35 percent di-, and 9 percent tri-germylphosphine after a 36-day reaction period.

The rate of condensations is increased significantly by having a small amount of water present with the GeH_3AsH_2.[81a] In a typical experiment,

TABLE IV

The Time Dependency for the Extent of Self-condensation of GeH_3PH_2

	Mole percent of germyl phosphines present		
Time (hr)	GeH_3PH_2	$(GeH_3)_2PH$	$(GeH_3)_3P$
0	100	0	0
12	86	14	0
26	77	22	1
35	74	23	3
46	68	27	5
71	74	19	7

TABLE V

The Time Dependency for the Extent of Condensation of GeH_3AsH_2 in the Presence of Water

	Mole percent of germyl arsines present		
Time (days)	GeH_3AsH_2	$(GeH_3)_2AsH$	$(GeH_3)_3As$
0	100	0	0
1	75	20	5
2	62	29	9
3	50	33	17
4	44	38	18
6	40	37	23

GeH_3AsH_2 (0.5 mmole) and water (0.05 mmole) were condensed into an nmr tube ($\cong$2 mm OD) that was then sealed and allowed to warm to room temperature. The extent of conversion under these conditions after a certain number of days as followed by nmr spectroscopy is given in Table V.

In the procedure for preparing $(GeH_3)_2AsH$ described in the literature, the water-catalyzed condensation was allowed to proceed until about 80 percent of the GeH_3AsH_2 had undergone condensation. The nmr tube was then cooled to $-196°$, opened under vacuum, and the products fractionated in the vacuum line. The $(GeH_3)_2AsH$ condenses in a trap at $-22°$.*

$(GeH_3)_2AsH$ prepared in this manner rapidly condenses further to give $(GeH_3)_3As$ and AsH_3.

D. Chlorophosphinogermane GeH_2ClPH_2

Drake, Goddard, and Riddle have extended the silylphosphine halogermane exchange reaction noted previously to include the interaction of SiH_3PH_2 with GeH_2Cl_2.[69] The compounds react at room temperature according to the equation

$$SiH_3PH_2 + GeH_2Cl_2 \longrightarrow GeH_2ClPH_2 + SiH_3Cl$$

The GeH_2ClPH_2 produced was identified by its ^1H nmr spectrum. An analogous exchange was noted for the interaction of CH_3GeHCl_2 with SiH_3PH_2, producing $CH_3GeHClPH_2$.[69] In both cases attempts to isolate the pure compounds in the vacuum line were not successful.

E. Germylphosphine-Borane Adduct $GeH_3PH_2BH_3$

Drake and Riddle have studied the adduct formed between GeH_3PH_2 and B_2H_6 by nmr spectroscopy.[36] In the experiment described in the literature,

* The $(GeH_3)_2AsH$ isolated from this reaction is reported to be at least 95 percent pure.

GeH_3PH_2 (0.4 mmole) was combined with B_2H_6 (0.2 mmole) and a small amount of tetramethylsilane in a nmr tube. The tube was sealed and inserted into the nmr spectrometer, the sample compartment of which was cooled to $-80°$. The spectrum of the mixture below $-20°$ was that of unreacted starting materials; at $-20°$, however, the features of the spectrum changed in accordance with the quantitative formation of the adduct $GeH_3PH_2BH_3$:

$$GeH_3PH_2 + \tfrac{1}{2}B_2H_6 \xrightarrow{-20°} GeH_3PH_2BH_3$$

The ^1H nmr spectrum for this adduct and the spectra obtained for the adducts $GeH_3PD_2BH_3$ and $GeD_3PH_2BH_3$ prepared in an identical manner can be analyzed readily in terms of first-order couplings.[36] The data leave no doubt whatsoever as to the chemical constitution and structures of the adducts.

F. Digermylcarbodiimide $GeH_3NCNGeH_3$*

Cradock and Ebsworth have reported the preparation of this extremely useful synthetic intermediate by the reaction of GeH_3F with $(CH_3)_3SiNCNSi(CH_3)_3$[68]:

$$2GeH_3F + (CH_3)_3SiNCNSi(CH_3)_3 \longrightarrow GeH_3NCNGeH + 2(CH_3)_3SiF$$

In a typical preparation, $(CH_3)_3SiNCNSi(CH_3)_3$† (5.69 mmoles) and GeH_3F (11.82 mmoles) are combined and allowed to react for 5 min at room temperature. The $GeH_3NCNGeH_3$ (5.41 mmoles, 95 percent yield) is purified easily in the vacuum line by its condensation in a trap at $-45°$.

IX. PREPARATION OF GERMYL COMPOUNDS CONTAINING GERMANIUM BOUND TO THE LOWER GROUP IVA ELEMENTS

Derivatives of the germanes that contain the group IVA elements, other than germanium, include organogermanes and the ternary silicon-germanium and germanium-tin hydrides. Organogermanes have been reviewed recently

* It has been concluded from an analysis of the vibrational spectra that the compound is a carbodiimide (*a*) rather than a cyanamide (*b*).[68]:

$GeH_3NCNGeH_3$ (*a*) NCN$(GeH_3)_2$ (*b*)

† Prepared by the 12-hr room-temperature reaction of $(CH_3)_3SiCl$ with Ag_2CN_2 in an all-glass ampul.[68,131]

by Hooton in this series[10] and by Glockling[8] and are not considered here. The inorganic mixed hydrides are considered only briefly at this point, owing to the recent coverage of the subject by Norman and Jolly.[11]

A. Mixed Germanium-Silicon and Germanium-Tin Hydrides

1. Germylsilane GeH_3SiH_3

Several methods are now available for the synthesis of GeH_3SiH_3. Spanier and MacDiarmid first reported the synthesis of the compound as a result of subjecting a mixture of GeH_4 and SiH_4 to an ozonizer silent electric discharge*[88]:

$$SiH_4 + GeH_4 \xrightarrow{\text{electrical discharge}} GeH_3SiH_3 + \text{other products}$$

Royen and Rocktäschel have prepared the compound and other mixed silicon-germanium hydrides by the action of 5*N* hydrochloric acid on a CaSiGe alloy.[126,132] Timms, Simpson, and Phillips have prepared the compound by the hydrolysis of an MgSiGe alloy.[133]

In all these procedures the yield of GeH_3SiH_3 is relatively low. For example, the yield of GeH_3SiH_3 reported by Spanier and MacDiarmid for the electric-discharge method is about 2 percent.[88] The hydrolysis procedures using CaSiGe or Mg_4SiGe alloys produce 3 percent or 0.6 percent GeH_3SiH_3, respectively.[126,133]

Silylgermane has also been prepared somewhat more conveniently by a coupling reaction, either involving $KSiH_3$ and GeH_3Cl as described by Varma and Cox[73] or $NaGeH_3$ and SiH_3Br as described by Dutton and Onyszchuk.[72] The Varma-Cox procedure calls for allowing GeH_3Cl to react with $KSiH_3$† in 1,2-dimethoxyethane at room temperature for 15 min. Purification of the GeH_3SiH_3 can be achieved by low-temperature fractionations. The reported yield by this method is approximately 20 percent. The Dutton-Onyszchuk method involves the condensation of SiH_3Br on ammonia-free $NaGeH_3$‡ and allowing the reaction to proceed at −78° for 16 hr, then at −45° for 1 hr. Before separating the products, the reactants are warmed from −23 to +1° over a 3 hr period. The yield of GeH_3SiH_3 by this method is reported to be 7.8 percent.[72]

* Details for the construction and use of a typical ozonizer discharge apparatus have been given by Gokhale, Drake, and Jolly.[13]

† Prepared by the reaction of Si_2H_6 with KH:

$$Si_2H_6 + KH \longrightarrow KSiH_3 + SiH_4$$

Other methods based on the reaction of SiH_4 with potassium sand are also available.[134]

‡ Readily prepared from the reaction of GeH_4 with sodium in liquid ammonia.[25] See also Section XI-A-l-a.

A variation of the Dutton-Onyszchuk procedure has been used by Lannon, Weiss, and Nixon for the preparation of SiH_3GeD_3 and SiD_3GeH_3.[135] The compounds were prepared in low yields for infrared studies by the interaction of $NaGeD_3$ with SiH_3I and $NaGeH_3$ with SiD_3I, respectively. Small amounts of disilyl ether formed in the reaction can be removed from the samples by passing the mixtures through an 18-ft column packed with Dow Corning 702 fluid on firebrick.[135]

2. *Other More Complex Germysilanes* $Si_xGe_yH_{2x+2y+2}$

A number of silicon-germanium hydrides that are more complex than SiH_3GeH_3 have been prepared by several methods. All the more complex derivatives must be purified by gas-chromatographic techniques as described by Phillips and coworkers.[136,137] The synthetic methods used thus far include the following:

1. The hydrolysis of various alloys containing silicon and germanium or SiO–GeO mixtures; for example,

$$Mg_{20}Si_9Ge \xrightarrow{H_2O} SiGeH_6\ (3\%),\ n\text{-}Si_3GeH_{10}\ (1.5\%),\ n\text{-}Si_4GeH_{12}\ (0.3\%),\ n\text{-}Si_5GeH_{14}\ (0.07\%) \quad \text{(Ref. 133)}$$

$$Mg_4SiGe \xrightarrow{H_2O} SiGeH_6\ (0.6\%),\ SiGe_2H_8\ (0.4\%),\ n\text{-}SiGe_3H_{10}\ (0.03\%),\ n\text{-}Si_2Ge_2H_{10}\ (0.2\%),\ n\text{-}Si_3Ge_2H_{12}\ (0.02\%) \quad \text{(Ref. 133)}$$

$$Ca(Ge_{0.5}Si_{0.5}) \xrightarrow{H_2O} SiGeH_6\ (3\%),\ Si_2GeH_8\ (0.7\%),\ SiGe_2H_8\ (0.2\%) \quad \text{(Ref. 126)}$$

2. The pyrolysis of a germane with a silane at 350 to 370°,[133] for example,

$$Si_2H_6 + Ge_2H_6 \xrightarrow{340°\ (0.5\ sec)} Si_2GeH_8\ (10\%)$$

$$Si_3H_8 + Ge_2H_6\ (1{:}2\ \text{ratio}) \xrightarrow{340°\ (0.5\ sec)} n\text{-}Si_3GeH_{10}\ (3\%) + n\text{-}Si_3Ge_2H_{12}\ (0.5\%)$$

$$Ge_3H_8 + \text{iso-}Si_4H_{10} \xrightarrow{360°\ (0.4\ sec)} [SiSi(Si)SiGe]H_{12}$$

3. Subjecting mixtures of GeH_4 and SiH_4 (or Si_2H_6) to an ozonizer silent electric discharge; for example,

$$GeH_4 + SiH_4 + Si_2H_6 \xrightarrow{\text{electrical discharge}} \text{neo-}Si_5H_{12},\ \text{neo-}Si_4GeH_{12},\ \text{neo-}Si_3Ge_2H_{12},\ \text{neo-}Si_2Ge_3H_{12},\ \text{neo-}SiGe_4H_{12},\ \text{iso-}SiGe_4H_{12},\ \text{iso-}Si_4Ge_2H_{14},\ \text{neo-}Ge_5H_{12},\ n\text{-}SiGe_4H_{12},\ n\text{-}Si_4Ge_2H_{14},\ \text{iso-}Si_3Ge_3H_{14},\ n\text{-}Si_3Ge_3H_{14},\ \text{a branched } Si_6H_{14},\ \text{and a branched } Si_7H_{16} \quad \text{(Ref. 87)}$$

The high-molecular-weight silicon-germanium hydrides have been characterized mainly by their chlorination and pyrolysis products, their ability to be adsorbed on molecular sieves, and their retention times on specific gas-chromatographic columns.

Mackay, Hosfield, and Stobart have carried out more extensive studies on the characterization of the hydrides Si_2GeH_8 and $SiGe_2H_8$ produced by the action of a silent electric discharge on SiH_4–GeH_4 mixtures.[138] By using a mixture of SiH_4:GeH_4 in a 1:4 ratio, the relative yields of compounds produced in a typical synthesis were $SiGeH_6$ (trace), Ge_2H_6(40), Si_3H_8 (trace), Si_2GeH_8(7), $SiGe_2H_8$(31), and Ge_3H_8(100). A silane-rich mixture also produces compounds that have a high germanium content. The Si_2GeH_8 fraction was found to contain both the symmetric and asymmetric position isomers, $SiH_3GeH_2SiH_3$ and $SiH_3SiH_2GeH_3$, in about equimolar ratios, and the $SiGe_2H_8$ fraction was almost pure $SiH_3GeH_2GeH_3$.[138]

The Si_2GeH_8 position isomers produced in the SiH_4–GeH_4 reaction could not be separated from one another by gas-chromatographic techniques, even at $-40°$; the unsymmetrical isomer $SiH_3SiH_2GeH_3$, however, was obtained almost free of the symmetric isomer by subjecting a mixture of Si_2H_6 and GeH_4 (1:2 ratio) to an electrical discharge. The compounds discussed were characterized by ^1H nmr, Raman, infrared, and mass spectroscopic studies.[138]

3. *Methyl(germyl)silanes* $(CH_3)_nSi(GeH_3)_{4-n}$ *and the Related Compound* $(CH_3)_3GeGeH_3$

Dutton and Onyszchuk have prepared a series of methyl(germyl)silanes of the type $(CH_3)_nSi(GeH_3)_{4-n}$, where $n = 0, 1, 2,$ or 3 by the reaction of ammonia-free $NaGeH_3$ with the appropriate methylchlorosilane[72]:

$$(CH_3)_3SiCl + NaGeH_3 \longrightarrow (CH_3)_3SiGeH_3 + NaCl$$

$$(CH_3)_2SiCl_2 + 2NaGeH_3 \longrightarrow (CH_3)_2Si(GeH_3)_2 + 2NaCl$$

$$CH_3SiCl_3 + 3NaGeH_3 \longrightarrow CH_3Si(GeH_3)_3 + 3NaCl$$

$$SiCl_4 + 4NaGeH_3 \longrightarrow (GeH_3)_4Si + 4NaCl$$

In each case the appropriate methylchlorosilane was condensed onto ammonia-free $NaGeH_3$ (see Section XI-A-1-a), and the mixture allowed to slowly warm over a 3 to 4 hr period from -23 to about 0°. Some pertinent information about the reactions is given in Table VI. The compounds can be purified in the vacuum line, although for $(CH_3)_3SiGeH_3$ the final purification was by gas chromatography.

Amberger and Mühlhofer have also reported the preparation of $(CH_3)_3SiGeH_3$ by an analogous reaction that used $(CH_3)_3SiCl$ and $KGeH_3$ in a monoglyme or diglyme solvent.[50]

TABLE VI

Synthesis of Methyl(germyl)silanes by the Reaction of Methylchlorosilanes with Sodium Germyl[72]

Compound	mmoles halosilane/ mmoles $NaGeH_3$	Product yield (%)	Reaction time (hr)	Reaction temperature (range, °C)
$(CH_3)_3SiGeH_3$[a]	9.1/10.4	12	4	−23 to 0
$(CH_3)_2Si(GeH_3)_2$	5.5/7.2	28	3	−23 to +2
$CH_3Si(GeH_3)_3$	9.1/12.0	34	4.5	−23 to +9
$(GeH_3)_4Si$	10.1/13.7	[b]	4	−23 to −1

[a] This compound can also be prepared from $(CH_3)_3SiF$ and $NaGeH_3$.

[b] This compound is prepared in a very low yield.

The related compound, 1,1,1-trimethyldigermane, $(CH_3)_3GeGeH_3$, has been prepared by an identical method[72]:

$$(CH_3)_3GeF + NaGeH_3 \longrightarrow (CH_3)_3GeGeH_3 + NaF$$

Trimethylfluorogermane (2.6 mmoles) was condensed onto a large excess of ammonia-free $NaGeH_3$ (21.3 mmoles), and the mixture allowed to react with slow warming over the range of −20° to −4° for a 2-hr period, followed by keeping the mixture at 0° for an additional 6 hr. The resulting $(CH_3)_3GeGeH_3$ (0.96 mmole, 36.4 percent yield) is purified by its condensation in a trap at −78°. The compound can also be prepared from $(CH_3)_3GeBr$ and ammonia-free $NaGeH_3$, although the yield is much lower (8.4 percent).[72] Mackay and coworkers have also prepared $(CH_3)_3GeGeH_3$ and its GeD_3 analog by a coupling reaction involving $KGeH_3$ or $KGeD_3$ and $(CH_3)_3GeX$.[138a]

4. *Germylfluorosilanes* $GeH_3(SiF_2)_nH$

Solan and Timms have reported the synthesis of germylfluorosilanes of the type $GeH_3(SiF_2)_nH$, where $n = 1$, 2, and 3, by the interaction of silicon difluoride with GeH_4.[139,140] The compounds are prepared by co-condensing a low-pressure gaseous mixture containing SiF_2, SiF_4, and GeH_4 in a roughly 4:2:3 mole ratio at −196, followed by allowing the frozen mixture to warm while continually pumping away the reaction products into another trap. The reaction also takes place by first condensing SiF_2 and SiF_4 (2:1 ratio) at −196°, followed by condensing GeH_4 onto the frozen mixture. The yield of volatile reaction products by this latter procedure is lower than when the reactants are co-condensed.

The products produced in the reaction are of similar volatilities and cannot be separated completely by an efficient low-pressure distillation column. Attempts have been made to separate the products by using gas chromatography; under the conditions used, however, the samples undergo complete decomposition. Three reasonably pure fractions of germylfluorosilanes have been obtained from the reaction products by tedious distillations using a special 3-ft long, low-temperature distillation column. The main component of each fraction as identified by infrared and mass spectroscopy proved to be of the following composition: GeH_3SiF_2H (main fraction), $GeH_3Si_2F_4H$, and $GeH_3Si_3F_6H$. The volatility and amount of the fractions decrease in the order $GeH_3SiF_2H > GeH_3Si_2F_4H > GeH_3Si_3F_6H$. The structures of the first two compounds have been confirmed as being GeH_3SiF_2H and $GeH_3SiF_2SiF_2H$ by their nmr spectra; it was not possible to obtain an nmr spectrum of $GeH_3Si_3F_6H$, however, owing to the unstable nature of the compound. It is presumed to be of the type $GeH_3(SiF_2)_3H$. A tentative identification of GeH_3SiF_3 as an impurity in the GeH_3SiF_2H fraction was also made in this study.

B. Germylstannane GeH_3SnH_3

This highly unstable compound has been prepared by Wiberg, Amberger, and Cambensi by the low-temperature $LiAlH_4$ reduction of $Cl_3GeSnCl_3$ or the corresponding hexaacetate derivative,[141] for example,

$$Cl_3GeSnCl_3 + LiAlH_4 \xrightarrow{\text{ether}\,(-80^\circ)} GeH_3SnH_3 + LiCl + AlCl_3$$

The compound is stable only in dilute etherial solutions and readily decomposes in more concentrated solutions or above −80°C:

$$GeH_3SnH_3 \longrightarrow GeH_4 + Sn + H_2$$

Attempts to prepare mixed hydrides of germanium and tin by the acid hydrolysis of an MgGeSn alloy have been unsuccessful.[133]

X. PREPARATION OF GERMYL COMPOUNDS CONTAINING GERMANIUM BOUND TO THE GROUP IIIA ELEMENTS

Very few compounds that contain germyl groups bound to the Group IIIA elements are known.* Amberger and coworkers[84] have been able to characterize partially the complex $K[(GeH_3)Ga(CH_3)_2Cl]$, formed by the reaction of $KGeH_3$ with $(CH_3)_2GaCl$, and also $GeH_3Ga(CH_3)_2$ (containing $(CH_3)_2GaH$ impurity) a product of the decomposition of the complex:

$$KGeH_3 + (CH_3)_2GaCl + 1.5CH_3OCH_2CH_2OCH_3 \xrightarrow{-20^\circ} K[(GeH_3)Ga(CH_3)_2Cl]\cdot 1.5CH_3OCH_2CH_2OCH_3$$

$$K[(GeH_3)Ga(CH_3)_2Cl]\cdot 1.5CH_3OCH_2CH_2OCH_3 \xrightarrow{\text{toluene}\,(-15^\circ)} GeH_3Ga(CH_3)_2 + KCl + 1.5CH_3OCH_2CH_2OCH_3$$

* Preliminary attempts to prepare mixed hydrides that contain Ge—B bonds by cracking germanium hydrides with boron hydrides have been unsuccessful.[133]

Rustad and Jolly have reported the formation of an intriguing salt by the reaction of B_2H_6 with $KGeH_3$ [51]:

$$KGeH_3 + B_2H_6 \longrightarrow KGeH_3BH_3$$

The physical and chemical properties of the salt are consistent with formulating the compound as potassium germyltrihydroborate $K^+[GeH_3BH_3]^-$, the anion of which is a hydroborate ion having one of its hydrogen atoms replaced by a germyl group, or, equivalently stated, a germyl anion coordinated to a borane group.

A. Potassium Germylchlorodimethylgallate $K[(GeH_3)Ga(CH_3)_2Cl]$

In a typical preparation, $KGeH_3$ (30 mmoles) dissolved in 1,2-dimethoxyethane is added slowly to a solution of $(CH_3)_2GaCl$ (28.5 mmoles) in 1,2-dimethoxyethane cooled to $-20°$.[84] There is no KCl precipitate formed, nor should there be any evolution of gas during the addition under these conditions. The solvent is removed at about $-25°$, and the residue dried at 0 to 5° under vacuum. The material that remains is a yellow viscous oil that, from analysis, is the pure solvated complex $K[(GeH_3)Ga(CH_3)_2Cl] \cdot 1.5CH_3OC_2H_4OCH_3$ (26.8 mmoles, 94 percent yield, based on $(CH_3)_2GaCl$). The complex slowly decomposes above 10° and more rapidly above 30°.

B. Germyldimethylgallane $GeH_3Ga(CH_3)_2$

The solvated adduct (15 mmoles) prepared from $KGeH_3$ and $(CH_3)_2GaCl$, as described in the previous section, is stirred with 80 ml of toluene at $-15°$ for 2 hr.[84] The solvent is then removed by distillation, first at $-20°$, then by pumping on the products maintained at $-15°$, for 2 days. A yellow chlorine-containing oil results. This oil produces, on distillation at 25 to 30° under high vacuum, a colorless liquid analyzed to be a mixture of $GeH_3Ga(CH_3)_2$ (76 percent) and $(CH_3)_2GaH$ (24 percent) that cannot be separated further. At 25°, $GeH_3Ga(CH_3)_2$ slowly decomposes to form $(CH_3)_2GaH$, polymeric $(GeH_2)_x$, and hydrogen.

C. Potassium Germyltrihydroborate $K[GeH_3BH_3]$

The basic method for preparing $K[GeH_3BH_3]$ involves condensing B_2H_6 (0.55 mmole) onto frozen $KGeH_3$ in 1,2-dimethoxyethane at $-196°$, followed by keeping the reactants at $-78°$ for about 15 min with occasional shaking, then warming the mixture to about 25°.*[51] A precipitate of KBH_4 (0.134 mmole) is removed by filtration under vacuum, and the solvent is then pumped

* A description of the apparatus used in carrying out this synthesis is given in Ref. 51.

away at room temperature. The white residue that results gradually turns yellow on standing and evolves GeH_4 (0.103 mmole) and a trace of hydrogen. This residue is extracted for 30 min with diethyl ether and is filtered, again under vacuum. The solid $KGeH_3BH_3$ (0.813 mmole) is obtained by evaporating the ether from the filtrate.

By-products of the reaction are KBH_4 and a compound thought to be potassium digermanyltrihydroborate $KGe_2H_5BH_3$, formed in the secondary reaction:

$$2KGeH_3BH_3 \longrightarrow KBH_4 + KGe_2H_5BH_3$$

The tentatively identified digermanyl derivative is stable only dissolved in 1,2-dimethoxyethane. The formation of the compound is promoted by the presence of excess B_2H_6.

XI. PREPARATION OF ALKALI-METAL AND TRANSITION-METAL DERIVATIVES OF GERMANE

A. Alkali-metal Derivatives of the Type $MGeH_3$ (M = Alkali Metal)

Alkali-metal derivatives of GeH_4 of the type $MGeH_3$, where M = Li, Na, K, have proved to be very useful intermediates in the synthesis of various germyl compounds, particularly those containing specific organic groupings. The alkali-metal derivatives themselves are normally not isolated in a pure form but rather are prepared and studied in a suitable solvent. The compounds, or solutions of them, can be prepared by two general methods. One involves the interaction of GeH_4 with an alkali metal dissolved in an appropriate solvent, and the second makes use of the deprotonation of GeH_4 by KOH.

1. From an Alkali Metal and Germane

a. In liquid ammonia. The reaction of GeH_4 with alkali metals has been studied in a variety of solvents. Kraus and coworkers carried out some of the original reactions of this type in liquid ammonia. Basically the procedure described involves passing GeH_4 slowly through a solution of sodium[25] or potassium[48] in liquid ammonia:

$$GeH_4 + M(NH_3) \longrightarrow MGeH_3 + \tfrac{1}{2}H_2 \qquad M = Na, K$$

The reaction is more complex than that written above, because hydrogen gas in excess of that required by the equation is evolved, especially at high metal concentrations.[30,142] This can be attributed to a side reaction of the type

$$GeH_4 + (2 + x)NH_3 \longrightarrow Ge(NH)_2 \cdot xNH_3 + 4H_2$$

Emeléus and Mackay have carried out conductometric titrations of sodium-ammonia solutions with GeH_4 at $-63.5°$.[30] The characteristic blue color of the ammonia solution is discharged at a GeH_4/Na ratio of 0.5, leaving a pale yellow solution and a yellow-green precipitate. The precipitate dissolves almost completely on the addition of more hydride up to a 1:1 mole ratio and was at one time thought to be the disodium salt, Na_2GeH_2,[30] that is,

First step:

$$GeH_4 + 2Na \longrightarrow Na_2GeH_2 + H_2$$

which would react with additional GeH_4 according to

Second step:

$$Na_2GeH_2 + GeH_4 \longrightarrow 2NaGeH_3$$

The results of Rustad and Jolly's investigation of the reaction at $-77°$, however, indicate that the precipitate is actually $NaNH_2$, identified by its X-ray powder pattern.*[142]

The actual course of the reaction is thus given as:

First step:

$$GeH_4 + 2Na + NH_3 \longrightarrow NaGeH_3 + NaNH_2 + H_2$$

*Second step:***

$$NaNH_2 + GeH_4 \longrightarrow NaGeH_3 + NH_3$$

The mechanism proposed for the interaction of GeH_4 with metal-ammonia solutions that accounts for the formation of the amide ion and germanium imide is[142]

$$e^- + GeH_4 \longrightarrow GeH_3^- + H$$

$$H + e^- \longrightarrow H^-$$

$$H^- + NH_3 \longrightarrow NH_2^- + H_2$$

$$2H \longrightarrow H_2$$

$$GeH_4 + NH_2^- \longrightarrow GeH_3^- + NH_3$$

$$GeH_4 + NH_2^- \longrightarrow [GeH_3NH_2] + H^-$$

The proposed intermediate $[GeH_3NH_2]$ presumably undergoes ammonolysis to form $Ge(NH_2)_4$ or $Ge(NH)_2$ rapidly.

From a synthetic point of view it should be noted that in the preparation of the alkali-metal salts of GeH_4 by this method one should continue to pass GeH_4 through the metal-ammonia solution for a short period after the

* Potassium amide is sufficiently soluble in liquid ammonia so that it does not precipitate during the course of the reaction of GeH_4 with potassium-ammonia solutions.

** In a separate experiment, it has been shown that GeH_4 reacts with KNH_2 in liquid ammonia to form $KGeH_3$, mainly.[142,143]

characteristic blue color is discharged. Also, secondary reactions that involve the ammonolysis of GeH_4 can be minimized by keeping the concentration of the metal-ammonia solution low.

It is reported that both $NaGeH_3$ and $KGeH_3$ can be isolated as ammonia-free white solids by pumping away the ammonia solvent.[25,48] The sodium salt has been shown to crystallize first from liquid ammonia as $NaGeH_3 \cdot 6NH_3$.[25] The ammonia of crystallization can be removed in several stages, the following four solid phases having been detected: $NaGeH_3 \cdot 6NH_3$, $NaGeH_3 \cdot 4.5NH_3$, $NaGeH_3 \cdot 2NH_3$, and $NaGeH_3$.[25] In light of the recent work discussed previously, the solid alkali-metal salts obtained in this manner undoubtedly contain some $Ge(NH)_2$ or $Ge(NH)_2 \cdot xNH_3$. The solid alkali-metal salts obtained decompose slowly at liquid-ammonia temperatures and much more rapidly at room temperature. The potassium salt seems to be somewhat more stable than the sodium salt.[48]

Amberger and Boeters report the formation of $LiGeH_3$ by the reaction of GeH_4 with lithium in liquid ammonia.[49] By removing the excess ammonia, a pale gray solid material identified as $LiGeH_3 \cdot 2NH_3$ can be recovered:

$$GeH_4 + Li(xNH_3) \xrightarrow{-\frac{1}{2}H_2} LiGeH_3 \cdot xNH_3 \xrightarrow{-(x-2)NH_3} LiGeH_3 \cdot 2NH_3$$

The diammoniate is thermally stable at room temperature.

As mentioned previously, it is not always necessary to isolate the solid alkali-metal derivatives if they are being used as intermediates. Thus the synthesis of CH_3GeH_3, $C_2H_5GeH_3$, or n-$C_3H_7GeH_3$ can be achieved by introducing CH_3I, C_2H_5Br, or n-C_3H_7Br, respectively, directly into the liquid-ammonia reaction mixture.[48] Several investigators, however, have chosen to remove most, if not all, of the ammonia before proceeding with a particular coupling reaction, for example,

$$SiH_3CH_2Cl + NaGeH_3(s) \longrightarrow SiH_3CH_2GeH_3 + NaCl \qquad \text{(Ref. 83)}$$

$$CH_3OCH_2Cl + NaGeH_3(s) \longrightarrow CH_3OCH_2GeH_3 + NaCl \qquad \text{(Ref. 78)}$$

All the ammonia must be removed in coupling systems that involve silicon-halogen bonds or any other bonds that undergo ammonolysis:

$$SiH_3Br + NaGeH_3(s) \longrightarrow GeH_3SiH_3 + NaBr \qquad \text{(Ref. 72)}$$

$$(CH_3)_3SiCl + NaGeH_3(s) \longrightarrow GeH_3Si(CH_3)_3 + NaCl \qquad \text{(Ref. 72)}$$

$$(CH_3)_3GeF + NaGeH_3(s) \longrightarrow GeH_3Ge(CH_3)_3 + NaF \qquad \text{(Ref. 72)}$$

Reactions of alkali-metal derivatives of GeH_4 are not always so straightforward as these equations seem to imply. For example, the reactions of $NaGeH_3$ with CH_2Br_2 and C_6H_5Br do not produce the corresponding coupled organogermanes but rather proceed according to:[48,143]

$$CH_2Br_2 + 2NaGeH_3 + NH_3 \longrightarrow CH_3GeH_3 + [GeH_3NH_2] \text{ (not isolated)} + 2NaBr$$

$$C_6H_5Br + NaGeH_3 \longrightarrow C_6H_6 + (GeH_2) + NaBr$$

Emeléus and Mackay have studied the interaction of Ge_2H_6 and Ge_3H_8 with sodium in liquid ammonia by conductimetric titrations, coupled with measurements of hydrogen evolution.[30] The conductivity results are consistent with essentially complete cleavage of germanium-germanium bonds in this system, for example,*

$$Ge_2H_6 + 2Na \longrightarrow 2NaGeH_3$$

$$Ge_3H_8 + 4Na \longrightarrow 2NaGeH_3 + Na_2GeH_2$$

Studies of the hydrogen evolution noted, however, suggest that other reactions, such as

$$Ge_2H_6 + Na \longrightarrow Na_2Ge_2H_4 + H_2$$

are also occurring.[30] No alkali-metal derivatives of the higher germanes have yet been isolated or characterized.

b. In hexamethylphosphortriamide and organic ethers. Other solvents have been investigated recently for the synthesis of alkali-metal derivatives of GeH_4 by this method. Cradock, Gibbon, and Van Dyke have found that solutions of alkali metals in hexamethylphosphortriamide $[(CH_3)_2N]_3PO$ (HMPT) react rapidly with GeH_4, providing a convenient source of the GeH_3^- ion for use in other syntheses[52]:

$$GeH_4 + M(HMPT) \longrightarrow M^+ + GeH_3^- + \tfrac{1}{2}H_2 \qquad M = Li, Na, K$$

$$RX + GeH_3^- \xrightarrow{HMPT} RGeH_3 + X^- \qquad RX = CH_3I, C_2H_5Br, CH_3OCH_2Cl$$

The authors give details for a typical reaction and note that variations in technique can result in very low yields of compounds.[52] The HMPT solvent is first degassed with pumping at room temperature, then at $\simeq 110°$ (reflux temperature). The reaction vessel is then cooled to about 10°, potassium (2.3 mmoles) added, and the vessel reevacuated. The metal partially dissolves to give a characteristic blue color. Germane (2.76 mmoles) is then condensed into a side arm on the vessel and allowed to build up pressure slowly over the metal-HMPT solution. After about 15 min, occasionally shaking the reactants, all the potassium dissolves, and the solution acquires a pale yellow color. (An orange- or green-colored solution at this stage implies that something has gone wrong with the preparation. If this happens, the synthesis must be repeated from the beginning with fresh starting materials.) After removing the hydrogen formed and the excess GeH_4, the solution can be used to prepare a variety of organogermanes, as illustrated above.

* Kraus and Carney have used this reaction to conveniently convert the higher germanes to GeH_4.[25] Thus the mixture of germanes produced in the reaction of magnesium germanide with ammonium bromide in liquid ammonia are passed through a solution of sodium in liquid ammonia. The germanes are presumably converted to $NaGeH_3$ and Na_2GeH_2. The sodium salts are soluble in liquid ammonia and all are converted to GeH_4 on the addition of ammonium bromide.

Amberger and Mühlhofer have described the preparation of $KGeH_3$ by the interaction of a Na/K alloy (80:20 weight percent) with GeH_4 in a 1,2-dimethoxyethane (monoglyme) or bis(2-methoxyethyl) ether (diglyme) solvent.[50] The quantitative reaction is reported to require 12 hr, although Stobart has found that on a 5-mmole scale, GeH_4 is consumed in 30 to 40 min at room temperature with shaking. The latter investigator recommends removing hydrogen at about the halfway stage of the reaction[29]:

$$2GeH_4 + 2K \longrightarrow 2KGeH_3 + H_2$$

Rustad and Jolly likewise have prepared the compound by allowing excess GeH_4 to react with potassium in 1,2-dimethoxyethane at $-63.3°$ for 36 hr.[51] The synthesis of $KGeD_3$ has also been achieved by this method, although a slow hydrogen-deuterium exchange occurs with the solvent.[138a] As discussed previously, the solutions of $KGeH_3$ prepared in this manner have been used to synthesize various germyl compounds.[50,51]

Garrity and Ring have studied the reaction of potassium with Ge_2H_6 in 1,2-dimethoxyethane.[144] The following two equations describe the main reaction:

$$xGe_2H_6 + K \longrightarrow xGeH_4 + x(GeH_2)$$

$$GeH_4 + K \longrightarrow KGeH_3 + \tfrac{1}{2}H_2$$

The formation of $KGeH_3$ directly from Ge_2H_6 in this solvent does not seem to be important[144]:

$$Ge_2H_6 + 2K \not\longrightarrow 2KGeH_3$$

2. *Synthesis by the Deprotonation of Germane*

Alkali-metal hydroxides act as very strong bases in nonhydroxylic solvents and are capable of deprotonating GeH_4 and other weak acids.[145,146] Jolly and coworkers have used this deprotonation reaction for conveniently preparing the potassium salt of GeH_4 and a variety of other weak acids in high yields.[145,146,147] Convenient solvents used in the reaction are dimethylsulfoxide or 1,2-dimethoxyethane. Excess KOH is present in the systems in order to dry the solvent and remove the water formed in the reaction as the hydroxide hydrate. Thus the net reaction is given as

$$2KOH + GeH_4 \longrightarrow K^+ + GeH_3^- + KOH \cdot H_2O(s)$$

In a typical preparation, powdered KOH (2.8 g) and 10 ml of 1,2-dimethoxyethane are placed in a 50-ml round-bottomed flask that is then attached to the vacuum line. The flask is surrounded by a $-78°$ bath, and the mixture is degassed with pumping for about 5 min. After this is complete the reactor is cooled to $-196°$ and thoroughly evacuated. Germane (

mmoles) is condensed into the reactor, and the mixture is allowed to react, with vigorous shaking, at room temperature for ½ hr. After the removal of any noncondensable gas from the reactor the $KGeH_3$ solution is ready for whatever further reaction is contemplated. Full details of the use of $KGeH_3$ prepared by this method in the synthesis of CH_3GeH_3 have been given by Rustad, Birchall, and Jolly[148]:

$$GeH_3^- + CH_3I \longrightarrow CH_3GeH_3 + I^-$$

B. Germyl Derivatives of Transition-Metal Complexes

Although a wide variety of transition-metal complexes that contain germanium-metal bonds is known, most of them involve organogermanium substituents rather than the simple germyl or related Ge—H-containing groupings.[8] Much research activity concerning the preparations of the simple germanium hydride transition-metal complexes can be noted in the literature at present, and undoubtedly this area will be greatly expanded in the future. The procedures used thus far in preparing the known germanium–transition-metal complexes that contain germanium-hydrogen bonds include the following:

1. The reaction of the Ge—H bond(s) of GeH_4 or a germyl halide with a transition-metal hydride:

$$GeH_4 + 2HMn(CO)_5 \longrightarrow GeH_2[Mn(CO)_5]_2 + 2H_2$$

$$\text{trans-X-Pt}[(C_2H_5)_3P]_2H + GeH_3X \longrightarrow \text{trans-X-Pt}[(C_2H_5)_3P]_2GeH_2X + H_2$$

$$X = Cl, Br, I$$

2. Cleavage of a metal-metal bond in transition-metal complexes by GeH_4:

$$GeH_4 + Mn_2(CO)_{10} \longrightarrow GeH_2[Mn(CO)_5]_2 + H_2$$

3. The interaction of a germyl halide with a sodium salt of a transition-metal complex:

$$GeH_3Br + NaMn(CO)_5 \longrightarrow GeH_3Mn(CO)_5 + NaBr$$

$$GeH_3Br + NaCo(CO)_4 \longrightarrow GeH_3Co(CO)_4 + NaBr$$

4. The reduction of halogermanium transition-metal complexes:

$$GeCl_2[Fe(CO)_2(\pi\text{-}C_5H_5)]_2 + NaBH_4 \longrightarrow GeH_2[Fe(CO)_2(\pi\text{-}C_5H_5)]_2$$

5. Silyl-germyl exchange reactions:

$$\text{trans-ClPt}[(C_2H_5)_3P]_2SiH_2Cl + GeH_3Cl \longrightarrow \text{trans-ClPt}[(C_2H_5)_3P]_2GeH_2Cl + SiH_3Cl$$

Other procedures applicable to the preparation of organogermanium transition-metal complexes have been reviewed by Hooton[10] and Glockling.[8]

1. Germylpentacarbonylmanganese $GeH_3Mn(CO)_5$

Mackay and George report the preparation of $GeH_3Mn(CO)_5$ in 75 percent yield by the reaction of GeH_3Br with $NaMn(CO)_5$.[74] Germyl bromide is condensed onto $NaMn(CO)_5$ in ether, and the mixture is allowed to warm to room temperature. Fractionation of the volatile products after a 15-min reaction produces the pure $GeH_3Mn(CO)_5$ as a condensate in a trap at $-45°$.

2. Digermanylpentacarbonylmanganese $Ge_2H_5Mn(CO)_5$

Stobart reports the synthesis of the first transition-metal complex that contains the Ge_2H_5-grouping by the interaction of $NaMn(CO)_5$ with Ge_2H_5I, the latter prepared *in situ* at $-63°$.[75b] The $Ge_2H_5Mn(CO)_5$ that results is a liquid which, owing to its low vapor pressure, is difficult to manipulate in a vacuum-system.

3. Germylpentacarbonylrhenium $GeH_3Re(CO)_5$

Mackay and Stobart report the preparation of $GeH_3Re(CO)_5$ by the same method used for the synthesis of $GeH_3Mn(CO)_5$, except that tetrahydrofuran is used as a solvent and the reaction time is 4 hr.[75a]

4. Germyltetracarbonylcobalt $GeH_3Co(CO)_4$

The reaction of GeH_3Br with $NaCo(CO)_4$ has been used by Mackay and George to prepare $GeH_3Co(CO)_4$.[75] The procedure followed is identical to that described previously for the synthesis of $GeH_3Mn(CO)_5$. A 35 percent yield is reported. The deuterated analog can also be prepared from GeD_3Br and $NaCo(CO)_4$.[75]

5. Bis(pentacarbonylmanganese)germane $GeH_2[Mn(CO)_5]_2$

Massey, Park, and Stone report the synthesis of $GeH_2[Mn(CO)_5]_2$ by the room-temperature interaction of GeH_4 with $HMn(CO)_5$.[47] In a typical preparation, as described in the literature, a mixture of GeH_4 (4.57 mmoles) and $HMn(CO)_5$ (3.23 mmoles) in tetrahydrofuran (3.19 mmoles) was sealed under vacuum in 100-ml glass bulb. After 8 days at room temperature the vessel was opened, and the volatile products (hydrogen and a trace of CO) and unreacted starting materials were removed. The crystalline material remaining in the vessel was sublimed (80° at 10^{-3} torr), yielding 600 mg of pale yellow $GeH_2[Mn(CO)_5]_2$ (95 percent yield based on the amount of $HMn(CO)_5$ consumed). There was no evidence whatsoever of the formation of $GeH_3Mn(CO)_5$ or $GeH[Mn(CO)_5]_3$ in this particular reaction.[47]

The formation of $GeH_2[Mn(CO)_5]_2$ has also been noted in the reaction of GeH_4 with $Mn_2(CO)_{10}$ at 140°.[47] It is difficult, however, to separate the compound from unreacted $Mn_2(CO)_{10}$, and this reaction also produces appreciable quantities of CO. There is no evidence of the formation of any manganese pentacarbonyl–substituted germanes in the reactions of GeH_4 or $KGeH_3$ with $ClMn(CO)_5$ or $BrMn(CO)_5$.[47]

6. *Bis(dicarbonyl-π-cyclopentadienyliron)germane* $GeH_2[Fe(CO)_2(\pi\text{-}C_5H_5)]_2$

Stone and coworkers have prepared $GeH_2[Fe(CO)_2(\pi\text{-}C_5H_5)]_2$ in 55 percent yield by the $NaBH_4$ reduction of $GeCl_2[Fe(CO)_2(\pi\text{-}C_5H_5)]$.*[149] As an illustration a solution of $NaBH_4$ (10.6 mmoles) in tetrahydrofuran (10 ml) is added with methanol (3 ml) to a well-stirred solution of $GeCl_2[Fe(CO)_2\text{-}(\pi\text{-}C_5H_5)]_2$ in tetrahydrofuran (100 ml) at 0°. After the solvent is removed, the residue is extracted with diethyl ether. The pure $GeH_2[Fe(CO)_2(\pi\text{-}C_5H_5)]_2$ (480 mg) is recovered by evaporating the ether and subliming the resulting solid.

7. *Platinum-Substituted Germyl Complexes*

The interaction of germyl halides with compounds of the type trans-X-$Pt[(C_2H_5)_3P]_2H$, where X = Cl, Br, I, in benzene at room temperature leads to the formation of platinum-substituted germyl halides[47a]:

$$\text{trans-X-Pt}[(C_2H_5)_3P]_2H + GeH_3X \longrightarrow \text{trans-X-Pt}[(C_2H_5)_3P]_2GeH_2X + H_2 \qquad X = \text{Cl, Br, I}$$

Germane reacts with the platinum hydride complexes in a similar manner, producing trans-X-$Pt[(C_2H_5)_3P]_2GeH_3$.[47a] When the halogen of the germyl compound and platinum hydride are different, a halogen exchange reaction leads to the formation of the product in which the heavier halogen is bound to platinum.[47a] Few specific details about the reactions are available. Characterization of the products (white or pale yellow easily hydrolyzable crystalline compounds) has been by analysis and spectral methods.[47a]

The compound *trans*-$ClPt[(C_2H_5)_3P]_2SiH_2Cl$ undergoes reaction with GeH_3Cl to give a number of products, including *trans*-$ClPt[(C_2H_5)_3P]_2GeH_2Cl$ and SiH_3Cl.[47a] The result implies that there has been a formal exchange of a Pt—Si with a Ge—H bond. Although numerous exchanges are known to occur between silyl and germyl compounds (see Section IV-B-2), this is the first one observed in which the Ge—H bond is involved. In all other silyl-germyl exchange reactions that have been reported, there is no net change in the number of Si—H and Ge—H bonds.

* This compound is prepared by the reaction of $GeCl_4$ with dicarbonyl-π-cyclopentadienyliron dimer and sodium in tetrahydrofuran.[149]

The formation of six-coordinate platinum complexes has been noted in the reaction of excess GeH_3Cl with *trans*-X-Pt$[(C_2H_5)_3P]_2H$.[47b] The complexes tentatively identified by their ^1H nmr spectra are:

$$\text{(A)}\quad ClGeH_2\text{—}\overset{H}{\underset{GeHCl_2}{Pt}}\text{—}GeH_2Cl,\ \ P(C_2H_5)_3,\ (C_2H_5)_3P$$

$$\text{(B)}\quad Cl_2GeH\text{—}\overset{H}{\underset{GeHCl_2}{Pt}}\text{—}GeH_2Cl,\ \ P(C_2H_5)_3,\ (C_2H_5)_3P$$

$$\text{(C)}\quad ClGeH_2\text{—}\overset{H}{\underset{GeH_2Cl}{Pt}}\text{—}GeH_2Cl,\ \ P(C_2H_5)_3,\ (C_2H_5)_3P$$

$$\text{(D)}\quad ClGeH_2\text{—}\overset{H}{\underset{GeH_2Cl}{Pt}}\text{—}GeHCl_2,\ \ P(C_2H_5)_3,\ (C_2H_5)_3P$$

The formation of these compounds is thought to proceed through the addition of GeH_3Cl to the H-coordinate platinum complex followed by the elimination of hydrogen and further addition:

$$Cl\text{—}Pt\text{—}H\ [P(C_2H_5)_3,\ (C_2H_5)_3P] \xrightarrow{GeH_3Cl} Cl\text{—}\overset{H}{\underset{GeH_2Cl}{Pt}}\text{—}H\ [P(C_2H_5)_3,\ (C_2H_5)_3P]$$

$$\xrightarrow[+GeH_3Cl]{-H_2} Cl\text{—}\overset{H}{\underset{GeH_2Cl}{Pt}}\text{—}GeH_2Cl\ [P(C_2H_5)_3,\ (C_2H_5)_3P]$$

$$\xrightarrow{-HCl} ClGeH_2\text{—}Pt\text{—}GeH_2Cl\ [P(C_2H_5)_3,\ (C_2H_5)_3P]\quad \text{(I)}$$

$$\xrightarrow{-GeH_2Cl_2} H\text{—}Pt\text{—}GeH_2Cl\ [P(C_2H_5)_3,\ (C_2H_5)_3P]\quad \text{(II)}$$

Compounds such as C or D would result from the addition of GeH_3Cl or of GeH_2Cl_2 to compound (I), or by a similar scheme involving further addition to (II) followed by the elimination of hydrogen. Hydrogen chloride and hydrogen were observed as products of the reaction.

XII. SOME SELECTED PHYSICAL PROPERTIES AND STRUCTURES OF GERMYL AND RELATED KINDS OF COMPOUNDS

A. Physical Properties

A few physical properties of most of the known germyl and related kinds of compounds are summarized in Table VII. Melting points are normally those which are obtained by using a Stock magnetic-plunger apparatus.[150] Most boiling points listed are values extrapolated from vapor-pressure equations derived from low-pressure measurements. The actual equations and other thermodynamic information derived from them are available in the corresponding reference listed. For many of the derivatives, vapor pressures at a particular temperature are listed. These data are of particular importance for checking the identity and purity of a sample. The ultimate sample identification is usually made by examining the compound's infrared, ^{1}H nmr, or mass spectra, references to which are also included in the table.

B. Structure Studies

Although information about the structures of germyl compounds is still rather scanty, enough data are available to illustrate the structural differences that exist within an analogous series of CH_3-, SiH_3-, and GeH_3-derivatives.* Comparative studies of the three series of group IVA compounds will undoubtedly contribute much to the understanding of how the bonding in a particular linkage changes as one proceeds down a group in the periodic table. Of particular concern at present is the evaluation of the degree to which the lower group IVA elements can use their vacant *d* orbitals in π bonding with adjacent atoms or groupings containing filled orbitals of the appropriate symmetry.

Both silicon and germanium have vacant *d* orbitals, which are clearly of the appropriate symmetry for overlap of the π type with filled *p* orbitals of attached atoms or groups.[188] The extent of the $(p \rightarrow d)\pi$ interaction, however, is subject to some debate. Most of the experimental results of structure and related studies now available for silyl and germyl compounds can be understood by considering that the vacant *d* orbitals of both silicon and germanium do participate in π bonding with certain donor atoms and that for the first-row donor atoms the overlap is more important in the case of silicon than for germanium.

This general conclusion is based largely on the observed structural parameters of several analogous series of CH_3-, SiH_3-, and GeH_3- compounds,

* No attempt is made to list all the structural parameters known for germyl and related compounds in this review. References are given in Table VII to structure studies in which this information is available.

TABLE VII

Some Selected Physical Properties of Germyl and Related Kinds of Compounds

	Compound	mp (°C)	bp (°C)	Vapor pressure (torr/temp, °C)	Reference	Reference ir, Raman	Reference nmr	Reference other studies[a]
Halogen derivatives	GeH_3F	−22	15.6	100/−22.8°	55	28, 92	38	MW (151)
	GeH_2F_2	...	...	18.6/0°	55, 97	97, 98, 152	38	
	GeH_3Cl	−52	28	72.5/−22.8°	46	18, 28	38	MW (153)
	GeH_2Cl_2	−68	69.5	38.7/0°	46	97, 98, 152, 154	38	
	$GeHCl_3$	−71	75.2	26.5/0.7°	101	155	107a	
	GeH_3Br	−32	52	87/0°	46	28, 92	38	MW (156)
	GeH_2Br_2	−15	89	6.1/0°	46	41, 97, 98, 152	38	
	GeH_2BrI	...	...	...	60	...	60	
	$GeHI_2Cl \cdot 2(C_2H_5)_2O$	...	...	...	111	...	107a	
	GeH_3I	−15	...	20/0°		28, 157	38	
	GeH_2I_2	45–47	...	<0.5/20°	53	53	53	
	Ge_2H_5Cl	...	88	13.4/5.3°	42	37, 42	42, 158	M (42)
	Ge_2H_5Br	...	...	...	42	37, 42	42, 158	M (42)
	Ge_2H_5I	~−17	...	...	42	37, 42	42, 158	M (42)
	Ge_3H_7I	...	...	nil/0°	9			
Halogenoid derivatives	GeH_3CN	...	...	5.8/0°	58	57	...	SIR (57, 158a), MW (113)
	GeH_3NCO	−44.0	71.5	27.7/0°	23	116, 115	115	MW (114), SIR (116)
	GeH_3NCS	18.6	150	3.1/20°	23	117	115	SIR (117)
	GeH_3N_3	−31	...	20/0°	61	61	61	SIR (61)
Group VIA derivatives	$(GeH_3)_2O$	...	...	66/0°	59	59, 159	81, 62	ED (160), M (68), SIR (59, 159)
	GeH_3OCH_3	−44.5	24.3	52/−23.2°	78	78, 159	78, 89	M (78)
	$GeH_3OC_6H_5$	...	...	...	79	79	79	M (79)

	GeH_3SCH_3	−97	87	33.5/0°	80, 81	159, 81	81	M (81)
	$GeH_3SC_6H_5$	...	...	...	79	79	79	M (79)
	GeH_3SH	...	...	...	...	...	119, 81a	
	GeH_3SSiH_3	...	...	...	88a	...	88a	
	$(GeH_3)_2Se$	−41	...	2–3/20°	62, 120	62, 120	62, 120	SIR (62, 120)
	$(GeH_3Se)_2GeH_2$	...	...	...	...	...	120	
	GeH_3SeH	...	...	...	...	...	119, 120	
	GeH_3SeSiH_3	...	...	...	88a	...	88a	
	$(GeH_3)_2Te$	−75	...	nil/20°	62	62	62	SIR (62)
	GeH_3TeH	...	...	...	...	...	119	
Group VA derivatives	$(GeH_3)_3N$	...	...	...	56, 82	56, 82	56, 82	SIR (56) ED (201)
	$(GeH_3)_3P$	−84	...	2/20°	63, 64	64	64	SIR (64)
	$(GeH_3)_3As$	−84	...	...	65	65	65	SIR (65)
	$(GeH_3)_3Sb$	−60	...	...	65	65	65	SIR (65)
	GeH_3PH_2	−133	51	...	36, 86, 126, 125, 77, 67, 69	86, 161a	162, 163, 36	
	GeH_3AsH_2	−115	67	...	86, 126, 125, 67, 76, 69	86, 36, 163a	162, 163	
	$(GeH_3)_2PH$	...	...	...	81a	...	81a	
	$(GeH_3)_2AsH$	...	...	...	81a	...	81a	
	GeH_2ClPH_2	...	...	...	69	...	69	
	GeH_2BrPH_2	...	...	...	69	...	69	
	$GeH_3PH_2BH_3$	...	...	...	36	...	36	
	$GeH_3NCNGeH_3$	10	...	0.6/0°	68	68	68	SIR (68)
Parent hydrides and selected group IVA derivatives	GeH_4	−165.9	−88.51	182/−112°	32, 164, 165	166, 167, 168	169, 38	M (170)
	GeD_4	−166.2	−89.2	186/−112°	171, 35	172, 167, 168	...	
	GeH_xD_{4-x} ($x = 1–3$)	...	...	...	...	172	...	
	Ge_2H_6	−109	29	236.5/0°	24, 173	174, 175	38, 163, 40	M (176)
	Ge_2D_6	−107.9	28.4	...	171	174	...	

(continued)

TABLE VII (*continued*)

Some Selected Physical Properties of Germyl and Related Kinds of Compounds

Compound	mp (°C)	bp (°C)	Vapor pressure (torr/temp, °C)	Reference	Reference ir, Raman	Reference nmr	Reference other studies[a]
Ge_3H_8	−101.8 (−105.6)	110 (est)	13.9/0°	177, 133, 24	40, 9	40	M (176)
Ge_3D_8	−100.3	110 (est)	...	171	...	...	
$(GeH_3)_2GeHD$	...	234 (est)	...	9	9	...	
n-Ge_4H_{10}	...	177 (est)	...	26, 9b	40, 9b	40, 9b	9b (M)
iso-Ge_4H_{10}	...		...	40, 9b	40, 9b	40, 9b	9b (M)
n-Ge_5H_{12}	...	234 (est)	...	26, 9b	40, 9b	9b	9b, 40 (M)
iso-Ge_5H_{12}	...		...	40, 9b	40, 9b	9b	9b, 40 (M)
neo-Ge_5H_{12}(?)	...	...	...	9b	9b	9b	
CH_3GeH_3	−158	−35.1	67.7/−78.5°	178, 179	180	38, 181	
$(CH_3)_2GeH_2$	−149	−0.6	...	178, 179	182	8, 181, 181b	
$C_2H_5GeH_3$	...	9.2	...	48	183	181	
n-$C_3H_7GeH_3$	...	30	...	184	...	...	
$CH_2{=}CHGeH_3$	...	−3.5	101.5/−45.9°	185	185	186	
$CH_3OCH_2GeH_3$	−121.6	44.4	105.3/0°	78	78	78	M (78)
$CH_3Ge_2H_5$	...	54.7	...	70	70	70	M (70)
$C_2H_5Ge_2H_5$	...	88.6	...	70	70	70	M (70)
$(CH_3)_3GeGeH_3$	−89.6	74.4	24.6/0°	72	72, 138a	72, 138a	
1,1-$(CH_3)_2Ge_2H_4$	...	73.4	...	71	71	71	M (71)
1,2-$(CH_3)_2Ge_2H_4$	...	88.8	...	71	71	71	M (71)
1,1,2-$(CH_3)_3Ge_2H_3$	...	95.4	...	71	71	71	M (71)
$CH_3GeH_2GeH_2GeH_3$	...	201	14.3/20.4°	9a	9a	9a	M (9a)
$GeH_3GeH(CH_3)GeH_3$	...	158	15.6/19.5°	9a	9a	9a	M (9a)
GeH_3SiH_3	−119.7	7.0	7.0/−78.5°	88	88, 135	158	MW (187)
GeD_3SiH_3	...	...	...	135	135	...	
GeH_3SiD_3	...	...	...	135	135	...	
$SiH_3GeH_2GeH_3$	−108.5	...	19.3/0°	133, 138	138	138	M (138)
$SiH_3SiH_2GeH_3$	−113.4		39.6/0°	133, 138	138	138	M (138)

	$GeH_3SiF_2SiF_2H$	−4	...	...	139	139	139	M (139)
	$(CH_3)_3SiGeH_3$	−77.4	73.6	...	72	72, 138a	72, 138a	
	$(CH_3)_2Si(GeH_3)_2$	−93.1	116.3	...	72	72	72	
	$CH_3Si(GeH_3)_3$	−101.0	154.6	...	72	72	72	
	$(GeH_3)_4Si$	−53.3	...	...	72	72	72	
	$SiH_3CH_2GeH_3$	−134.8	30	...	83	...	83	
	$GeH_3CH_2SiH_2Cl$	−129.2	104	...	83	...	83	
	$GeH_3CH_2SiHCl_2$	...	...	3.7/0°	83	...	83	
	$(SiH_3CH_2)_2GeH_2$	...	...	8.6/0°	83	...	83	
	GeH_3SnH_3	...	...	...	141	141	...	
Group IIIA derivatives	$K[(GeH_3Ga(CH_3)_2Cl]$	−20 to −25	...	...	84	...	...	
	$GeH_3Ga(CH_3)_2$	...	...	...	84	...	...	
	$K[GeH_3BH_3]$	98–99	...	...	51	51	51	
Anion derived from alkali metals	GeH_3^-	...	...	...	See text	181	181b	
Transition-metal derivatives	$GeH_3Mn(CO)_5$	23.8	...	5.9/32° 13.5/47°	74	74	74	M (74)
	$GeH_3Co(CO)_4$	−36 to −38	...	2.98/0.2° 8.4/16.2°	75	75	75	M (75)
	$GeH_2[Mn(CO)_5]_2$	87–88	...	...	47	47	47	
	$GeH_2[Fe(CO)_2(\pi\text{-}C_5H_5)]_2$	110 (decomp)	...	...	149	149	149	
	trans-X-Pt$[(C_2H_5)_3P]_2GeH_3$	...	...	...	47a	...	...	
	trans-X-Pt$[(C_2H_5)_3P]_2GeH_2X$ (X = Cl, Br, I)	...	...	...	47a	...	...	
	$H[(C_2H_5)_3P]_2Pt(GeH_2Cl)_3$	...	...	...	47b	...	47b	
	$H[(C_2H_5)_3P]_2Pt(GeH_2Cl)_2GeHCl_2$	...	...	...	47b	...	47b	
	$H[(C_2H_5)_3P]_2Pt(GeHCl_2)_2GeH_2Cl$	...	...	...	47b	...	47b	
	$GeH_3Re(CO)_4$	−53 to −54	...	0.5/20°	75a	75a	75a	
	$Ge_2H_5Mn(CO)_5$	<0	...	...	75b	75b	75b	M (75b)
	$Ge_2H_5Co(CO)_4$	(unstable *in vacuo*)	...	...	29	...	...	

[a] M = mass spectrum, MW = microwave spectrum, SIR = structural data from infrared spectrum, ED = electron diffraction

particularly the ether, amine, and halogenoid derivatives. It is assumed that carbon does not participate in π bonding of this kind, and that a qualitative measure of the extent of π bonding in silicon and germanium compounds with first-row donor elements is reflected by changes in bond angles about the donor atom. Steric differences and effects due to the electronegativity differences of silicon and germanium are not considered important. A lone pair of electrons localized on an atom in a molecule does have an effect on the structure of the molecule, and if the lone pair is delocalized from the atom to an acceptor atom, some structural change is expected, just on the basis of electrostatic interactions. As an illustration, the Si—O—Si bond angle in $(SiH_3)_2O$ (144.1 $\pm$ 0.8°, electron diffraction)[189] is considerably wider than the C—O—C bond angle of $(CH_3)_2O$ (111.7° $\pm$ 0.3, microwave).[190] The angle found for $(CH_3)_2O$ is approximately that expected, considering the two sigma bonds and two lone pairs of electrons about the oxygen ($\simeq sp^3$ hybrid) atom of the molecule. The wider valence angle of the silicon ether can be attributed to the partial delocalization of the two electron pairs on oxygen into the 3*d* orbitals of silicon as a result of the proposed $(p \rightarrow d)\pi$ interaction.[1–3,5–7]

The conclusion that the 4*d* orbitals of germanium can participate in $(p \rightarrow d)\pi$ bonding with first-row elements but to a lesser extent than the 3*d* orbitals of silicon is based, in part, on the observation that the Ge—O—Ge bond angle of $(GeH_3)_2O$ (126.5 $\pm$ 0.4°, electron diffraction)[160] is wider than the C—O—C angle of $(CH_3)_2O$ but not so wide as the Si—O—Si angle of $(SiH_3)_2O$. A similar trend is noted by comparing structures of $(CH_3)_3N$, $(SiH_3)_3N$, and $(GeH_3)_3N$. The heavy atom skeleton of the $(CH_3)_3N$ molecule is nonplanar as a result of the three sigma bonds and one lone pair of electrons about the nitrogen, although the heavy atom skeleton of $(SiH_3)_3N$ is planar (Si—N—Si bond angle = 119.6 $\pm$ 1°, electron diffraction),[191] with three σ bonds about the nitrogen, the lone pair of electrons being delocalized through the three silicon atoms by way of the $(p \rightarrow d)\pi$ bond. Infrared studies of $(GeH_3)_3N$ indicate that this molecule has a nonplanar heavy atom skeleton, with a GeNGe bond angle estimated to be 116° $\pm$ 2°.[56*]

The E—N—C angles, where E = C, Si, Ge, in the heavy atom skeletons of CH_3NCS (CNC = 142°, microwave,[192] SiH_3NCS (SiNC = 180°, microwave),†[193] and GeH_3NCS (GeNC = 156°, infrared-Raman)[117] are also in accordance with the relative importance of $(p \rightarrow d)\pi$ bonding involving

* *Note added in proof:* A recent electron diffraction study of $(GeH_3)_3N$ indicates that the molecule has a planar structure with Ge—N bond lengths of 1.836 $\pm$ 0.005 Å.[201]

† Sheldrick has recently reported an angle of 159° for the SiNC angle of SiH_3NCS determined by electron diffraction.[193a] It is suggested that this is due to the excitation of a low-frequency skeletal-bending mode for a molecule not in the vibrational ground state, resulting in the deviance from the 180° microwave result.

silicon and germanium with first-row donor atoms, as discussed above. The E—N—C angles of the corresponding isocyanates likewise agree with the trend; the magnitude of the GeNC angle in GeH_3NCO, however, is presently not accurately known. Estimates range from 170° by infrared methods[116] to 140 to 154° by microwave techniques.[114] The CNC and SiNC angles of CH_3NCO and SiH_3NCO are 140° (electron diffraction) and 180° (microwave), respectively.[194,195]

Effects of the relative extent of $(p \rightarrow d)\pi$ bonding in silyl and germyl compounds should also be apparent in examining the calculated and experimental bond lengths of the specific linkages in which this kind of bonding is thought to occur. The calculated values are subject to some uncertainty, but nevertheless approximate values can be obtained. The calculated and observed values of the Ge—O distance in $(GeH_3)_2O$ are 1.827 and 1.766 ± 0.004 Å, respectively.[160] The 0.061 Å shortening in the germanium series is about half of that observed for the Si—O bond shortening in $(SiH_3)_2O$.[189] The difference between the calculated and observed lengths of the Si—F and Ge—F bonds of SiH_3F and GeH_3F also show the same trend.[151]

The relative base strengths of the series of ethers GeH_3OCH_3 > $(CH_3)_2O$ > SiH_3OCH_3 also indicate that $(p \rightarrow d)\pi$ bonding is more important in the Si—O bond than in the Ge—O bond.[78]

The reason for the difference in π electron acceptor capability of silicon and germanium has not been established conclusively. It may be related in part to the presence of a radial node in the 4*d* orbitals of germanium but not silicon.[196]

The situation with regard to π bonding involving silicon and germanium with donor atoms of second- and lower-row elements is not very well understood. It is generally thought that π interactions involving either silicon or germanium are less important with the lower-row donor atoms than with the first-row donor atoms. It seems that structural parameters for these systems cannot be used to assess the presence or extent of π bonding with silicon or germanium. For example, the sulfur valence angles of $(CH_3)_2S$, $(SiH_3)_2S$, and $(GeH_3)_2S$ show little variation, being 98.9 ± 0.2° (microwave),[197] 97.4 ± 0.7° (electron diffraction),[198] and 98.9 ± 0.3° (electron diffraction),[160] respectively. Also, both $(SiH_3)_3P$ and $(GeH_3)_3P$ have pyramidal structures.[63,64,199] No appreciable bond shortenings have been observed in any of these systems.

Evidence for π bonding in cases involving second-row donor atoms with silicon or germanium is obtained mainly from base-strength studies. For example, evidence of $(p \rightarrow d)\pi$ bonding in the Si—S and Ge—S linkages is found in the observation that $(SiH_3)_2S$, $(GeH_3)_2S$, SiH_3SCH_3 and GeH_3SCH_3, are weaker bases toward phenol than $(CH_3)_2S$.[80,81] Likewise the feeble base character of both $(SiH_3)_3P$ and $(GeH_3)_3P$ can be rationalized on the basis of some $(p \rightarrow d)\pi$ interaction in the Si—P and Ge—P bonds.

Cradock and Ebsworth have recently sought evidence for the existence of $(p \rightarrow d)\pi$ bonding in silicon and germanium compounds by studying the photo-electron spectra of silyl and germyl halides.[200] The results strongly suggest that some degree of $(p \rightarrow d)\pi$ bonding is present in both the Si—Cl and Ge—Cl linkages, and that the interaction is less in the case of germanium than with silicon.

Acknowledgements

The author gratefully acknowledges his colleagues and students for providing helpful suggestions and constructive criticisms of the material presented in this chapter. Special thanks are extended to John Drake and Stephen Stobart for sending manuscripts of their recent work prior to publication. I am also indebted to Mrs. Kathryn Severn for her patience and care in typing the manuscript.

References

1. F. G. A. Stone, *Hydrogen Compounds of the Group IV Elements*, Prentice-Hall, Englewood Cliffs, N.J., 1962.
2. K. M. Mackay, *Hydrogen Compounds of the Elements*, Spon, London, 1966.
3. A. G. MacDiarmid, *Advan. Inorg. Chem. Radiochem.*, **3**, 207 (1961); *Quart. Rev. (London)*, **10**, 208 (1956).
4. A. G. MacDiarmid, "Halogen and Halogenoid Derivatives of the Silanes," in *Preparative Inorganic Reactions*, W. L. Jolly (ed.), Wiley-Interscience, New York, 1964, vol. I, p. 165.
5. E. A. V. Ebsworth, *Volatile Silicon Compounds*, Pergamon, New York, 1963.
6. B. J. Aylett, *Advan. Inorg. Chem. Radiochem.*, **11**, 249 (1969).
7. C. H. Van Dyke, "The Silanes," in *Kirk-Othmer Encyclopedia of Chemical Technology*, A. Standen (ed.), Wiley, New York, 1969, vol. 18, p. 172.
8. F. Glockling, *The Chemistry of Germanium*, Academic, New York, 1969.
9. K. M. Mackay and P. Robinson, *J. Chem. Soc.*, **1965**, 5121.

9a. S. T. Hosfield and K. M. Mackay, *J. Organometal. Chem.*, **24**, 107 (1970).

9b. K. M. Mackay and K. J. Sutton, *J. Chem. Soc. (A)*, **1968**, 2312.

10. K. A. Hooton, "Organogermanium Compounds," in *Preparative Inorganic Reactions*, W. L. Jolly (ed.), Wiley-Interscience, New York, 1968, vol. 4, p. 85.
11. W. L. Jolly and A. D. Norman, "Hydrides of Groups IV and V," in *Preparative Inorganic Reactions*, W. L. Jolly (ed.), Wiley-Interscience, New York, 1968, vol. 4, p. 1.
12. W. L. Jolly and J. E. Drake, *Inorg. Syn.*, **7**, 34 (1963).
13. S. D. Gokhale, J. E. Drake and W. L. Jolly, *J. Inorg. Nucl. Chem.*, **27**, 1911 (1965).
14. W. L. Jolly, *Synthetic Inorganic Chemistry*, Prentice-Hall, Englewood Cliffs, N.J., 1960, chap. 10; *The Synthesis and Characterization of Inorganic Compounds*, Prentice-Hall, Englewood Cliffs, N.J., 1970, pp. 139–181.

15. R. T. Sanderson, *Vacuum Manipulation of Volatile Compounds*, Wiley, New York, 1948.
16. D. F. Shriver, *The Manipulation of Air-Sensitive Compounds*, McGraw-Hill, New York, 1969.
17. S. Dushman and J. M. Lafferty, *Scientific Foundations of Vacuum Technique*, Wiley, New York, 1949.
18. R. C. Lord and C. M. Steese, *J. Chem. Phys.*, **22**, 542 (1954).
19. C. M. Steese, M. S. thesis, Massachusetts Institute of Technology, 1953.
20. D. L. Morrison and A. P. Hagen, *Inorg. Syn.*, **13** in press.
21. W. L. Jolly, *J. Am. Chem. Soc.*, **83**, 335 (1961).
22. T. S. Piper and M. K. Wilson, *J. Inorg. Nucl. Chem.*, **4**, 22 (1957).
23. T. N. Srivastava, J. E. Griffiths, and M. Onyszchuk, *Can. J. Chem.*, **40**, 739 (1962).
23a. P. Royen and C. Rocktäschel, *Z. Anorg. Allgem. Chem.*, **346**, 279 (1966).
24. L. M. Dennis, R. B. Corey, and R. W. Moore, *J. Am. Chem. Soc.*, **46**, 657 (1924).
25. C. A. Kraus and E. S. Carney, *J. Am. Chem. Soc.*, **56**, 765 (1934).
26. E. Amberger, *Angew. Chem.*, **71**, 372 (1959).
27. K. Borer and C. S. G. Phillips, *Proc. Chem. Soc.*, 189 (1959).
28. D. E. Freeman, K. H. Rhee, and M. K. Wilson, *J. Chem. Phys.*, **39**, 2908 (1963).
29. S. R. Stobart, private communication.
30. H. J. Emeléus and K. M. Mackay, *J. Chem. Soc.*, **1961**, 2676.
31. J. M. Bellama and A. G. MacDiarmid, *Inorg. Chem.*, **7**, 2070 (1968).
32. A. E. Finholt, A. C. Bond, Jr., K. E. Wilzbach, and H. I. Schlesinger, *J. Am. Chem. Soc.*, **69**, 2692 (1947).
33. E. D. Macklen, *J. Chem. Soc.*, **1959**, 1984.
34. S. Sujishi and J. N. Keith, *J. Am. Chem. Soc.*, **80**, 4138 (1958).
35. A. D. Norman, J. Webster, and W. L. Jolly, *Inorg. Syn.*, **11**, 176 (1969).
36. J. E. Drake and C. Riddle, *J. Chem. Soc. (A)*, **1968**, 1675.
37. K. M. Mackay, P. Robinson, and R. D. George, *Inorg. Chim. Acta*, **1**, 236 (1967).
38. E. A. V. Ebsworth, S. G. Frankiss, and A. G. Robiette, *J. Mol. Spectry.*, **12**, 299 (1964).
39. E. D. Macklen, *J. Chem. Soc.*, **1959**, 1989.
39a. A. D. Zorin, I. A. Frolov, P. N. Galkin, and I. N. Skachkova, *Zh. Neorg. Khim.*, **15**, 2032 (1970); *Chem. Abstr.*, **73**, 115846 (1970).
40. J. E. Drake and W. L. Jolly, *Proc. Chem. Soc.*, 379 (1961); *J. Chem. Soc.*, **1962**, 2807.
41. A. L. Beach and J. E. Griffiths, *Can. J. Chem.*, **44**, 743 (1966).
42. K. M. Mackay, P. Robinson, E. J. Spanier, and A. G. MacDiarmid, *J. Inorg. Nucl. Chem.*, **28**, 1377 (1966).
43. K. M. Mackay and P. J. Roebuck, *J. Chem. Soc.*, **1964**, 1195.
44. C. H. Van Dyke and J. T. Wang, unpublished results.
45. S. Sujishi and W. Ando, unpublished results.
46. L. M. Dennis and P. R. Judy, *J. Am. Chem. Soc.*, **51**, 2321 (1929).
46a. W. A. Guillory and C. E. Smith, *J. Chem. Phys.*, **53**, 1661 (1970).
46b. G. K. Barker and J. E. Drake, unpublished results.
47. A. G. Massey, A. J. Park, and F. G. A. Stone, *J. Am. Chem. Soc.*, **85**, 2021 (1963).
47a. J. E. Bentham, S. Cradock, and E. A. V. Ebsworth, *Chem. Comm.*, **1969**, 528.
47b. J. E. Bentham and E. A. V. Ebsworth, *Inorg. Nucl. Chem. Letters*, **6**, 671 (1970).
48. G. K. Teal and C. A. Kraus, *J. Am. Chem. Soc.*, **72**, 4706 (1950).
49. E. Amberger and H. D. Boeters, *Angew. Chem. Intern. Ed.*, **2**, 686 (1963).
50. E. Amberger and E. Mühlhofer, *J. Organometal. Chem. (Amsterdam)*, **12**, 55 (1968).

51. D. S. Rustad and W. L. Jolly, *Inorg. Chem.*, **7**, 213 (1968).
52. S. Cradock, G. A. Gibbon, and C. H. Van Dyke, *Inorg. Chem.*, **6**, 1751 (1967).
53. S. Cradock and E. A. V. Ebsworth, *J. Chem. Soc.* (*A*), **1967**, 12.
54. S. Cradock and A. G. Robiette, unpublished observations as quoted in Ref. 53.
55. T. N. Srivastava and M. Onyszchuk, *Proc. Chem. Soc.*, **1961**, 205.
56. D. W. H. Rankin, *J. Chem. Soc.*, (*A*), **1969**, 1926.
57. T. D. Goldfarb, *J. Chem. Phys.*, **37**, 642 (1962).
58. S. Sujishi and J. N. Keith, Abstr. *134th Meeting A.C.S.*, *Inorg. Chem. Div.*, 1958, p. 44N.
59. T. D. Goldfarb and S. Sujishi, *J. Am. Chem. Soc.*, **86**, 1679 (1964).
59a. D. W. H. Rankin, unpublished results.
59b. S. Cradock, unpublished results.
60. S. Cradock and E. A. V. Ebsworth, *J. Chem. Soc.* (*A*), **1967**, 1226.
61. S. Cradock and E. A. V. Ebsworth, *J. Chem. Soc.* (*A*), **1968**, 1420.
62. S. Cradock, E. A. V. Ebsworth, and D. W. H. Rankin, *J. Chem. Soc.* (*A*), **1969**, 1628.
63. S. Cradock, G. Davidson, E. A. V. Ebsworth, and L. A. Woodward, *Chem. Comm.*, **21**, 515 (1965).
64. S. Cradock, E. A. V. Ebsworth, G. Davidson, and L. A. Woodward, *J. Chem. Soc.* (*A*), **1967**, 1229.
65. E. A. V. Ebsworth, D. W. H. Rankin, and G. M. Sheldrick, *J. Chem. Soc.* (*A*), **1968**, 2828.
66. E. A. V. Ebsworth, A. G. Lee, and G. M. Sheldrick, *J. Chem. Soc.* (*A*), **1968**, 2294.
67. J. E. Drake, N. Goddard, and J. Simpson, *Inorg. Nucl. Chem. Letters*, **4**, 361 (1968).
68. S. Cradock and E. A. V. Ebsworth, *J. Chem. Soc.* (*A*), **1968**, 1423.
69. J. E. Drake, N. Goddard, and C. Riddle, *J. Chem. Soc.* (*A*), **1969**, 2704; C. Riddle, Ph.D. Thesis, Southampton (1969).
70. K. M. Mackay, R. D. George, P. Robinson, and R. Watt, *J. Chem. Soc.* (*A*), **1968**, 1920.
71. R. D. George and K. M. Mackay, *J. Chem. Soc.* (*A*), **1969**, 2122.
72. W. A. Dutton and M. Onyszchuk, *Inorg. Chem.*, **7**, 1735 (1968).
73. R. Varma and A. P. Cox, *Angew. Chem.*, **76**, 649 (1964).
74. K. M. Mackay and R. D. George, *Inorg. Nucl. Chem. Letters*, **5**, 797 (1969).
75. K. M. Mackay and R. D. George, *Inorg. Nucl. Chem. Letters*, **6**, 289 (1970).
75a. K. M. Mackay and S. R. Stobart, *Inorg. Nucl. Chem. Letters*, **6**, 687 (1970).
75b. S. R. Stobart, *Chem. Comm.*, **1970**, 999.
76. J. W. Anderson and J. E. Drake, *Inorg. Nucl. Chem. Letters*, **5**, 887 (1969); *J. Chem. Soc.* (*A*), **1970**, 3131.
77. D. C. Wingleth and A. D. Norman, *Chem. Comm.*, **1967**, 1218.
78. G. A. Gibbon, J. T. Wang, and C. H. Van Dyke, *Inorg. Chem.*, **6**, 1989 (1967),
79. C. Glidewell and D. W. H. Rankin, *J. Chem. Soc.* (*A*), **1969**, 753.
80. J. T. Wang and C. H. Van Dyke, *Chem. Comm.*, **1967**, 612.
81. J. T. Wang and C. H. Van Dyke, *Inorg. Chem.*, **7**, 1319 (1968).
81a. J. E. Drake and C. Riddle, *J. Chem. Soc.* (A), **1968**, 2709.
82. D. W. H. Rankin, *Chem. Comm.*, **1969**, 194.
82a. M. Onyszchuk, private communication.
83. G. A. Gibbon, E. W. Kifer, and C. H. Van Dyke, *Inorg. Nucl. Chem. Letters*, **6**, 617 (1970).
84. E. Amberger, W. Stoeger, and J. Hönigschmid, *J. Organometal. Chem.* (*Amsterdam*), **18**, 77 (1969).

85. W. L. Jolly, "The Use of Electrical Discharges in Chemical Synthesis," in *Technique of Inorganic Chemistry*, H. B. Jonassen and A. Weissberger (eds.), Wiley-Interscience, New York, 1963.
86. J. E. Drake and W. L. Jolly, *Chem. Ind.*, **1962**, 1470.
87. T. D. Andrews and C. S. G. Phillips, *J. Chem. Soc.* (*A*), **1966**, 46.
88. E. J. Spanier and A. G. MacDiarmid, *Inorg. Chem.*, **2**, 215 (1963).
88a. J. E. Drake and C. Riddle, *J. Chem. Soc.* (*A*), **1970**, 3134.
89. G. A. Gibbon, Y. Rousseau, C. H. Van Dyke, and G. J. Mains, *Inorg. Chem.*, **5**, 114 (1966).
90. Y. Rousseau and G. J. Mains, *J. Phys. Chem.*, **70**, 3158 (1966).
91. R. Varma, Abstracts, Fourth International Conference on Organometallic Chemistry, Univ. of Bristol, 1969, Paper M-10.
92. J. E. Griffiths, T. N. Srivastava, and M. Onyszchuk, *Can. J. Chem.*, **40**, 579 (1962).
93. J. E. Griffiths and A. L. Beach, *Can. J. Chem.*, **44**, 1227 (1966).
94. C. H. Van Dyke, unpublished results.
95. J. E. Bulkowski and C. H. Van Dyke, unpublished results.
96. F. A. Anderson, B. Bak, and A. Hillebert, *Acta Chem. Scand.*, **7**, 236 (1953).
97. T. N. Srivastava, J. E. Griffiths, and M. Onyszchuk, *Can. J. Chem.*, **41**, 2101 (1963).
98. E. A. V. Ebsworth and A. G. Robiette, *Spectrochim. Acta*, **20**, 1639 (1964).
99. C. Winkler, *J. Prakt. Chem.*, **142**, (N.S. 34) 177 (1886).
100. C. Winkler, *J. Prakt. Chem.*, **144**, (N.S. 36) 188 (1887).
101. L. M. Dennis, W. R. Orndoff, and D. L. Tabern, *J. Phys. Chem.*, **30**, 1049 (1926).
102. C. W. Moulton and J. G. Miller, *J. Am. Chem. Soc.*, **78**, 2702 (1956).
103. L. M. Dennis and R. E. Hulse, *J. Am. Chem. Soc.*, **52**, 3553 (1930).
104. L. S. Foster, *Inorg. Syn.*, **2**, 102 (1946).
105. O. M. Nefedov and S. P. Kolesnikov, *Izv. Akad. Nauk SSSR* (*Engl. Transl.*), **1963**, 1910.
106. T. K. Gar and V. F. Mironov, *Izv. Akad. Nauk SSSR* (*Engl. Transl.*), **1965**, 827.
107. V. F. Mironov and T. K. Gar, *Organometal. Rev.* (*A*), **3**, 311 (1968).
107a. O. M. Nefedov, S. P. Kolesnikov, and W. I. Schejtschenko, *Angew. Chem.*, **76**, 498 (1964).
108. G. M. Brewer and L. M. Dennis, *J. Phys. Chem.*, **31**, 1526 (1927).
109. V. F. Mironov and T. K. Gar, *Izv. Akad. Nauk SSSR* (*Engl. Transl.*), **1965**, 740.
110. P. Mazerolles and G. Mannel, *Bull. Soc. Chim. France*, **1967**, 2511.
111. O. M. Nefedov and S. P. Kolesnikov, *Izv. Akad. Nauk SSSR* (*Engl. Transl.*), **1966**, 187.
112. T. D. Goldfarb and B. P. Zafonte, *J. Chem. Phys.* **41**, 3653 (1964).
113. R. Varma and K. S. Buckton, *J. Chem. Phys.*, **46**, 1565 (1967).
114. K. R. Ramaprasad, R. Varma, and R. Nelson, *J. Am. Chem. Soc.*, **90**, 6248 (1968).
115. K. M. Mackay and S. R. Stobart, *Spectrochim. Acta*, **26A**, 373 (1970).
115a. K. F. Chew, W. Derbyshire, N. Logan, A. H. Norburg, and A. I. P. Sinha, *Chem. Comm.*, **1970**, 1708.
116. J. E. Griffiths, *J. Chem. Phys.*, **48**, 278 (1968); J. E. Griffiths and A. L. Beach, *Chem. Comm.*, **19**, 437 (1965).
117. G. Davidson, L. A. Woodward, K. M. Mackay, and P. Robinson, *Spectrochim. Acta*, **23A**, 2383 (1967).
118. J. S. Thayer and R. West, *Inorg. Chem.*, **3**, 889 (1964).
119. C. Glidewell, D. W. H. Rankin, and G. M. Sheldrick, *Trans. Faraday Soc.*, **65**, 1409 (1969).
120. J. E. Drake and C. Riddle, *J. Chem. Soc.* (*A*), **1969**, 1573.

121. L. M. Dennis and R. W. Work, *J. Am. Chem. Soc.*, **55**, 4486 (1933).
122. S. Sujishi and T. D. Goldfarb, Abstr. *140th Meeting A.C.S.*, *Inorg. Chem. Div.*, 1958, p. 35N.
123. S. Sujishi, *Abstr. XVIIth Intern. Congr. Pure Appl. Chem.*, 1959, p. 53.
124. A. H. Blatt, *Organic Syntheses*, Wiley, New York, 1943, coll. vol. (II), p. 345.
125. P. Royen, C. Rocktäschel, and W. Mosch, *Angew. Chem. Intern. Ed.*, **3**, 703 (1964).
126. P. Royen and C. Rocktäschel, *Z. Anorg. Allgem. Chem.*, **346**, 290 (1966).
127. E. Amberger and H. Boeters, *Angew. Chem. Intern. Ed.*, **2**, 52 (1962).
128. G. Davidson, E. A. V. Ebsworth, G. M. Sheldrick, and L. A. Woodward, *Spectrochim. Acta*, **22**, 67 (1966).
129. E. Amberger and H. D. Boeters, *Ber.*, **97**, 1999.
130. A. E. Finholt, C. Helling, V. Imhof, L. Nielsen, and E. Jacobson, *Inorg. Chem.*, **2**, 504 (1963).
131. J. Pump and U. Wannagat, *Ann. Chem.*, **652**, 21 (1962).
132. P. Royen and C. Rocktäschel, *Angew. Chem.*, **76**, 302 (1964).
133. P. L. Timms, C. C. Simpson, and C. S. G. Phillips, *J. Chem. Soc.*, **1964**, 1467.
134. R. C. Kennedy, L. P. Freeman, A. P. Fox, and M. A. Ring, *J. Inorg. Nucl. Chem.*, **28**, 1373 (1966).
135. J. A. Lannon, G. S. Weiss, and E. R. Nixon, *Spectrochim. Acta*, **26A**, 221 (1970).
136. C. S. G. Phillips and P. L. Timms, *Anal. Chem.*, **35**, 505 (1963).
137. C. S. G. Phillips, P. Powell, J. A. Semlyen, and P. L. Timms, *Z. Anal. Chem.*, **167**, 202 (1963).
138. K. M. Mackay, S. T. Hosfield, and S. R. Stobart, *J. Chem. Soc.* (*A*), **1969**, 2937.
138a. R. D. George, K. M. Mackay, and S. R. Stobart, *J. Chem. Soc.* (*A*), **1970**, 3250.
139. D. Solan and P. L. Timms, *Inorg. Chem.*, **7**, 2157 (1968).
140. D. Solan, *U.S. Clearinghouse Fed. Sci. Tech. Inform.*, *PB Rep.*, **1969**, PB-187819; *U.S. Govt. Res. Develop. Rep.*, **70**, 61 (1970); *Chem. Abstr.*, **72**, 96190 (1970).
141. E. Wiberg, E. Amberger, and H. Cambensi, *Z. Anorg. Chem.*, **351**, 164 (1967).
142. D. S. Rustad and W. L. Jolly, *Inorg. Chem.*, **6**, 1986 (1967).
143. S. N. Glarum and C. A. Kraus, *J. Am. Chem. Soc.*, **72**, 5398 (1950).
144. S. P. Garrity and M. A. Ring, *Inorg. Nucl. Chem. Letters*, **4**, 77 (1968).
145. W. L. Jolly, *J. Chem. Educ.*, **44**, 304 (1967).
146. W. L. Jolly, *Inorg. Chem.*, **6**, 1435 (1967).
147. W. L. Jolly, *Inorg. Syn.*, **11**, 113 (1968).
148. D. S. Rustad, T. Birchall, and W. L. Jolly, *Inorg. Syn.*, **11**, 128 (1968).
149. N. Flitcroft, D. A. Harbourne, I. Paul, P. M. Tucker, and F. G. A. Stone, *J. Chem. Soc.* (*A*), **1966**, 1130.
150. A. Stock, *Hydrides of Boron and Silicon*, Cornell University Press, Ithaca, N.Y., 1933.
151. J. E. Griffiths and K. B. McAfee, *Proc. Chem. Soc.*, **1961**, 456.
152. J. E. Drake and C. Riddle, *J. Chem. Soc.* (*A*), **1969**, 2114.
153. B. P. Dailey, J. M. Mays, and C. H. Townes, *Phys. Rev.*, **76**, 136 (1949).
154. J. E. Drake, C. Riddle, and D. E. Rogers, *J. Chem. Soc.* (*A*), **1969**, 910.
155. L. P. Lindeman and M. K. Wilson, *Spectrochim. Acta*, **9**, 47 (1957).
156. A. H. Sharbaugh, B. S. Pritchard, V. G. Thomas, J. M. Mays, and B. P. Dailey, *Phys. Rev.*, **79**, 189 (1950).
157. J. E. Griffiths, *Can. J. Chem.*, **45**, 2639 (1967).
158. E. J. Spanier and A. G. MacDiarmid, *J. Inorg. Nucl. Chem.*, **31**, 2976 (1969).
158a. K. Venkateswarlu and P. Bhamambal, *Acta Phys. Pol.* (*A*), **37**, 661 (1970); *Chem. Abstr.*, **73**, 19860 (1970).

159. S. Cradock, *J. Chem. Soc.* (*A*), **1968**, 1426.
160. C. Glidewell, D. W. H. Rankin, A. G. Robiette, G. M. Sheidrick, B. Beagley, and S. Cradock, *J. Chem. Soc.* (*A*), **1970**, 315; G. Glidewell, D. W. H. Rankin, A. G. Robiette, G. M. Sheldrick, S. Cradock, E. A. V. Ebsworth, and D. Beagley, *Inorg. Nucl. Chem. Letters*, **5**, 417 (1969).
161. T. N. Srivastava and M. Onyszchuk, *Can. J. Chem.*, **41**, 1244 (1963).
161a. K. M. Mackay, K. J. Sutton, S. R. Stobart, J. E. Drake, and C. Riddle, *Spectrochim. Acta*, **25A**, 925 (1969).
162. J. E. Drake and W. L. Jolly, *J. Chem. Phys.*, **38**, 1033 (1963).
163. J. E. Drake and W. L. Jolly, *Lawrence Radiation Lab. Rept.* UCRL-10422, University of California, August, 1962.
163a. J. E. Drake, C. Riddle, K. M. Mackay, S. R. Stobart, and K. J. Sutton, *Spectrochim. Acta*, **25A**, 941 (1969).
164. K. Clusius and G. Faber, *Z. Physik. Chem.*, **51B**, 352 (1942).
165. G. G. Devyatykh and I. A. Frolov, *Zh. Neorgan. Khim.*, **8**, 265 (1963); *Chem. Abstr.*, **59**, 2179h (1963).
166. J. W. Straley, C. H. Tindal, and H. H. Nielsen, *Phys. Rev.*, **62**, 161 (1942).
167. I. W. Levin, *J. Chem. Phys.*, **42**, 1244 (1965).
168. D. C. McKean and A. A. Chalmers, *Spectrochim. Acta*, **23A**, 777 (1967).
169. H. Dreeskamp, *Z. Naturforsch.*, **19a**, 139 (1964).
170. F. E. Saalfeld and H. J. Svec, *Inorg. Chem.*, **2**, 46 (1963).
171. A. H. Zeitmann and G. C. Fitzgibbon, *J. Am. Chem. Soc.*, **76**, 2021 (1954).
172. L. P. Lindeman and M. K. Wilson, *Z. Physik. Chem.*, **9**, 29 (1956).
173. S. R. Gunn and L. G. Green, *J. Phys. Chem.*, **65**, 779 (1961).
174. V. A. Crawford, K. H. Rhee, and M. K. Wilson, *J. Chem. Phys.*, **37**, 2377 (1962).
175. D. A. Dows and R. M. Hexter, *J. Chem. Phys.*, **24**, 1029 (1956).
176. F. E. Saalfeld and H. J. Svec, *Inorg. Chem.*, **2**, 50 (1963).
177. S. R. Gunn and L. G. Green, *J. Phys. Chem.*, **68**, 946 (1964).
178. E. Amberger and H. Boeters, *Angew. Chem.*, **73**, 114 (1961).
179. J. E. Griffiths, *Inorg. Chem.*, **2**, 375 (1963).
180. J. E. Griffiths, *J. Chem. Phys.*, **38**, 2879 (1963).
181. T. Birchall and W. L. Jolly, *Inorg. Chem.*, **5**, 2177 (1966).
181a. C. Schumann and H. Dreeskamp, *J. Magnetic Resonance*, **3**, 204 (1970).
181b. T. Birchall and I. Drummond, *J. Chem. Soc.*, (*A*), **1970**, 1859.
182. D. F. Van der Vondel and G. P. Van der Kelen, *Bull. Soc. Chim. Belge*, **74**, 467 (1965).
183. K. M. Mackay and R. Watt, *Spectrochim. Acta*, **23A**, 2761 (1967).
184. O. H. Johnson and L. V. Jones, *J. Org. Chem.*, **17**, 1172 (1952).
185. F. E. Brinckman and F. G. A. Stone, *J. Inorg. Nucl. Chem.*, **11**, 24 (1959).
186. C. H. Van Dyke and R. Neilson, unpublished results.
187. A. P. Cox and R. Varma, *J. Chem. Phys.*, **46**, 2007 (1967).
188. D. P. Craig, A. Maccoll, R. S. Nyholm, L. E. Orgel, and L. E. Sutton, *J. Chem. Soc.*, **1954**, 332.
189. A. Almenningen, O. Bastiansen, V. Ewing, K. Hedberg, and M. Traetteberg, *Acta Chem. Scand.*, **17**, 2455 (1963).
190. U. Blukis, P. H. Kasai, and R. J. Myers, *J. Chem. Phys.*, **38**, 2753 (1963).
191. K. Hedberg, *J. Am. Chem. Soc.*, **77**, 6491 (1955).
192. C. I. Beard and B. P. Dailey, *J. Am. Chem. Soc.*, **71**, 929 (1949).
193. D. R. Jenkins, R. Kewley, and T. M. Sugden, *Proc. Chem. Soc.*, **1960**, 220; *Trans. Faraday Soc.*, **58**, 1284 (1962).

193a. G. M. Sheldrick, *Abstr. Chem. Soc./R.I.C. Meeting*, Paper 7.6, Edinburgh, Scotland (1970).
194. R. F. Curl, Jr., V. M. Rao, K. V. L. N. Sastry, and J. A. Hodgeson, *J. Chem. Phys.*, **39**, 3335 (1963).
195. M. C. L. Gerry, J. C. Thompson, and T. M. Sugden, *Nature*, **211**, 846 (1966).
196. D. S. Urch, *J. Inorg. Nucl. Chem.*, **25**, 771 (1963).
197. L. Pierce and M. Hayashi, *J. Chem. Phys.*, **35**, 479 (1961).
198. A. Almenningen, K. Hedberg, and R. Seip, *Acta Chem. Scand.*, **17**, 2264 (1963).
199. B. Beagley, A. G. Robiette, and G. M. Sheldrick, *Chem. Comm.*, **1967**, 601; *J. Chem. Soc.* (*A*), **1968**, 3002.
200. S. Cradock and E. A. V. Ebsworth, *Chem. Comm.*, **1971**, 57.
201. C. Glidewell, D. W. H. Rankin, and A. G. Robiette, *J. Chem. Soc.* (*A*), 1970 (2935).

AUTHOR INDEX

Numbers in parentheses are reference numbers and show that an author's work is referred to although his name is not mentioned in the text. Numbers in *italics* indicate the pages on which the full references appear.

A

B

H

L

T

U

SUBJECT INDEX

CUMULATIVE INDEX, VOLUMES 1–7